TOMS RIVER

TOMS RIVER

A STORY OF SCIENCE
AND SALVATION

DAN FAGIN

BANTAM BOOKS • NEW YORK

Published in the United States by Bantam Books, an imprint of
The Random House Publishing Group, a division of Random House, Inc., New York.

BANTAM BOOKS and the rooster colophon are registered trademarks of
Random House, Inc.

Library of Congress Cataloging-in-Publication Data

Fagin, Dan.
Toms River : a story of science and salvation / Dan Fagin.
p. cm.
Includes bibliographical references and index.
ISBN 978-0-553-80653-3
eBook ISBN 978-0-345-53861-1
1. Drinking water—Contamination—Health aspects—Toms River Region.
2. Cancer—Toms River Region.   3. Groundwater—Pollution—Health aspects—
Toms River Region.   4. Water quality—New Jersey—Toms River Watershed.
5. Toms River Watershed (N.J.)—Environmental conditions.   I. Title.
RA592.N5F34 2013
363.7209749'48—dc23   2012017030

Printed in the United States of America on acid-free paper

www.bantamdell.com

2 4 6 8 9 7 5 3 1

First Edition

For Alison, Anna, and Lily, always.
And for Lois Levin Roisman, 1938–2008.

Every Tree
carries the snow with its own grace
bends to the breeze with its own sway
etches the clouds with its own stroke
bows to the ice with its own resolve
rights its trunk clenched by its own roots
drenches itself in its own desire
and creates its own spring.

—Lois Levin Roisman

Contents

≈

III. COUNTING

IV. CAUSES

Prologue: Marking Time

≈

On the rare occasions when Michael Gillick needed to know what day it was, he could check his pillbox. It was the size of a small briefcase and had seven compartments, one for each day of the week. Each compartment was subdivided into sections, for the five times each day that Michael took his pills: seven o'clock in the morning, noon, 3:30, 8:30, and eleven at night. (He set his cell phone alarm to the times, to make sure he did not forget.) Once a week, Michael or his mother would refill the compartments, in a careful ritual that was the pharmacopoeial equivalent of turning the hourglass.

For a typical week, he counted out 138 pills: tiny pink morphine tablets for pain, yellow steroids to normalize his immune system, white phenobarbitals for seizures, and blue oval-shaped antihistamines for nausea and dizziness. There was also Prevacid for heartburn, Corgard for high blood pressure, and a yogurt pill for indigestion. Three times a day, Michael took a powerful blood pressure medication called Regitine. Years earlier, the drug's manufacturer—the company's name at the time was Ciba-Geigy—had stopped making Regitine in pill form, but Michael had secured a large stockpile and had been working his way through it ever since.

Michael lived with his parents in a ranch-style house on a shady street in the comfortable Brookside Heights section of Toms River, New Jersey. He did not get out much. He loved movies, but a trip to the theater was an ordeal because he was extremely small. Strangers

would point and say, "Oh, what a cutie!" Once, when he was fourteen, he stepped out into the lobby to look for a bathroom, and a
woman demanded to know why he was wandering off without his
mother. He had tried dating, but it did not go well. When Michael
was sixteen, he developed a mad crush on the girl who delivered the
newspaper. He would watch her from his bedroom window every
morning. But when he finally got up the courage to try to speak to her,
he kept his eyes on the floor. Later, he realized why: He did not want
to watch her watching him.

Born in 1979, Michael was now a man. He stood four feet six
inches tall and weighed about one hundred pounds.

If Michael held a job or attended school, there would be other
ways to measure the passage of time besides his pillbox. Michael had
tried one semester of community college. He was definitely smart
enough—he ended up with straight A's—but everyone in the Gillick
family agreed that school put too much stress on his damaged body;
a job would be even worse. So Michael stayed home instead. Most
days he would sleep until noon, watch soap operas in the afternoon,
exercise (he liked lifting light weights), mess around a bit on the gui-
tar, and then play video games or watch professional wrestling for
hours and hours after dinner.

Michael was a night owl, but he had his reasons. Staying up until
three o'clock and watching television was far superior to trying to
sleep. There were two nightmares he could not shake. In the first, he
watched through his bedroom window as the family dog ran out into
the street and was hit by a passing car. The second nightmare, far
worse, came from the horror movies and fantasy video games Michael loved. It featured a hideous, blood-soaked man brandishing a
huge knife in a howling thunderstorm. "I'll always be with you, and
in the end, I'll come for you," the man would tell Michael as lightning
crashed. Then the ghoul would kill Michael's parents and older
brother, one by one, as Michael watched. After one especially horrific
night when Michael was young, his mother asked a police sketch art-
ist to come to the house and draw the imaginary man based on Michael's description, but it didn't help much. The gruesome killer was
true to his word: He kept coming back. His name was "Sir Kan," and

only later did Michael and his mother invert the name's syllables and figure out its significance.

The towering fact of Michael Gillick's life was that he had cancer. He had *always* had cancer. When he was three months old, Michael was diagnosed with neuroblastoma, a fast-spreading cancer of the nervous system. By that time, the disease was already so far advanced that it was apparent that he had been afflicted even while still inside his mother's womb. The doctors had told Linda and Raymond (Rusty) Gillick that Michael had only a fifty-fifty chance of reaching his first birthday. They missed their guess by decades, but survival came at a terrible price. Tumors cost Michael the full use of his left eye and ear, ruined his balance, and shifted the location of his internal organs. Steroid drugs stunted his growth and bloated his face, while chemotherapy weakened his heart and lungs, destroyed the lining of his stomach, and dissolved his bones to the point that walking was painful. When he was younger, Michael's body was so sensitive that he would scream if anyone so much as touched him. Now he mainly just felt exhausted, breathless, and nauseated—and that was on a pretty good day. On the bad days he could hardly move. And while his pharmaceutical regimen seemed to be holding his neuroblastoma in check, no doctor ever dared tell him that he had beaten it.

Michael had no memory of living any other way, so he tended not to wallow in his problems. When he was younger, he had at times doubted his Catholic faith, but he had since come to believe deeply in an afterlife and in a god that was both just and merciful—even if there was precious little evidence of that in the circumstances of his own life. He salted his Catholicism with New Age ideas about the healing power of crystals (he sometimes wore one around his neck) and tried hard to avoid being morose. "Every night you're still alive, that's been a good day—even if it's a bad TV day. Tonight is wrestling, so it's a good day," he would say. "People take things for granted, and they shouldn't." Michael always made a special effort to sound upbeat when his mother, who ran a cancer support group, asked him to talk to a newly diagnosed child.

Only his parents and a few close friends knew the secret behind Michael's stoicism: He was waiting. Michael had long ago resigned himself to the fact that he would never be healthy, but there was something he wanted almost as much—sometimes, even more. What Michael Gillick wanted was justice. For as long as he could remember and with a certainty he could never fully explain—"I just feel it, here," he would say, tapping his scarred chest—Michael was absolutely convinced that something, some*one*, was responsible for giving him cancer and making his life so painful. And thanks to a remarkable sequence of events in his hometown of Toms River, events in which Michael and his mother played a significant part, he was now equally certain he knew who had done it.

"When I first heard about what might have caused my cancer, when I was young, I said, 'I want to live and fight, so I can see them punished,'" he recalled years later. "I said, 'I won't die until I get retribution.' I didn't know the word *retribution* at the time, so I probably said 'revenge.' That's what I want. We're still waiting for it, and we're not going away. As far as I'm concerned, there's a lot more we're going to find out, and when we do, it's going to blow people's minds."

Many of his neighbors didn't believe him, and it was easy to understand why. Michael's convictions about the cause of his illness threatened almost everything that the people of Toms River believed about themselves, their town, and even their country. With its strip malls and package stores, its subdivisions and ball fields, Toms River was no different than thousands of other towns. It had grown very fast—as fast as any community in the United States for a while—but growth was the engine that created its wealth. If the detritus of Toms River's prosperity had been quietly buried, dumped, or burned within the town's borders, then many residents seemed to regard that as a necessary if unpleasant tradeoff, like rush-hour traffic or crowded beaches in July. Besides, environmental risk was everywhere. The choices that the people of Toms River had made over the decades—to defer to authority, to focus on the here and now, to grow at almost any cost—were hardly unique. If Michael Gillick was right, then all of those choices were wrong—and not just in Toms River.

Michael had been waiting for a very long time, and he was willing to keep waiting. In bleak hospital wards as far away as New York City and Philadelphia, he and his mother had met dozens of other young people from Toms River with cancer—far too many to be a coincidence, he was certain. Many of those friends were gone now, gone forever, but Michael was still here, waiting. He had sat through hundreds of committee meetings and press conferences and strategy sessions in lawyers' offices. He had waited for the results of scientific investigations that seemed to drag on forever, including the big one—the one that was supposed to prove that he and his mother were wrong, that they were just being emotional, hysterical. The so-called experts had gotten a surprise then, hadn't they?

Michael and Linda Gillick had started out knowing nothing, and now, more than thirty years later, they knew almost everything. Along with many other people, some of whom they had never even met, the Gillicks had helped to uncover the secret history of Toms River: a dark chronicle of dumpers at midnight and deceptions in broad daylight, of corporate avarice and government neglect. They had fought the fears and delusions of their own neighbors, and they had been vindicated. Now Michael felt he was closer than ever to achieving his final goal. It was just a matter of biding his time, and then the whole truth would come out at last. He could wait a little longer for that.

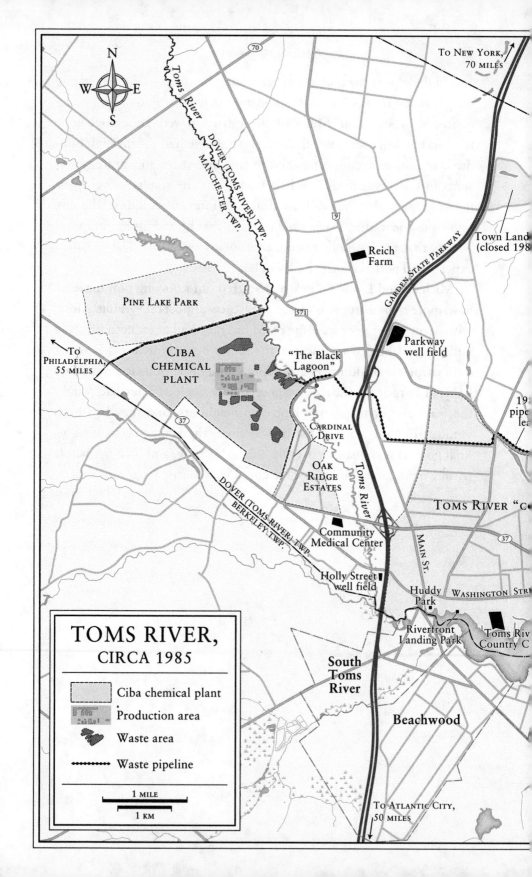

TOMS RIVER,
CIRCA 1985

Ciba chemical plant

Production area

Waste area

Waste pipeline

1 MILE

1 KM

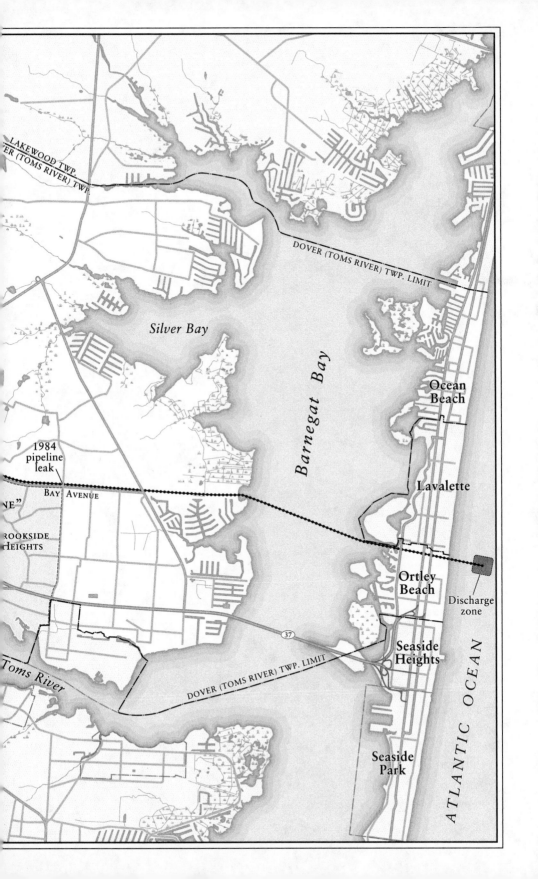

PART I

THE ICE CREAM FACTORY

CHAPTER ONE

≈

Pirates

Who Tom was, if he ever was, is the first unsolved mystery of Toms River. He may have been an adventurer named Captain William Tom who helped chase the Dutch out of New Amsterdam in 1664 and then prospered as the British Crown's tax collector in the wildlands to the south, in the newly established province of New Jersey. Or he may have been an ancient Indian named Old Tom who lived on the cliffs near the mouth of the river and spied on merchant ships during the Revolutionary War on behalf of the British or the Americans, depending on which side paid the larger bribe.

The people of Toms River, in their infinite capacity for self-invention, prefer a different origin story, one that features neither taxes nor bribery. Despite some doubt about its veracity, the story is enshrined on park plaques, in local histories, and even in a bit of doggerel known grandly as the township's "Old Epic Poem."[1] According to this version, a man named Thomas Luker came alone to the dense pine forests of central New Jersey in about 1700 and settled near the bay, on the northern side of a small river that would bear his name. He lived peacefully among the natives, took the name Tom Pumha ("white friend" in the Lenape language), and married the chief's daughter, Princess Ann.

Today, just up Main Street from the spot where their wigwam supposedly stood, is a bank that used to be known as First National Bank of Toms River. For decades, First National fueled the town's frenzied growth with easy credit before it finally imploded in 1991 under the weight of hundreds of millions of dollars in bad real estate loans and a Depression-style run on its assets by frantic depositors. There was nothing apocryphal about its spectacular collapse, which was the largest bank failure in New Jersey history and a harbinger of even more jarring local crises to come, but no plaques or epic poems commemorate the event. In Toms River, history has always been a fungible commodity.

Before the chemical industry came to town in the 1950s and the supercharged growth began, the most exciting thing that had ever happened in Toms River was the American Revolution. In the years before the war, because of its quirky geography, the village had been a haven for small-time piracy. Cranberry Inlet, a narrow passage between the Atlantic Ocean and Barnegat Bay, was one of the few places on the New Jersey coast where ships could safely wait out storms. But captains who brought their ships in through the inlet to seek shelter in the bay became sitting ducks for the local riffraff, who could slip out of Toms River in whaleboats, attack the ships, and steal their cargo before scurrying back to hide in the shoals. Scavenging shipwrecks was another lucrative pastime. If too few boats ran aground on their own, enterprising locals occasionally moved things along by posting lights in unfamiliar places on the beach to confuse ship pilots looking for the inlet.

With the coming of the Revolutionary War, such underhanded tactics suddenly were not only legal, they were regarded as acts of patriotism. The men of Toms River pursued British shipping with gusto and cooperated with American privateers who seized Loyalist ships and sold their contents at auction in the town square. The British struck back in 1781 by torching the town's salt works. After losing still more supply ships, the Redcoats returned the following year to burn the entire town, including all fifteen houses, the rebuilt salt works, and the local tavern. Holed up in a small stockade in what is now Huddy Park in downtown Toms River, an outnumbered force of twenty-five rebel militiamen led by Captain Joshua Huddy tried unsuccessfully to hold off the attackers. An account of the raid in a Tory newspaper described the subsequent rout: "The Town, as it is called, consisting of about a dozen houses, in which none but a piratical set of banditti resided, together with a grist and saw-mill, were with the blockhouse burned to the ground, and an iron cannon spiked and thrown into the river."[2] Huddy was captured, held in irons on a prison ship for two months, and then hanged without a trial. His execution was a major diplomatic incident that enraged General George Washington; the uproar even led to a brief suspension of the Paris peace talks that ended the war.

In the thirty years that followed, the population of Dover Township (the town's official name until 2006, though almost everyone called it Toms River) quadrupled and its "banditti" went straight, more or less.[3] They prospered as merchants in a bustling port that featured two inns and was a busy stop on the coastal stagecoach route. Unfortunately for the town, a storm in 1812 sealed Cranberry Inlet, and with it the community's chief source of income and main connection to the outside world. (Exactly 200 years later, Hurricane Sandy would devastate parts of Toms River and the shoreline communities, destroying more than 400 Ocean County homes and causing major damage to more than 1,100 others. Sandy came close to reopening Cranberry Inlet but did not quite succeed because so many hardened structures had been built in Ortley Beach during the real estate boom of the 1960s and 1970s.) With the closure of the inlet in 1812, Toms River was once again unimportant, and it would stay that way for the next 140 years. Its population stagnated at less than three thousand for a century before edging upward starting in the early 1900s with the arrival of the railroad and summer tourists from Philadelphia and New York. Toms River was the sleepy center of what was, literally and figuratively, a backwater county. The 1920 census of Ocean County recorded about twenty-two thousand people in a county of almost nine hundred square miles, a third of it under water. Most were still farmers or tradesmen, with a sprinkling of wealthier landowners.

Secure in their insularity, the town burghers hunted in the pine-lands, fished in the bay, and sailed on the river. The same families lived in the same comfortable homes from generation to generation, perched comfortably atop a hierarchy that was as rigidly defined as it was unchanging. The most powerful family, the Mathises, lived in a white mansion on Main Street. A mariner turned automobile dealer, Thomas A. "Captain Tom" Mathis and his son William Steelman "Steets" Mathis ran the all-powerful Ocean County Republican Committee for fifty years, exercising iron control over patronage in town and county government from World War I to the mid-1960s. For much of that time, the father or the son (they took turns) represented Ocean County in the New Jersey State Senate.[4]

Everything in Toms River had its place, as did everyone. Anything that mattered had been settled long ago. The pirate days were over.

The very big idea that would transform Toms River and reshape the global economy was born in 1856 in the attic laboratory of a precocious eighteen-year-old chemistry student named William Henry Perkin, who lived with his family in London's East End. It was Easter vacation, and Perkin was using the time off to work on some coal tar experiments suggested by his mentor at the Royal College of Chemistry, August Wilhelm von Hofmann.

No one in the world knew more about the chemical properties of coal tar than Hofmann, and coal tar was a very important compound to know about. It was, arguably, the first large-scale industrial waste. By the mid-1800s, coal gas and solid coke had replaced candles, animal oils, and wood as the most important sources of light, heat, and cooking fuel in many European and American cities. Both coal gas and coke were derived from burning coal at high temperatures in the absence of oxygen, a process that left behind a thick, smelly brown liquid that was called coal tar because it resembled the pine tar used to waterproof wooden ships. But undistilled coal tar was not a very good sealant and was noxious, too, and thus very difficult to get rid of. Burning it produced hazardous black smoke, and burying it killed any nearby vegetation. The two most common disposal practices for coal tar, dumping it into open pits or waterways, were obviously unsavory. But Hofmann, a Hessian expatriate who was an endlessly patient experimenter, was convinced that coal tar could be turned into something useful. He had already established a track record of doing so at the Royal College of Chemistry, where he was the founding director. Knowing that the various components of coal tar vaporized at different temperatures as it was heated, Hofmann spent years separating its many ingredients. In the 1840s, his work had helped to launch the timber "pickling" industry, in which railway ties and telegraph poles were protected from decay by dipping them in creosote, made from coal tar. But the timber picklers were not interested in the lighter and most volatile components of coal tar, which were still nothing but toxic waste—more toxic, in fact, than undistilled coal tar. So Hofmann and his students kept experimenting.

One of those students was young William Perkin. Hofmann had him working on a project that involved breaking down some key components of coal tar to their nitrogen bases, the amines.[5] Hofmann knew that quinine, the only effective treatment for malaria and thus vital to the British Empire, was also an amine, with a chemical structure very similar to that of several coal tar components, including naphtha. He also knew that bark from Peruvian cinchona trees was the only source of quinine, which is why the medicine was costly and very difficult to obtain. But what if the miracle drug could be synthesized from naphtha or some other unwanted ingredient of coal tar? Hofmann did not think it could, but he considered it a suitable project for his promising teenage protégé.

Perkin eagerly accepted the challenge; like his mentor Hofmann, he was an obsessive experimenter. Perkin set to work during his Easter vacation, while Hofmann was in Germany. Laboring in a small, simple lab on the top floor of his family's home, Perkin decided to experiment with toluene, a toxic component of coal tar that would later play a major role in Toms River. Perkin isolated a derivative called allyl-toluidine, then tried to transform it into quinine by oxidizing it in a mixture with potassium dichromate and sulfuric acid. When he was finished, his test tube contained a reddish-black powder, not the clear medicine he was hoping to see. So Perkin tried again, this time choosing a simpler amine called aniline, which was derived from benzene, another coal tar component that would become notorious later. Once again, he mixed it with potassium dichromate and sulfuric acid, and again the experiment flopped. This time, a black, gooey substance was at the bottom of his test tube, and it certainly was not quinine.

When Perkin washed the black goo out of the test tube, however, he saw something that intrigued him: a bright purple residue on the glass. The color was vivid, and it clung stubbornly to the glass. Even more interestingly, when he treated the gunk with alcohol, its purple color transferred flawlessly to a cotton cloth he used to clean his test tubes. Perkin had stumbled upon the molecular magic of aniline. Benzene, toluene, and other components of coal tar were colorless because they absorbed ultraviolet light undetectable by the human eye. But if those aromatic hydrocarbons were treated with an acid to create aniline or another amine, after some additional steps the newly synthesized molecules very efficiently absorbed light particles from specific wavelengths in the visible spectrum. The young chemist did not know *why* the resulting color was so vivid; the ability of molecules to absorb photons at specific wavelengths based on the structure of their shared electron bonds would not be worked out for another fifty years. He did not even know exactly what he had created; the precise molecular structure of his new chemical would not be deduced until the 1990s. But Perkin did not need anything more than his own eyes to know that what was at the bottom of his test tube might prove very useful, especially after its color transferred so flawlessly onto the cotton cloth. A few months earlier, Perkin and a fellow student had tried to synthesize a textile dye and failed; now he had somehow succeeded while trying to create a medicine for malaria. As Perkin knew, whoever created the first artificial dye capable of staining silk, cotton, and other fabrics with a beautiful color might get very rich. Perhaps, the teenager thought, his failed experiment might not be a failure after all.

Dyes were a very big business, and always had been. The human impulse to drape our bodies in color is primal; ancient cultures from India to the Americas colored their clothes and skin with dyes extracted from wood, animals, and flowering plants.[6] The most celebrated hue of the ancient world, by far, was Tyrian purple. It could be produced only from the milky mucosal secretions of several species of sea snails, or whelks, especially one in the Eastern Mediterranean known as the spiny dye-murex. The reddish purple dye was prized because it was both dazzling in hue and vanishingly scarce. Each murex typically produced only a few drops of dye—and only when freshly caught. It was a color of legendary origin, supposedly discovered by Heracles (Hercules, to the Romans). According to Greek myth, the great hero saw that his dog's mouth was stained purple after chewing shells on the Levantine shore. Heracles considered the hue to be so magnificent that he presented a purple robe to the king of Phoenicia, who promptly declared the color to be a symbol of royalty and made Tyre the ancient world's center of murex dye production. And that is why, on the Ides of March in the year 44 B.C., Julius Caesar was wearing his ceremonial robe of Tyrian purple when he was slain by Brutus in the senate house of Rome. It is also why, thirteen years later at the Battle of Actium, the sails of Cleopatra's royal barge were dyed vivid purple.

With the decline of the Roman Empire, the elaborate system of murex cultivation and dye production established by the Romans disappeared, and so did the purple hue itself. A millennium of grays, browns, and blacks followed. A new dye industry finally arose in the late Middle Ages, allowing Catholic cardinals to cloak themselves in scarlet drawn from the shells of tiny kermes insects and tapestry makers to weave with vivid reds from dyewood trees native to India and Brazil.[7] There were purples, too, mostly from lichens, but they were pale and faded quickly. The deep reddish purple of Caesar and Heracles, hue of power and wealth, monarch of colors, was no longer in the dye maker's palette. It was gone, sustained only in legend.

And then, suddenly, there it was, clinging tenaciously to the glass walls of eighteen-year-old William Henry Perkin's test tubes, without a sea snail in sight. Within six months, Perkin had patented his dye-making process and resigned from the Royal College of Chemistry (over the objections of his mentor, Hofmann, who thought he was being reckless) to devote himself to the manufacture of the dye he first called Tyrian purple. He later switched to an appellation that would go down in history as the first commercial product of the synthetic chemical industry: Perkin's mauve, or mauveine. At first, Perkin and his brother, Thomas, made their dye in William's top-floor workshop. Then they switched to the garden behind the family home, and finally to a factory on the outskirts of London alongside the Grand Junction Canal. Luckily for the Perkin brothers, light purple happened to be *très chic* in the salons of Paris and London in 1857 and 1858. Mauve, as the French called it, was the favorite hue of both Empress Eugénie of France and her close friend Queen Victoria of England. Perkin's new dye was not only brighter than the mauves his French competitors laboriously produced from lichen, it was also much cheaper. Thanks to Perkin, any fashionable woman could afford to wear Eugénie's favorite color, and by 1858 almost all of them did. The dye houses of Europe took notice, creating their own crash research programs in aniline chemistry and sending delegations to London to negotiate access to Perkin's manufacturing secrets.

Two rival dye makers from Basel, Switzerland, were among the closest observers of Perkin's success. Johann Rudolf Geigy-Merian was among the fourth generation of Geigys in the dyewood business in Basel; his great-grandfather Johann Rudolf Geigy-Gemuseus had founded the firm one hundred years earlier in 1758. His competitor Alexander Clavel was a relative newcomer to Basel and was not even Swiss. Clavel was a Frenchman who resettled in Basel because that city, situated strategically on the Rhine River between Germany and France, was a thriving center of the textile trade. Geigy-Merian and Clavel shared a fascination with Perkin's breakthrough in aniline chemistry and the cheaper, brighter dyes it produced. Their enthusiasm quickened with the discovery, in 1858, of the second great aniline dye. It was a bright red called fuchsine that could be produced even more cheaply than Perkin's mauveine.

To Geigy and Clavel, there seemed to be no reason not to try to out-Perkin Perkin, especially because the young Englishman had failed to secure patents in any countries except his own. Even if he had, it would not have mattered, since Switzerland did not enforce patents and would not recognize any chemical process as protectable intellectual property for another fifty years. (The resentful French

called Switzerland *le pays de contre-facteurs,* the land of counterfeit-
ers, while the even angrier Germans called it *der Räuber-Staat,* the
nation of pirates.) Geigy and Clavel did not bother trying to negotiate
with Perkin; he had discussed his methods with enough people that
they were now effectively in the public domain—in patent-free Switzerland, at least. By the end of 1859, Geigy and Clavel had each established his own thriving aniline dye manufacturing operation in Basel,
within a few miles of each other on canals near the Rhine. In doing
so, they set their firms on course to become two of the largest chemical manufacturers in the world—and eventual partners in a sprawling
manufacturing operation in a small New Jersey town that had its own
history of piracy: Toms River.

Over the next ten years of frenetic activity along the Rhine, in Germany as well as Switzerland, the production of aniline dyes—purples,
reds, and blacks first, then every color in the rainbow—transformed
one small family firm after another into international colossi. By
1870, thanks to the new synthetic dyes, most of the companies that
would dominate the chemical industry for the next century and a half
had established themselves as global players. The list included Geigy,
Bayer, Hoechst, Agfa (an acronym for *Aktiengesellschaft für Anilin-
fabrikation,* or the Corporation for Aniline Production), and the biggest of all, BASF, which stood for *Badische Anilin- und Soda-Fabrik,*
or the Baden Aniline and Soda Factory. Alexander Clavel's company
prospered, too, especially after he sold it in 1873. Eleven years later,
the company took the name *Gesellschaft für Chemische Industrie im
Basel,* Society for Chemical Industry in Basel, or Ciba for short. The
third great Basel dye maker, Sandoz, jumped into the game soon afterward, in 1886.

The companies' success began with the appropriation of Perkin's
big idea, but it did not end there. An even more important decision
was to follow the instinct of his mentor, Hofmann, by pulling apart
coal tar and finding uses for all of its constituent parts, not just aniline. After the aniline dyes, derived from benzene, came magentas
made from toluene, reds from anthracene, pinks from phenol, and
indigos from naphthalene. These were all hydrocarbons, the abundant and inexpensive building blocks of organic chemistry. Hydrocar-

bons proved extremely useful to the new world of chemical fabrication for the same reason that hydrogen and carbon are vital to the chemistry of life. When atoms of hydrogen and carbon form molecules, they tend to arrange themselves into durable structures of rings and long chains in which the atoms bond strongly via shared electrons. About four billion years ago, the strength of those hydrogen-carbon bonds allowed increasingly complex molecules—amino acids, DNA, and proteins—to evolve from the primordial soup, making life possible. Now, upon the stable platform of the hydrocarbon polymers in coal tar, chemists began to build a galaxy of new materials that were stronger, more attractive, and cheaper than what nature provided.

Dyes came first, soon followed by paints, solvents, aspirin, sweeteners, laxatives, detergents, inks, anesthetics, cosmetics, adhesives, photographic materials, roofing, resins, and the first primitive plastics—all synthetic and all derived from coal tar, the fountainhead of commercial chemistry. (Coal tar shampoos and soaps came too—and are still available in very diluted form as approved treatments for psoriasis and head lice.) Germany's Ruhr Valley, with its vast deposits of bituminous coal, became the industrial heartland of Europe and thus the world. The British satirical magazine *Punch*, which back in 1859 had lampooned "mauve measles" as a fashion epidemic that should be treated with a "dose of ridicule," by 1888 was singing the praises of aniline chemistry, with only a tinge of sarcasm:

Beautiful Tar, the outcome bright
Of the black coal and the yellow gas-light,
Of modern products most wondrous far,
Tar of the gas-works, beautiful Tar! . . .

Oil, and ointment, and wax, and wine,
And the lovely colours called aniline;
You can make anything from a salve to a star,
If you only know how to, from black Coal-tar.[8]

When the chemical manufacturers finally did expand beyond coal tar chemistry at the end of the nineteenth century, they did so by adapting their manufacturing protocols to petroleum and other raw materials, thereby producing an even larger array of tremendously successful products, from acetone to X-ray plates. Ciba even acquired its own shale oil deposits in the Alps as a new feedstock. By the time the three huge Basel-based chemical makers (Ciba, Geigy, and Sandoz) had formed an alliance to make dyes and other products in the United States—first in Cincinnati, Ohio, in 1920 and then in Toms River in 1952—the industry had proved itself capable of synthesizing almost any material. It was a phenomenally profitable business, as long as no one paid too much attention to what the manufacturing process left behind.

Johann Rudolf Geigy-Merian and Alexander Clavel brought aniline chemistry to the banks of the Rhine, but they were not the first to have to face its consequences. That distinction belongs to a little-known Geigy manager named Johann Jakob Müller-Pack, who in 1860 leased one of Geigy's factory sites and formed his own company to make aniline dyes on a grand scale. The story of what happened next is uncannily similar to what would happen in Toms River more than a century later.[9] Müller-Pack's motivation in launching his own manufacturing company was obvious: By 1860, it was clear that fuchsine, the red aniline dye, would be an even bigger moneymaker than Perkin's mauve. Fuchsine was not just an excellent magenta dye, it was also an intermediate in the production of many colors. As with mauve, those dyes were produced by mixing aniline with oxidizing agents. However, instead of using sulfuric acid as an oxidizer, as Perkin did, fuchsine manufacturers used arsenic acid. This colorless acid was as toxic as arsenic itself, the fabled murder weapon of Renaissance nobility.

Fuchsine production required large quantities of arsenic acid, and much of it came out as waste at the end because the dye manufacturing process was so inefficient. As one aniline chemist later wrote: "In the action of arsenic acid . . . on aniline, only forty percent of soluble, useful coloring matter is formed from the aniline consumed; the rest of the aniline goes over into resinous masses, insoluble in water or in diluted acids. Their nature has not yet been exactly determined in science, their quantity, however, amounts to many times as much as the quantity of magenta formed."[10] In other words, this astonishingly profitable new industry generated far more toxic waste than useful product, and no one had any idea what was actually in that waste or how to get rid of it. This was still true a century later in Toms River, where Ciba and Geigy were still using the same crude disposal method Müller-Pack had selected back in 1860: dumping untreated, unidentified waste into open pits and unlined lagoons on the factory property.

Müller-Pack was selling fuchsine as fast as he could make it, so in 1862 the Geigy family built a second factory for aniline production and rented this one to him also. The new factory was larger and required even more arsenic acid: 200 kilograms per day, or 441 pounds. That was too much for a lagoon to handle alone (even one that was unlined and leaked like a sieve), so this time Müller-Pack adopted an additional disposal method that would become all too familiar a hundred years later in New Jersey: He discharged his arsenic-laced wastewater into the nearest waterway—in this case, a canal beside the plant that led to the Rhine. On the outskirts of London, Perkin was doing the same thing in the canal next to *his* factory, though on a smaller scale and with less arsenic. Even so, the pollution was apparent enough that his neighbors could tell what color Perkin was making that day by looking at the waters of the canal.[11]

The Swiss factories were not isolated in the countryside; they were in the middle of a busy city with a long tradition of close attention to public health. In May of 1863, a chemist who worked for the city of Basel, Friedrich Goppelsröder, inspected the two Müller-Pack factories, as well as Alexander Clavel's, and concluded that working conditions were dangerous and the disposal practices unsanitary. (Goppelsröder did not tell the companies in advance that he was coming—a sharp contrast to what would happen more than a century later in Toms River.) Nine months after Goppelsröder's inspections, the city council ordered Müller-Pack to stop dumping into the canal and banned Clavel from making any aniline dyes at all. Clavel ignored the order for a while and then built a new factory (still used by Ciba almost 150 years later) that was outside the city limits and, most importantly, next to the Rhine. As an official company history put it,

"The immediate proximity of the river as a direct outlet for effluents and also for the disposal of rubbish had become a vital necessity of colour manufacture."[12] But when Müller-Pack, stuck inside the city limits, proposed discharging his factory's waste into a tributary of the Rhine, the city rejected his idea because the tributary was too dry. Desperate to resume full production, he then suggested putting all of his waste in barrels and emptying them directly into the river. This scheme, too, was rejected.

By then, Müller-Pack had even bigger problems: His waste was making some of his factory's neighbors seriously ill. The citizens of Basel got their water from shallow wells, and some of those wells were very close to a factory where Müller-Pack had been dumping waste into an unlined lagoon for two years. In 1863, a railway worker became sick after drinking contaminated water from a nearby well. The following year, a gardener and maid who worked in a home next to the plant fell ill after drinking tea made from water pumped from a tainted well. The owner of the home was a wealthy man, and he summoned Goppelsröder to investigate. The chemist analyzed the well water and reported to city health officials that arsenic levels were "so high that the water must be designated as poisoned, which thus clearly explains the attacks of vomiting, etc."[13] He also noticed that the water was yellowish and had an "indefinable, peculiar, and somewhat repulsive odor"—one that was indisputably awful but difficult to describe (similarly vague terms would be used a century later in Toms River). Goppelsröder then tested the factory's lagoon and soil and even the sediment at the bottom of a nearby canal, finding contamination everywhere he looked. Based on his report, the city in 1864 ordered the older Müller-Pack plant closed. The city also sued Müller-Pack on behalf of the poison victims (by then there were seven). In March of 1865, after eight months in court, he was found guilty of gross negligence. Müller-Pack was ordered to pay a large fine and compensate the victims as well as nearby landowners for the loss of their property values. He even had to deliver clean drinking water to the neighborhood. The humiliation and expense were too much to bear, so a few months after the guilty verdict, he moved to Paris.

Aniline dyes were still a booming business, however, and the

Geigys, as the landowners, were not about to let Müller-Pack's factories sit idle. They took over dye production, and soon Johann Rudolf Geigy-Merian convinced the city to let him deal with the waste by building a pipeline, six thousand feet long, to the Rhine. When the pipeline proved inadequate, Geigy workers began to make clandestine nighttime visits to Basel's Middle Bridge to dump barrels of waste into the fast-moving current in the center of the river.[14] Since the Rhine flowed north and Basel was a border city, Geigy's waste, and Clavel's, too, became Germany's problem, not Switzerland's. From that point on, chemical manufacturers all over the world would follow the same strategy for getting rid of their waste. They would dump on their own property first, since that was always the cheapest alternative, and then if the authorities foreclosed that option, they would instead discharge their liquid waste into the largest and fastest-flowing body of water available. It was no coincidence that the great chemical companies of Switzerland and Germany built many of their factories beside one of the widest and swiftest rivers in Europe and that Perkin made sure that his much smaller factory was next to a canal that led to the Thames.[15]

Even the mighty Rhine, however, could not sufficiently dilute all the hydrocarbon waste that the dye companies were pouring into it. In 1882, a chemistry professor at the University of Basel placed fish in cages at various points in the Rhine to prove that they were being harmed by dye waste—perhaps the first example of a controlled experiment involving wildlife and industrial pollution.[16] By the 1890s, Geigy, Bayer, Ciba, BASF, and others were dumping benzene, toluene, naphthalene, nitrobenzene, and other toxic distillates of coal tar into the river at volumes that would have made Müller-Pack blush. Meanwhile, in the countryside just outside of Basel, the neighbors of Ciba's huge dye works continued to complain bitterly about the "disagreeable steam or vapours escaping into the atmosphere" that had destroyed the gardens of their country homes. No one was in a position to make the companies stop. The chemical industry was a crucial component of rising German power and Swiss prosperity. When hundreds of Basel residents downwind from Ciba's smokestacks tried to block a planned factory expansion in 1900, their protest was rejected on the grounds that "pure Alpine air could not be expected in an industrial area."[17]

The rapidly expanding factories, meanwhile, were becoming extremely dangerous places to work. They were booming in every sense of the word, since explosions were a constant threat. So many powerful acids and volatile solvents were used in the dye manufacturing process that Ciba engineers developed an ingenious potential solution: the first wearable respirator, a breathing apparatus designed to protect laborers as they mixed vaporous chemicals by hand. The device was so hot, heavy, and bulky, however, that workers shunned it. Instead, most laborers simply held cloths over their faces—an action that provided almost no protection. Supervisors generally did not insist that the respirators be used because their employees worked much faster without them. Washup rules also went unenforced; not only did Basel's creeks turn bright blue and red with dumped dye waste, the city's aniline workers added to the polychromatic spectacle by walking the streets "with their hands and sometimes faces and necks colored in all hues of the spectrum."[18] Accidental poisonings were frequent, with the most common symptoms being convulsions, bloody urine, and skin discoloration. As a result, extremely high rates of worker attrition were considered normal in the dye industry, with one American commentator noting in 1925 that superintendents in aniline factories "considered that their duty had been properly performed if they were able to get out the required production without more than ten percent of their men continuously on leave and if such men as were left were able to at least stand up."[19]

Even more ominously, physicians were noticing a new kind of illness they called "aniline tumors." The man who coined the term was a Frankfurt surgeon named Ludwig Wilhelm Carl Rehn. In 1895, he diagnosed bladder cancer in three of the forty-five dye workers he examined who were engaged in the production of fuchsine magenta. By 1906, he had documented thirty-eight similarly stricken workers in Frankfurt, and other doctors in Switzerland and Germany were making similar observations. Within four years, a leading Swiss medical professor was

calling bladder cancer in aniline factories "the most noticeable occupational disease that made a most terrible impression on all who came in
contact with it because of its awfulness and malignancy."[20]

The growing evidence of harm did nothing to slow the industry's
growth. Like its competitors, Ciba expanded all over Europe after the
turn of the century, building factories in Poland, Russia, France, and
England. By 1913, Ciba had almost three thousand employees, most
of them making products for export or working overseas. The German companies grew even faster. In fact, dyes and pharmaceuticals
were the two biggest sources of export revenue for Switzerland and
Germany until World War I, when many of the German factories
switched over to making explosives and poison gas for the Kaiser's
armies. The neutral Swiss eventually picked up the slack and thrived.
In 1917, Ciba's revenues topped fifty million Swiss francs (equivalent
to about US $180 million today)—and 30 percent of it was profit.[21]

In the aftermath of the lost war, and the forced abrogation of
many of their prized patents, Germany's chemical companies embarked on a new survival strategy. The formerly fierce competitors
began to work very closely with each other to try to stay ahead of
newly strengthened foreign competitors, especially in the United
States. Their efforts would climax in the 1925 merger of BASF,
Bayer, Hoechst, Agfa, and others into the conglomerate *Interessen-
Gemeinschaft Farbenindustrie* (Community of Interest of the Dye Industry), or I.G. Farben, which would gain infamy during World War II
as the patent holder of Zyklon B, the cyanide-based poison gas used
at Auschwitz and other Nazi death camps. In the years after the First
World War, however, the German companies were still admired for
their technical prowess.

The Germans' new cartel strategy had a predictable impact across
the border in Basel, the manufacturing hub of *der Räuber-Staat*, with
its long tradition of appropriating foreign ideas. Swiss profits were
falling as German companies reentered world markets, so Ciba, Geigy,
and Sandoz, the three largest Swiss dye manufacturers, decided to
form a "community of interest" of their own. It was a partnership,
not a merger (the mergers would not come until 1971 and 1996), and
it was aimed in part at breaking into the biggest market in the world.

American tariffs were high. The only way around them was to buy or build a plant in the United States to make products for the American market. In 1920, the three Swiss companies did just that, buying two old factories in Cincinnati, Ohio.

The dye makers of Basel were nothing if not consistent. If the Ohio Valley was America's Ruhr, its industrial heartland, then the Ohio River was its Rhine and Cincinnati its Basel. The Ohio was wide and deep with a brisk current, and Cincinnati was full of factories, which meant that the newly named Cincinnati Chemical Works would not stand out. Best of all, the factories the Swiss bought were already hooked up to the city's sewer system, which "treated" waste only in the loosest sense. As in Basel, the city pipes simply channeled it into the creeks and canals that emptied into the concealing waters of the Ohio. At the time, no one seemed bothered that the river was also the principal source of drinking water for more than seven hundred thousand people who lived in Cincinnati and in fourteen other cities farther downstream. After more than a half-century of dumping hazardous chemicals into the Rhine, the Swiss companies were not interested in changing the way they did business now that they had arrived in America.

Subsequent events unfolded like a movie sequel. By the mid-1920s, the Cincinnati Chemical Works was generating steady profits for the Swiss, who responded by expanding into resins and specialty chemicals as well as dyes. Both factories—one in the city's Norwood neighborhood and the other nearby in St. Bernard—grew quickly, and so did their smokestack emissions and wastewater discharges. The growth reached a fever pitch during World War II when the Cincinnati Chemical Works made dyes for military uniforms and smoke grenades and also became the country's biggest producer of DDT (dichlorodiphenyltrichloroethane), the "miracle chemical" whose potent insecticidal properties had been discovered in 1939 by a Geigy chemist named Paul Hermann Müller. Amid the prosperity, few people paid attention to the appearance of what seemed to be an unusual number of bladder cancer cases among the St. Bernard plant's dye workers, who were handling the same chemicals that had triggered bladder tumors back in Europe.

As in Basel, however, there were limits to what the local government was willing to tolerate. Cincinnati municipal officials, and business leaders, too, were embarrassed by the condition of the Ohio and the ruination of the small streams that carried so much toxic sewage into the river. The St. Bernard plant, for example, was responsible for nearly one million gallons of wastewater per day. Exactly what was in that wastewater remained as much of a mystery as it had been in Basel, but company chemists knew that it contained high volumes of sulfuric acid, which along with nitrobenzene had replaced arsenical acid as a major pollutant in almost all types of dye manufacture.[22] Once it reached the Ohio River, the acidic wastewater from the two Cincinnati Chemical Works factories mixed with effluents from dozens of others enterprises that were, in some cases, even more noxious. Collectively, all that waste made the stretch of river near Cincinnati the most polluted section of the entire thousand-mile Ohio.[23] When long-overdue testing confirmed that disease-carrying bacteria were thriving in the foul mixture of untreated sewage and chemical waste, the fouling of one of America's great rivers became a regional scandal and the subject of four congressional hearings between 1936 and 1945.

The Swiss owners of the Cincinnati Chemical Works, in which Ciba was the senior partner, had been through all this before in Basel and elsewhere: the talk of cancer among employees, the pollution complaints from neighbors, and the government crackdowns that would inevitably follow. The Swiss could see what was coming, and they reacted in time-honored fashion: They made plans to skip town. By the time the City of Cincinnati finally built three large sewage treatment plants in the 1950s and passed a law requiring manufacturers to either pre-treat their waste or pay huge fees to the city, Ciba had already shifted most of its production elsewhere.

Instead of moving to another big, boisterous city like Basel or Cincinnati, Ciba found a much more remote location, a sleepy town where hierarchies were respected and authority trusted, a place where the Swiss could do coal tar chemistry on a grand scale without interference from outsiders, where the river was theirs for the taking. The new property was virgin territory, deep in the New Jersey pinelands, virtually untouched but for a single tumbledown farm along the river. That farm was known as Luker Farm, and its former owners claimed to be descended from a legend named Tom Luker and his Indian princess bride, who long ago had shared a wigwam a few miles away, alongside the same river, according to the old story.

Two hundred and fifty years later, something new was coming to Tom's river.

CHAPTER TWO

≈≈

Insensible Things

The monument to coal tar chemistry that Ciba raised in the New Jersey pinelands was like no manufacturing complex the company had built in Basel, Cincinnati, or anywhere else. For one thing, the property it purchased in June of 1949 was immense. The irregularly shaped parcel (it was vaguely triangular) consisted of almost two square miles, with a meandering mile-long stretch of the Toms River forming part of the eastern border. Except for the old farm near the river, Luker Farm, it consisted almost entirely of dense pine and oak forest. At 1,350 acres, the site was large enough to hold more than three hundred factories the size of the original Ciba plant that Alexander Clavel had built back in 1864 on the outskirts of Basel.

Normally, in order to reduce infrastructure costs, factory buildings would be situated as close as possible to existing roads and utility lines. But after almost a century of conflicts with neighbors, the Swiss executives directing the project took the opposite approach in New Jersey. They decided to clear thirty-five acres right in the middle of the vast property, leaving forested buffers of a half-mile or more between the buildings and the outside world. No matter how large the complex grew—and by the 1970s there would be twenty-two buildings, five waste lagoons, and more than a dozen dumps on the property—

passersby would see nothing but the front gate and a solid wall of pine and oak. A careful observer might notice, in the middle distance, the top of the water tower and the roofs of the production buildings peeking over the concealing blanket of green. From the windows of a moving car, however, they would be practically invisible. For everyone who did not work there, the plant would be out of sight and out of mind.

The first three buildings were large and low-slung, totaling four million cubic feet (a bit less than the ocean liner *Titanic*). The Swiss had correspondingly expansive expectations for them.[1] The buildings cost $18 million (roughly $150 million today), but Ciba executives thought that the potential payoff was worth the investment. There was just one purpose for all the construction: to produce thousands of pounds of vat dyes every day, around the clock, as cheaply as possible. Ciba had been making vat dyes in Basel since 1907 and in Cincinnati for almost as long, but never at the scale the company envisioned in Toms River.

Vat dyes were an innovative product from a familiar source: coal tar. Their chief building block, a yellowish brown powder called anthraquinone, could be derived from three coal tar components: anthracene, naphthalene, and benzene. There was nothing new about that, of course. Chemists had been making dyes from the ubiquitous tar since William Henry Perkin first did it in his parents' attic in 1856. What was different about the anthraquinone vat dyes was their durability. Vat dyes got their name not because they were made in vats (they were not), but because of their peculiar chemistry. The dyes would not dissolve in water, yet in the presence of a catalyst could perform the nifty trick of converting to a soluble salt, known as a vat, while coloring a fabric. When the fabric was dried, the dye would revert to being insoluble. That made vat dyes incredibly useful because they bound tightly to fabrics in just one immersion yet were highly resistant to water and sunlight. Almost anything they dyed *stayed* dyed—even cotton, the bane of dyers ever since the invention of the modern cotton gin in 1793 popularized the fabric.[2]

Military uniforms had been the dominant use for vat dyes during World War II—many khakis, browns, olives, and blues were anthraquinone-based colors—and now that the war was over, Ciba was sure that civilian demand would soar. Toms River, company managers decided, should become the center of vat dye production in the United States, challenging bigger rivals like DuPont and Allied Chemical. When the plant opened in 1952, it was capable of producing thirty-five colors and four million pounds of dye per year, about 10 percent of all vat dye production in the United States. That was enough to generate about $6 million a year in sales for Ciba, or $50 million in today's dollars. Moreover, the company was careful to design the road grid and the rest of the plant infrastructure so that production could be doubled or even tripled if demand grew as much as Ciba hoped. Its predecessor companies, starting with Alexander Clavel's, had been in the dye business longer than its competitors, but Ciba usually lagged behind BASF in Europe and DuPont in America in both innovation and production. Vat dyes were an opportunity to close the gap, and Toms River was the place to do it.

There was, and still is, only one major drawback to the production of anthraquinone vat dyes. They are, in the words of dye industry historian Robert J. Baptista, "undoubtedly the class of dyestuffs that involves more hazardous materials, and more hazardous wastes, than any other." The durability of the finished dyes was hard won. Building those rugged, long-chained molecules required a multistep manufacturing process that was complex and very dangerous. Almost every chemical required for the various reactions was either explosive or poisonous, or both. The list of necessary raw materials included prodigious amounts of hazardous solvents (nitrobenzene, naphthalene, and benzene among them), muriatic and sulfuric acids, lye, ammonia, and heavy metals such as mercury, chromium, vanadium, copper, and the seemingly inescapable arsenic. Nitrobenzene was a special problem. It was toxic when inhaled or touched, but it was used in so many different stages of vat dye production that company memos described it as the "blood stream" of the operation.[3] Performing all of the various steps of manufacture not only required huge volumes of hazardous materials, it also took up a lot of space—far more than was available at the two cramped factories in Cincinnati, where Ciba could make vat dyes only in small batches. (Those two Ohio plants continued to operate in the 1950s, though at lower production levels because of the shift to Toms River.) In the wilds of central New Jersey, however, there was enough room for buildings so cavernous that they could accommodate the entire production process under one roof. So that is what the company built in Toms River: a mini-city optimized for the manufacture of a single product.

For all the care the Swiss lavished on its efficient design, however, the mini-city in the pinelands had one incongruous characteristic: It wasted much more than it produced. Dye manufacture had always been a waste-intensive business; that was obvious from nearly a century of dumping into the Rhine, the Ohio, and other rivers. But even by the dismal standards of dye manufacture, vat dyes were a new low. A maxim of the chemical industry is that every step added to the manufacturing process decreases efficiency and increases waste. Making anthraquinone vat dyes required more steps than any other type of dye process. Various combinations of solvents, acids, and metals were boiled in huge kettles, run through high-pressure autoclaves, squeezed through filters, stirred in tanks, and dried in ovens and vacuum chambers. At each step, the volatility of the required ingredients ensured that some of the chemicals would not react properly, creating unwanted byproducts. Some unreacted chemicals could be reused, and a few byproducts could be sold, but most were useless waste. The process was so inefficient that at Toms River, making brown vat dye, a typical example, required five and a half pounds of raw material to produce one pound of finished dye, with almost all of the remainder discarded as waste.[4] In other words, the Toms River plant's production capacity when it opened in 1952 was not really four million pounds per year. It was actually four million pounds of dye and approximately eighteen million pounds of hazardous chemical waste.[5]

Shipping out the finished dye would be simple enough: Trucks and freight cars (a spur from the Central Railroad of New Jersey ran right up to the production buildings) would cart it off to textile plants in the Carolinas and New England. But Ciba was not about to give the same treatment to the much greater quantities of toxic waste it would produce at Toms River. To cart it all to an off-site landfill would be very costly, and the company's plans to catch up to DuPont and its other competitors left no room for unnecessary expense. The dye would leave Toms River, but the waste would stay.

Four hundred and twenty years before the dye makers of Basel came to Toms River, a cantankerous and mostly self-taught physician with unorthodox beliefs about the uses of chemistry and the nature of illness left the same Swiss city in the dead of night. He was fleeing for his life. Despised by his peers, the itinerant healer was certain that his ideas would triumph in the end, though he surely never conceived of the central role they would someday play in the defilement and belated redemption of a small town on the other side of the ocean. He was poor but carried a noble name, Theophrastus Bombastus von Hohenheim, which he further embellished several times before settling on a pen name that reflected his supreme self-confidence: Paracelsus, which was Latin for "surpassing Celsus." (An encyclopedist who lived in the first century, Celsus wrote a celebrated compendium of Roman medicine.)

Long before Basel was the cradle of the chemical industry, it was a center of learning. Interrupting a lifetime spent wandering Central Europe, the thirty-three-year-old Paracelsus was summoned there in early 1527 to treat an eminent patient, the printer Johann Froben. At the time, Froben was sharing lodgings with the even more illustrious Erasmus, the humanist theologian, who was also ailing. Paracelsus temporarily cured them both, and in gratitude the two luminaries secured an appointment for him as city physician and lecturer at the University of Basel. He quickly wore out his welcome. Forever scowling and prone to crude insults, especially when drinking wine to excess, Paracelsus never missed an opportunity to alienate those with whom he disagreed and to prophesy their ultimate humiliation.[6] At the St. John's Day bonfire on June 24, 1527, Paracelsus told his students to burn the writings of Avicenna, a Persian healer who lived five hundred years earlier and was still revered by Renaissance physicians, including those at the University of Basel.[7] Soon after, Paracelsus's patron Froben fell ill again and died, and Paracelsus became enmeshed in a lawsuit with a sick cleric who had not paid his bill. Upon losing the suit in early 1528, Paracelsus insulted the judge and was obliged to flee to avoid imprisonment or worse. He never returned and died in 1541 while treating another church official (despite having vowed never to do so again), this time in Salzburg.

What made Paracelsus so despised during his single year in Basel, besides his irascible demeanor and unlimited self-regard, was the contempt he showed for conventional medical practice. He had no patience for the Greek, Roman, and Islamic texts that comprised the canon of Renaissance medicine. "Let me tell you this, the stubble on my chin knows more than you and all your scribes, my shoe buckles are more learned than your Galen and your Avicenna, and my beard has more experience than all your high colleges," he wrote.[8] He reserved his most cutting scorn for the classical assertion that disease was caused by an imbalance of four bodily fluids, or "humors": blood, yellow bile, black bile, and phlegm, corresponding to the earthly elements of air, fire, earth, and water. In a brief manifesto he distributed in Basel in the summer of 1527, Paracelsus declared that his mission was to correct the mistakes of his fellow physicians through the promotion of new treatments of his own devising: "I did not compile them from excerpts of Hippocrates and Galen. In ceaseless toil I created them anew upon the foundation of experience, the supreme teacher of all things. If I want to prove anything I shall do so not by quoting authorities, but by experiment and by reasoning thereupon."[9]

In those few sentences, printed on a one-page handbill passed out on the streets of Basel on June 5, 1527, Paracelsus articulated a new kind of science, based not on received doctrine but on observation and experimentation. In truth, Paracelsus did not always live up to his words. His evidence-based medicine also included liberal doses of mysticism and alchemy, and some of its key tenets—including the humor-esque notion that all things are composed of mercury, salt, and sulfur (terms he did not use in their modern context)—were as wrong as the classical concepts they sought to replace. But it is also true that by rejecting dogma, by closely observing patients and taking detailed case histories from them and by experimenting with various treatments and keeping track of the results, Paracelsus became the most famous physician in Europe, shunned by peers but legendary for his cures. The spread of his ideas, especially after his death, helped to spur the scientific revolution of the late Renaissance. More than three centuries after his abortive attempt in Basel, his ideas even helped to reform mainstream medicine.

By questioning the presumed wisdom of the ancients, Paracelsus also helped set into motion a centuries-long intellectual discourse that would give rise to the chemical industry and also to the new branches of medicine—toxicology and epidemiology—that would eventually come to grips with the industry's lethal consequences in places like Toms River. These other legacies of Paracelsus stemmed from his belief that illness was chemical in nature and therefore so were its cures. In his manuscripts he referred to "alchemy," not "chemistry," but he meant it in a different way than the medieval alchemists who sought to transmute lead into gold, and his remedies were different than the bloodlettings and herbal concoctions of ancient humoristic medicine. To Paracelsus, the process of chemistry, the separation of the pure from the impure, was both a sacred act and a practical one, found in both the divine creation of the world (a separation from chaos) and the medicines Paracelsus distilled in his glass vessels. William Henry Perkin was working squarely within the Paracelsian tradition in 1856 when he separated the components of coal tar and accidentally launched the synthetic chemical industry. So was Perkin's antipode, the Basel chemist Friedrich Goppelsröder, who seven years later was the first person to investigate, and document, the terrible toll the industry was taking.

Paracelsus's ideas about pollution and disease are known because he wrote a book generally considered to be the first full-length text on occupational health: *Von der Bergsucht und anderen Bergkrankheiten,* or *On the Miners' Sickness and Other Miners' Diseases.*[10] As a boy, Paracelsus apprenticed to a smelter and thus grew up hearing the miners tell tales of the mountain gnomes who guarded their ore with poisonous vapors. Later, he returned to the mines as a physician keenly interested in learning the secrets of metallurgy and in treating workers who inhaled the noxious fumes generated by the mining and processing of ore. As with all of Paracelsus's books, long passages of *Von der Bergsucht* are indecipherable to modern readers, and his explanations for many diseases are far off the mark, including his confident assurance that the positions of the stars and moon are responsible for the shivering and teeth-chattering symptoms of mercury poisoning. Despite these errors, he capably described the symptoms of arsenic and mercury poisoning in the book and was the first to observe that pollutants can lodge in lung tissue and cause long-term disease. The "wasting disease" Paracelsus described was almost certainly lung cancer, and he even noticed that bystanders, not just industrial workers, were at risk if they spent enough time near smelters and mines.[11]

After Paracelsus's unhappy life ended in 1541, a succession of unconventional scientists, many of them fellow outsiders, would refine his ideas and contribute their own to the new science of environmental medicine. The most influential was the Italian physician Bernardino Ramazzini. Unlike his colleagues, who thought that such undignified excursions were beneath them, Ramazzini insisted on inspecting work sites himself and compiling elaborate occupational histories of his patients. Published in 1700, his *De Morbis Artificum Diatriba,* or *Diseases of Workers,* included startlingly accurate descriptions of diseases associated with fifty-two occupations, from lead poisoning of potters and mercury poisoning of mirror makers to the hunched spines and overtaxed minds of sedentary "learned men." Ramazzini's penetrating observations set the agenda for all the industrial hygienists who followed. The prescient solutions he proposed—including bathing, exercise, ventilation, gloves, masks, and the isolation of hazardous materials—eventually formed the basis of the public health regulations that would so annoy Ciba in Basel, Cincinnati, and finally Toms River.

As they forged the first solid links between illness and chemical contaminants, Paracelsus and Ramazzini began to build a new kind of medical science, one that rejected inherited dogma and instead focused on the specific experiences of individual patients—what they breathed, touched, and consumed. Freed from orthodoxy, they detected patterns that others missed, including lung damage in miners and memory loss in mirror makers. Precisely what was causing those maladies was something at which they could only guess, and many of their guesses turned out to be wrong. But what mattered was this: Disease could no longer be explained away as the uncontrollable consequence of capricious deities, jealous mountain gnomes, or humoral imbalances. And the modern equivalents of those excuses—"it's just bad luck," or "you must have bad genes" (the people of Toms River would hear both)—would not hold water, either, at least for the vast majority of diseases that were not wholly hereditary. After Paracelsus and Ramazzini, it was clear that many afflictions were a consequence of human action. Risk could be reduced by precaution, or it could be promoted by recklessness in pursuit of treasure. After the rise of the synthetic chemical industry, those choices would become starker, but they were present even in the quarries and workshops of the preindustrialized world. Paracelsus understood this, writing, "We must also have gold and silver, also other metals, iron, tin, copper, lead and mercury. If we wish to have these, we must risk both life and body in a struggle with many enemies that oppose us."[12]

By the time the huge new plant in Toms River opened for business in 1952, both Paracelsian traditions—robust chemical experimentation and careful attention to the health consequences of those experiments—were firmly established in the industrialized world. Only one of them, however, turned a consistent profit.

If the people of Toms River had any concerns about the chemical industry's arrival in their town in 1952, there is no record of it. No Paracelsus of the pinelands passed out angry handbills on Washington Street or tossed Ciba's promotional brochures into a bonfire on the beach. Instead, the general attitude, according to people who remember, was euphoric. "It was a very big deal when Ciba came to Toms River. It was *the* place to work in Ocean County," recalled Roden Lightbody, a future mayor whose brother worked at the plant. Ocean County had not thrived in the years immediately after World War II; the only big change was an influx of Jewish refugees recognizable by the concentration camp numbers tattooed on their arms. The newcomers were liberals, often socialists, which created tension with the town's conservative establishment. (Julius and Ethel Rosenberg's two young sons, ages ten and six, were living with family friends in Toms River in 1953 when their parents were executed as Soviet spies, but the boys had to move back to New York six months later when the

school superintendent acceded to complaints from parents and expelled them because they were not legal residents of New Jersey.)[13]
The immigrants were not wealthy. Many took up egg farming because
land was cheap and markets in New York and Philadelphia were near.
The closest big newspaper, the *Asbury Park Press,* began carrying two
full pages of "poultry news" on Saturday, as well as meeting an-
nouncements for B'nai B'rith, the Jewish service organization. So
many egg farms lined the old stagecoach road, now Route 9, that the
thoroughfare looked like it was permanently dusted in snow. It was
actually a coating of white feathers.

Ciba's arrival meant that poultry was no longer the only growth
industry in Ocean County. The initial 200 employees included 120
transferees from Cincinnati, but there would be hundreds of additional openings as dye production expanded. The annual payroll by
1954 would be $1.5 million. Perhaps just as importantly, the opening
of the plant meant that, for the first time since the American Revolution, Toms River would matter to the outside world. Its dyes would
ship all over the globe. The dedication ceremony on June 4, 1953,
after the plant had been operating for more than six months, was one
of the biggest spectacles Ocean County had seen in years—certainly
the biggest since the 1937 explosion of the airship *Hindenburg* at
Lakehurst Naval Air Station, just five miles to the west. More than five
hundred people attended. As another blimp from Lakehurst flew
overhead, a brass band played and a member of the Board of Chosen
Freeholders (as the county legislature was called) presented a scroll to
Ciba president Robert Kaeppeli. The main speaker was New Jersey
governor Alfred E. Driscoll, a Republican who later ran a pharmaceutical company. Driscoll told the crowd that Ciba was taking a risk in
building a huge factory and thus should be allowed to "reap the rewards" without excessive taxation or government interference—
a message that no doubt resonated in conservative Toms River.[14]

The Ciba executive who spoke that day was Harry B. Marshall,
the head of Ciba's United States subsidiary. (Kaeppeli, who was Swiss,
was relegated to scroll-receiving.) Marshall told the crowd that the
plant would *improve* the water quality of the Toms River, not pollute
it. "One of the most important aspects of this plant has been the care

taken to make it an asset to the community," he said, according to an admiring account of his speech in a local newspaper.[15] "When effluent water is released into the Toms River it will be clear and pure and in no way contaminate the stream or harbors which provide livelihood and pleasure to residents and guests of the area." By all accounts, the assembled crowd responded warmly. A booklet Ciba produced in 1953 included glossy pictures of the new waste-handling system with a caption explaining that "the purified effluent, clear, neutral and harmless to fish life, is discharged into the Toms River."[16]

There was only one small discordant note among the credulous chorus that greeted the factory's opening. On March 24, 1953, ten weeks before the dedication ceremony, a brief item appeared in the *Brooklyn Eagle*'s hunting and fishing column: "New Jersey faces another pollution peril as another chemical concern is now erecting a plant on Toms River which is geared to extract one million gallons from the river daily and discharge it back again, contaminated with sulphuric acid and pharmaceutical wastes. This would completely ruin the waters of the river and Barnegat Bay."[17] It was just one paragraph in the middle of a column published on the back pages of the sports section of an out-of-town newspaper, but the Swiss took no chances. They responded by arranging a tour of the factory for local reporters and were rewarded with a laudatory headline on the front page of the Toms River–based *New Jersey Courier:* "Tour of Ciba Disposal Plant Dispels Rumors of Pollution."[18] For Ciba, it was an ideal situation. In Toms River, the company could operate as if the past had never happened and the future never would.

The solid waste produced at the newly opened factory was generally undiluted by water or any other benign material and was thus very hazardous, despite the affectionate terminology used to describe it. For example, "filter cake" was grainy material, soaked with solvents, captured by the filters used in various steps of the dye-making process. "Still bottoms" were gummy layers of solids that formed at the bottom of kettles and other reaction vessels at high temperatures. "Clarification residues" (apparently named by someone with less imagination) were chunks of unreacted chemicals floating in the liquid dye. A major task for workers on the factory floor was to remove all that waste, often by hand. Every few hours, or at least at the end of each eight-hour shift, the line workers would scrape still bottoms, strain off clarification residue, and wash or change the filters to get rid of filter cake. Then they would have to dump the waste they had just collected.

Where to put it? Ciba made the cheapest and most obvious choice, if not the safest one. The company would get rid of its solid waste in Toms River the same way it had in Basel (at first) and the same way its competitors did it at their own plants if they were lucky enough to have sufficient room. Ciba would dump it on the factory grounds—all of it, enough to fill more than a thousand fifty-five-gallon drums every year. (By the late 1970s, the annual total would be closer to ten thousand drums.) Drums were buried in pits and trenches or just dumped in unmarked clearings within the dense forest. Often, the waste would not be contained at all. A half-mile south of the production buildings, for instance, the company dug a pit (and later two more beside it) for arsenic acid, which was mixed with lime and dumped in what came to be known as the Acid Pits. Other dumps acquired nicknames like the Moon and the Smudge Pots; there were too many to name them all. They proliferated like a pox, eventually covering more than forty-eight acres in more than a dozen locations. The exact number of dumps will never be known because company officials did not bother to map the locations of the smallest ones, which were subsequently forgotten. With almost two square miles of woods at their disposal, there did not seem to be any reason to keep track. "Back in those days I don't think anybody ever really thought about landfills, period," said Jorge Winkler, a senior executive at the Toms River plant in the 1970s and 1980s. "You just put [waste] someplace that one thought was suitable, and suitable was basically any cheap land that was out of the way."

Liquid waste at the Toms River plant could not be handled quite so blithely; there was just too much of it. As in Basel and Cincinnati, it included large amounts of sulfuric acid, anthraquinone, and other acids and solvents. To reduce its toxicity, the liquid waste was diluted with water taken from the river, typically in a ratio of about two hundred gallons of water for each gallon of chemical waste. As a result, the volume of wastewater generated at the Toms River factory was tremendous: more than two million gallons per day by the mid-1950s and five million by the 1970s. Even compared to other dye factories, the Toms River discharges were a deluge. The extra steps required to produce anthraquinone vat dyes generated almost one thousand gallons of wastewater per pound of finished dye—more than ten times as much as any other type of dye manufacturing.[19]

The river was the obvious place to dump the wastewater, but the Toms was not a wide, fast-flowing river like the Rhine or the Ohio. It was, and still is, a serpentine, languid stream, with a current slow enough to encourage cedar swamps and beaver dams. The stretch beside the factory is less than fifty feet wide, and the river stays narrow until it reaches the heart of town two miles away. There, the Toms finally widens and becomes tidal as it nears Barnegat Bay. In 1952, fishing and small-scale tourism were the cornerstones of the struggling local economy, and the river rolled right past the pillared 1850 courthouse that was the seat of power for State Senator William Steelman "Steets" Mathis, the county's political boss and, not coincidentally, an avid fisherman.

To preserve good relations with the locals, the company needed the community's namesake river to continue to look, as Harry B. Marshall put it at the 1953 dedication ceremony, "clear and pure"—or at least as clear and pure as the Toms had always looked. On that point, Mother Nature had given Ciba a gift. Upstream from the factory property, cedars grew thickly along the river's banks, and their roots lent the water a tea-colored tinge. In addition to being tinted, the river water was also murky because the sandy soil and verdant swamps added so much organic material to the water. The Toms River was a flourishing and nearly pristine ecosystem before the factory started discharging chemical waste into it, but the river had never *looked* clean to the locals. Ciba capitalized on this incongruity. Some of the news articles in those early days included comments from company officials disparaging the quality of the natural river water. One even claimed that the corrosive acids in Ciba's wastewater would make the river water taste better to fish and humans alike because in its natural state the Toms was too alkaline.[20] By the mid-1960s, that claim would be laughable, but in 1953 it was taken at face value.

Even the canard that its wastewater was "cleaning" the river did not give the company free rein simply to channel all of its wastewater into the Toms, however. The river was just too narrow, too slow, and too important to the community's identity and economy. Even in a town as deferential as Toms River, dumping two million gallons of acid-laced waste every day into a small stream that cut right through the center of town would be a provocative act. The company needed to find a way to reduce its direct discharges into the river. Here, nature bestowed a second gift, one Ciba would repay with decades of torrential dumping. Underneath the topsoil at the factory site, as it turned out, the ground consisted of highly porous sand and gravel that was saturated with groundwater starting just a few feet below the surface. This meant that the company could dig shallow wells and withdraw millions of gallons of water every day to produce dye and dilute its waste. Just as importantly, wastewater dumped directly onto the sand would seep into the ground and seemingly disappear. Every gallon that vanished into the ground was a gallon that would not have to be dumped directly into the river via a large and very visible outfall pipe. The two great benefits of porous soil—as a source of clean water and as a dumping ground for liquid waste—seem contradictory: Why would the company risk contaminating its own water wells? The answer was that company officials thought they could both pump and dump without fouling their own nest if the wastewater lagoons were a few thousand feet away from Ciba's water wells. They were wrong about that, too.

Even before the plant was built, the company's intention to dump most of its wastewater onto the ground, in a place no outsider could see, was clear to everyone involved. Privacy was crucial, one Ciba executive wrote in a 1949 memo: "There is a psychological factor involved in locating the sewage treatment area at considerable distance from the river. . . . Such area is better located in the deep wooded area, well screened from any outside observation." The same memo suggests that Ciba knew that much of its liquid waste would disappear into the sand: "we expect considerable seepage into adjacent area from ditch, which will reduce the volume of effluent carried into the river."[21] Another memo from 1949 is even blunter: "At Toms River, all objectionable sewage can be disposed of by irrigation of the sandy soil with only a minimum treatment, without difficulty and at very moderate cost."[22]

The company was far less forthcoming in describing its plans to the state health department, which could have forced improvements but did not. In a 1949 letter to the state, a senior Ciba manager named Philip Kronowitt acknowledged that solvent-laced wastewater would be disposed of "by irrigation of a suitable acreage of sand beds." But he also asserted that the volume would be "very small, probably not over a few thousand gallons per month" and would occur "at such distance from the river and neighborhood development as to make harmful seepage impossible."[23] Both of those assertions would soon prove wildly inaccurate—by 1955, his staff was telling Kronowitt that more than one million gallons were seeping into the ground *every day*—but the state accepted them at face value and issued the operating permits Ciba needed to open in 1952.[24]

The wastewater "treatment" process the company built at Toms River took full advantage of the sponge-like characteristics of the pinelands soil.[25] It consisted of a series of basins in which the factory wastewater was mixed with treated sewage, blended with lime to reduce its acidity, and then sent to what was essentially a huge man-made lake with a surface area of about ten acres and a depth of ten feet.[26] This unlined "settling lagoon" had a theoretical capacity of twenty-nine million gallons, but because there was nothing underneath it except sand and gravel, it was more like a bottomless pit. At first, company engineers predicted that solid material in the waste would settle on the bottom of the lagoon and reduce the seepage, but their self-serving assessment was quickly proved wrong. The lagoon should have filled up in a couple of months, but instead took more than eighteen months to fill because so much wastewater was seeping down through its sandy bottom into the groundwater beneath. On days when it was not replenished with fresh effluent, the lagoon's water level dropped by more than five feet. Finally, in mid-1954, the lagoon filled enough for the wastewater to begin flowing over the top of a weir and into a rectangular basin into which air was bubbled. From there, the wastewater was sent to a chlorinating basin and then into a long ditch that led toward the river. This ditch emptied into a "fish pond" that the company announced it would stock with perch in order to demonstrate that the effluent was safe before it reached the river. The pond, in turn, emptied into the Toms River through a large outfall pipe.

For all its stages and basins, however, the system provided very little actual treatment of the factory's liquid waste. The added lime reduced the wastewater's acidity and chlorine killed pathogens, but beyond that the system relied solely on dilution, aeration, gravity, and time. The system did manage to remove some solids from the liquid waste, but those eventually ended up in groundwater too. This dried sludge, including muck periodically dredged from the settling lagoon, was dumped into an unlined ten-acre gully known as the Moon, located southeast of the lagoon. But the waste did not stay at the Moon; eventually, solvents and mercury from the sludge were found in the water table as well. Chemicals that did not seep into the ground went into the river, which is why the company quietly shelved its much-ballyhooed plans to stock the "fish pond" with perch. The tremendous volume of organic waste that remained in the effluent consumed all of the dissolved oxygen in the pond water. Nothing that needed oxygen—including perch—could have lived for long in that beleaguered pond, which emptied directly into the Toms River.

As ineffective as the system was, however, it was still an upgrade from the direct dumping Ciba was doing in Cincinnati and was typical for a large chemical factory in the 1950s.[27] "The kind of treatment Ciba used was so rudimentary that it didn't stand a chance of working, but that was true almost anywhere you looked at that time," said James Etzel, a professor emeritus of environmental engineering at Purdue University and an authority on industrial waste treatment. "Based on what was in the technical literature at the time, Ciba would have known that the materials they were using, and that were finding their way into their wastewater, were not the kinds of things you'd put in your drink. Their hope was that if you discharge the chemicals [into the ground] they would just go away. As a result, they saved a lot of money on waste treatment."

None of this would have mattered, at least not very much, if the people of Toms River had gotten their drinking water someplace else. But the same characteristics that made the factory site so perfect for the company—sandy soil, a high water table, and the nearby river— also made the surrounding area the best place in town to find fresh drinking water. The township's water provider, the Toms River Water Company, supplied the entire township from a shallow well two miles downstream from the chemical plant on Holly Street. In 1953, Toms River Water began using a second well at the same spot. The water company did not directly tap the river, but its two wells never ran dry because their constant pumping actually pulled river water out of the Toms. The water seeped through the riverbank and replenished the underground reservoir tapped by the two shallow wells, which were situated just a few hundred feet away.

Not everyone in town was a customer of Toms River Water. Many relied on their own backyard wells, including homeowners in the new Oak Ridge Estates subdivision, located about seven hundred feet east of the factory's sludge dump and a half-mile from the settling lagoon, both of which had no protective liner and therefore leaked copiously. So if you lived in Oak Ridge Estates and got your water from your own well, or if you lived anywhere in town and were a customer of the water company, it was quite possible that you would eventually drink whatever Ciba was dumping into the ground and river.

Many years later, the question of whether the company should have known, even in the 1950s, that its waste-handling practices were dangerous was examined in a hard-fought court battle. After two tri- als and eleven years of litigation, a New Jersey superior court judge ruled in 1998 that Toms River Chemical had followed typical prac- tices of the 1950s and 1960s and had not intended to contaminate the drinking water.[28] Citing the testimony of a Ciba-Geigy expert witness who was a former top New Jersey environmental official, the judge declared that the factory's wastewater treatment system at the time was one of the most advanced in the state among large manufactur- ers.[29] Therefore, the judge ruled, its insurers would have to reimburse Ciba-Geigy for the full cost of cleaning up the company's pollution between 1952 and 1984, when the insurance coverage ended. The total cost was reportedly more than $400 million.

But even if the pollution had not been deliberate, as the judge ruled, should the company have known better, even in the earliest years of the Toms River plant's operation? After all, Ciba had been struggling with water pollution issues for decades in Basel, Cincinnati, and elsewhere, and there had already been many lawsuits over groundwater contamination from liquid waste dumped on sandy soil, including in New Jersey.[30] By 1934, the standard textbook on chemical factory design was warning that "disposal of waste into a stream or river is no longer satisfactory" and that any factory that relies on "seepage through the ground" should make sure that water supplies are unaffected "in order to avoid trouble from neighboring plants or the local authorities."[31] "The risk to groundwater was very well known; any plant manager who was on top of his job should have been alert to these issues," said Louisiana State University anthropologist Craig Colten, an expert on the waste-handling practices of that era.[32]

Four centuries before Ciba decided to dump its waste into Toms River's water supply, Paracelsus tried to explain the dangers he had perceived in the poisons that could not be seen, touched, or sometimes even smelled. Writing with characteristic fervor in his revolutionary treatise on the diseases of miners, he declared, "For man should know this, that in the insensible things there is a uniform enmity. For the exhalations of such minerals also kill us in the same manner as the crocodile with his breath corrupts and kills man."[33] By 1952, there was every reason to believe him. Yet the men who ran the huge new chemical plant in Toms River were content to watch the daily dross of dye manufacture—by the mid-1950s, about two million gallons of wastewater and several large drums brimming with solid waste—disappear into the sand and the river. They did not worry about where it would go next, and neither did anyone else. In a different sense of the word, they, too, were insensible.

≈

CHAPTER THREE

First Fingerprints

On steamy summer afternoons, the water felt blissfully cool. The cedars grew thickly along the riverbanks, but there was a small beach for sunbathing. Better still, the swimming hole was deep in the woods and hard to reach, so adults and little kids stayed away. For a generation of Toms River teenagers in the 1950s, the shady pool was a shared secret of summer life, a place to take your girlfriend and sneak cigarettes and beer. "We called it the Black Lagoon," remembered George Woolley, who grew up in town but had only a vague idea of what was on the other side of the chain-link fence just upstream. "We were basically swimming in Ciba's effluent, but we were oblivious to that."

Don Bennett swam in the river too. He and Woolley were high school friends (Woolley ushered at Bennett's wedding), though they would later drift apart as they assumed prominent roles on opposite sides of the drama that would engulf their town. Woolley, who liked science, went to work at the chemical plant in 1964 and stayed there for thirty-two years. Bennett, who liked to write, became a newspaperman and worked for local papers for even longer. "We all swam there, despite the smell," Bennett remembered. "I was just a kid, but I knew there was something awful there. It was only a third of a mile downstream from where the company was discharging stuff into the river, so there wasn't much dilution. We still swam in it, sure we did. We were kids."

Three summers later, in 1962, George Woolley returned to the Black Lagoon on a whim, a twenty-year-old in search of nostalgia and a cool respite from a hot day. The pool was nothing like what he remembered. The water was dark and the stench foul. Against his better judgment, Woolley took a quick swim and swallowed some water, which tasted awful. Surfacing, he noticed purplish foam clinging to his body.

No one swam in the Black Lagoon anymore.

Once Toms River finally started to grow, it did not stop. Between 1950 and 1960, the population of Dover Township (its formal name) more than doubled to 17,000, while Ocean County's population grew almost as fast, to 108,000. The growth would accelerate further in the 1960s, when Ocean was one of the fastest-growing counties in the United States and all of Toms River seemed to be perennially under construction. Ciba stoked the fire by expanding steadily. In 1955, the company's two partners in Cincinnati, Geigy and Sandoz, bought minority stakes in the Toms River plant from Ciba. The three firms were envisioning big things from their mini-city in the pinelands, which they rechristened the Toms River–Cincinnati Chemical Corporation.[1] The factory had suffered through a few lean years in the mid-1950s because of weak demand for vat dyes, but by the end of the decade it was firing on all cylinders and was about to begin a major expansion. Already, with a payroll of almost five hundred in 1959, the chemical plant was the county's largest private employer, by far.

The completion of the Garden State Parkway in 1955 ended the regional isolation that had begun in 1812 when a hurricane sealed Cranberry Inlet and crippled the marine trades in Toms River. The huge highway project, which took eight years to finish and included 172 miles of meticulously engineered roadway, was an economic boon because it drastically reduced the travel time to Toms River from the big cities to the north.[2] During most of the town's long somnolence, the trip from New York could take a day or more. With the completion of the parkway, it took less than two hours. Weekend visitors raced down the highway to beach communities in Ocean County, so real estate prices soared. The Barnegat Peninsula and Long Beach Island went from attracting a few thousand tourists every summer to more than one hundred thousand. Some visitors stayed and looked for year-round housing in the towns they drove through on the way to the shore, including Toms River. Land prices rose as egg farmers sold out to developers. The character of the township began to change. The newcomers arrived in wave after wave, and Toms River gradually stopped being a place where everyone knew your family and your place in the social order. The residents of the new subdivisions were rootless, united only by a shared conviction that their town's rapidly rising property values and wholesome image must be zealously guarded, come what may.

It was a golden time for anyone lucky enough to land a job at the chemical plant. The company was on a hiring binge; the payroll hit one thousand in the summer of 1961 and kept rising. The pay was outstanding: almost three dollars an hour for an experienced equipment operator. John Talty, at age twenty, joined the company in 1960 after proving to his interviewer that he could calculate the sides of a right triangle using the Pythagorean theorem, a feat he vaguely recalled from high school geometry class. At $78 a week, his salary as an entry-level laboratory assistant was unheard of in Ocean County for someone without a college degree. Talty was ecstatic. "In 1960, it was the place to go if you lived around here," he remembered. Talty met his future wife in the lunchroom (she had been hired a week after him) and would work at the factory for thirty-six years until it finally closed. In a community where housing costs were already soaring to levels that were out of reach for many blue-collar workers, a job at the plant was a secure route to prosperity. John Talty, George Woolley, and thousands of others would repay the company with their unwavering loyalty, come what may.

Its payroll was not the only way the company spread money around the community. Its executives were the driving force behind a nine-year campaign to build Ocean County's first hospital. Community Medical Center opened in 1961 on a choice piece of land near the parkway, just down the road from the factory on Route 37. The following year, the company bought a run-down nine-hole public golf course on an old dairy farm near downtown, built a new clubhouse and swimming pool, and reopened the Toms River Country Club, whose sixty-seven acres quickly became the private playground of the town's business and political elite. Company executives sat on the boards of the First National Bank of Toms River, the Ocean County Boy Scouts Council, the Chamber of Commerce, and the Toms River Yacht Club and were reliable contributors to the local Republican Party (and occasionally the Democrats). The company's senior executives were, by all accounts, deeply involved in town affairs, even if they were not quite *of* the town. Many of the top managers were from Switzerland, and their German accents and advanced educations (many had doctorates in chemistry) stood out.

As smoothly as things were going in Toms River, Ciba was having trouble elsewhere. Since 1947, the company had made epoxy resins and adhesives in Kimberton, Pennsylvania, thirty miles northwest of Philadelphia. The factory was on top of a hill, the village in the valley below. Anything Ciba discharged would flow downhill toward the streams that supplied Kimberton with its drinking water, but that did not deter the company from digging a series of unlined seepage lagoons to serve as waste dumps.[3] By 1957, there were eight lagoons on the hill. That same year, engineers working for the local water company detected carbolic acid and salt, two of the principal waste products of resin manufacturing, in the streams at the bottom of the hill. Facing accusations that it had contaminated the water supply, Ciba agreed to dig new wells for the village far from the tainted streams. That was enough to stave off a lawsuit, but it did not mollify Kimberton residents, who did not hide their hostility.[4] In Cincinnati, meanwhile, the two factories Ciba had co-owned with Geigy and Sandoz since 1920 were increasingly unprofitable, in part because of the city's newfound insistence that the factories treat their waste before dumping it into the public sewer system and the Ohio River.

In 1958, Ciba made a portentous decision: It closed all three factories (two in Cincinnati and one in Kimberton) and shifted the work to a location where relations with the locals were far more amicable and where there was plenty of room for waste dumping. The newly renamed Toms River Chemical Corporation (Cincinnati was dropped from the title) would no longer be a place that made only vat dyes.

The new signature products at Toms River would be azo dyes, a family of dyestuffs that traced its chemical lineage back one hundred years to the fuchsine magenta that was the first competitor to William Perkin's mauve. Azo dyes were not quite as durable as vat dyes, but the colors were brighter and more varied, and the dyes were cheaper to produce. They required just three or four manufacturing steps, compared to seven or more for vat dyes. Azo dyes had been hugely popular since the 1880s and had been manufactured for decades in Cincinnati. Now manufacturing was moving to Toms River, to a cluster of new buildings north of the original vat dye manufacturing area.

By 1960, the azo complex was finished: four new production buildings plus two laboratories and two giant warehouses. Unlike vat dyes, azo dye production required phosgene gas, a colorless, sweet-smelling vapor that could kill on contact and was the most deadly gas used in World War I (first by the Germans and then by the Allies). The gas was so dangerous that Toms River Chemical eventually built an isolated, bunker-like building with reinforced concrete walls and special one-ton holding tanks whose steel walls were a quarter-inch thick, just to store phosgene. It was a place no employee ever wanted to visit, especially old-timers who had fought in the Great War. Nearby was another small building used only to store another ingredient of azo dyes, benzidine, which had already wreaked carcinogenic havoc at the Cincinnati Chemical Works, though very few workers at Toms River knew it.

A second new manufacturing cluster at Toms River Chemical, east of the original vat dye area, was just as dangerous. There, also in 1960, the company constructed another large factory building— Building 108—and an attached warehouse for the production and storage of an entirely new slate of products: resins, brighteners, insoluble pigments, and a variety of specialty chemicals, most of which had previously been made in Kimberton. In addition to producing vat and azo dyes to color fabric, the Toms River plant would now make products to color plastic, paper, carpet, leather, food, and even detergent, as well as super-strong adhesives and protective coatings. Almost all of those products required heavy use of highly volatile solvents, particularly toluene, xylene, trichloroethylene, and, most alarmingly, epichlorohydrin. The latter was a quadruple threat: flammable, very irritating to eyes and lungs, capable of burning through skin, and a likely carcinogen.[5] Between 1961 and 1965, Toms River Chemical manufactured its own supply of epichlorohydrin in a small structure next to another small building the company erected in 1957 to make anthraquinone, the keystone ingredient of vat dyes that was also highly flammable and a carcinogen, though its cancer-causing properties were not known at the time.[6]

In the early years of the plant, most of the men who went to work at Toms River Chemical, even the ones who transferred from Cincinnati, had only a vague understanding of the dangers posed by the chemicals they were handling. The same was true of the women, many of whom worked as secretaries in the adjacent office buildings and had to get used to the powerful smells. Ray Talty was working as a stockboy at a local A&P when he got a job at Toms River Chemical in 1962, following in the footsteps of his older brother, John, who had been hired two years earlier. Ray, like his brother, met his future wife at the factory. Jackie Talty gave up her job in 1971 to raise their children, but before she did she went through many pairs of nylon stockings. Any secretary sent on an errand to one of the production buildings, where the air was thick with solvent and acid vapors, would run the risk that her nylons would melt on her legs. "You'd get these holes in your stockings," Jackie Talty remembered. "If you complained to one of the bosses, they'd say you could put in a receipt for a new pair of stockings, and then they'd say, 'Be grateful you have a job.' I don't think anybody ever put in a receipt." Said Ray Talty: "Early on, we didn't really know much. In the sixties, if you said anything the supervisors could be pretty sarcastic. Some of them would say, 'What do you think this is, an ice cream factory?'"

Because the Talty brothers and George Woolley all worked in laboratories at Toms River Chemical and not in a production building, they had a chance to see the entire factory complex. They quickly learned which places to dread. Building 102, the main vat dye production area, "was all benzene, chlorobenzene—all these solvents in this really big dark building. It was the worst building that you could work in or even just walk through," said Ray Talty. Explosions were a constant risk: On December 22, 1960, for example, on the second floor of Building 102, a pressurized kettle containing an explosive mixture of eight thousand pounds of tar, dye, and nitrobenzene burst and crashed through the floor to the ground level, blowing out dozens of windows and hurling chunks of iron forty feet in all directions. Workers "took cover wherever they were, some lying flat on the floor," according to a company memo that blamed operator error for the accident.[7] George Woolley was especially worried about the building where resins and specialty chemicals were made. "Some of the stuff that came out of Building 108 was very bad: epichlorohydrin, ethylene oxide and also toluene and xylene, but the epichlorohydrin overshadowed everything, it was bad stuff," said Woolley. The phosgene gas was intimidating too. Even the labs could be dangerous places; the technicians were handling chemicals they knew almost nothing about. One day in the early 1960s, John Talty was accidentally soaked with dimethyl sulfate, a derivative of sulfuric acid that was later determined to be a likely carcinogen. The foreman sent him to the nurse, who told him to go home and take a shower. Afterward, while lathering up for a shave, he discovered that bleeding sores had opened up all over his face. Soon afterward, he decided to get involved in the factory's union; his brother and Woolley later joined him.

Lurking behind the more conspicuous fears about fires, explosions, spills, and melting stockings was something darker. "You would hear stories about people working in the plant who got cancer, even back in the sixties. It wasn't something people talked about back then, but it was there," said Woolley. Most were older workers who had transferred from Cincinnati or Kimberton and had thus been exposed for decades. A few were much younger, including Jackie Talty, who was diagnosed with breast cancer when she was twenty-six, four years after quitting the factory to have children. Another twenty years would pass before the first tentative efforts to find out whether the number of cancer cases among employees of Toms River Chemical was unusually high. No one wanted to ask during the go-go 1960s, when the community and the company were growing like crabgrass in full sun. It was one thing for managers and employees to acknowledge the risk of fires and spills; it was quite another to face the possibility that merely breathing the factory's air was a serious long-term health risk. "They were much more concerned about acute safety, about somebody getting burned or blown up, than they were about the exposures to toxic fumes day after day," said Woolley. "As long as you didn't drop dead on the spot, they didn't care."

If Paracelsus and Bernardino Ramazzini essentially invented occupational medicine, laying the foundations for all the environmental health research that would follow, it fell to others to turn their unconventional observations into a disciplined science. One of the first to do so was a physician with the mellifluous name of Percivall Pott, a short-statured dandy who always wore his powdered wig in public. He prospered as a surgeon catering to London's upper classes in the mid-eighteenth century, but Pott was also an innovator who developed new surgical instruments and dressings as well as improved treatments for ailments ranging from hernias and spine injuries to skull fractures. He had a deep distrust of conventional surgical practices, many of which were akin to sanctioned butchery. In 1756, when he was forty-three years old, Pott fractured an ankle after being thrown by his horse. He insisted on lying on the cold ground for hours so that he could direct the construction of an improvised stretcher made from a carriage door and the poles from a sedan chair. Having thus devised a safe way to reach the hospital, Pott managed to avoid the standard treatment of the time, which was amputation.[8]

Despite his cultivated airs, Pott had the social conscience of a man who had grown up poor. His father died when he was just a toddler, so Pott relied on patrons to finance his medical education. He later returned the favor by housing destitute students and by treating the poor as well as the wealthy (his famous patients included the writer Samuel Johnson and the painter Thomas Gainsborough). As chief surgeon at St. Bartholomew's Hospital, founded in 1123, Pott was moved by the dire circumstances of London's chimney sweep boys. As he would later write, "they are thrust up narrow, and sometimes hot chimneys, where they are bruised, burned and almost suffocated; and when they get to puberty, become liable to a most noisome, painful, and fatal disease."[9]

After the Great Fire in 1666 destroyed large parts of the city, London's chimneys were rebuilt as long and narrow flues with tight turns and sharp angles. The passages were so narrow—sometimes no more than nine by fourteen inches—that only young boys, typically four to seven years old, could fit inside.[10] Many were half-starved orphans exploited by avaricious masters, who justified the abuse on the grounds that dirty chimneys were a fire risk. Cleaning the flues was agonizing work; it was not uncommon for a "climbing boy" to get stuck and suffocate. Clothes increased the risk of becoming trapped, so sweeps in England generally worked in the nude, scraping their knees and elbows raw as they squirmed inside the brick labyrinths.[11] They labored in a hellish murk of dust and smoke and often had to maneuver while carrying a mortar and trowel so that they could fill in cracks between the bricks and chip away the hardened soot. Because the boys bathed extremely rarely—in some cases just once a year—the soot clung to their bodies and was ground into any exposed abrasions.

What appalled Percivall Pott even more than the nightmarish working conditions, however, was the "noisome, painful and fatal disease" that many former chimney sweeps developed after puberty: cancer of the scrotum. Exactly when and how Pott first made the connection between the cancer patients he was treating at St. Bartholomew's and their former occupation as chimney sweeps is unknown; he did not elaborate on that point in his extensive writings. We do not even know how many young men with scrotal tumors Pott treated before he was convinced. But it was a landmark moment nonetheless: the first medically documented identification of a cancer caused by a pollutant. Pott was not the first to speculate correctly about a cause of cancer; Ramazzini, for example, in 1700 rightly guessed that nuns had higher rates of breast cancer because of their childlessness, though he did not know why.[12] But Percivall Pott moved from guesswork to meticulous observation and documentation of cases.

In 1775, when Pott finally wrote up his conclusions in an essay that began with a tribute to Ramazzini, he evinced little doubt about the cause: "The disease, in these people, seems to derive its origin from the lodgment of soot in the rugae [skin folds] of the scrotum." James Earle, in a biography of his father-in-law he published thirty-three years later, was even more specific: "This species of cancer, which Mr. Pott has so accurately described, appears to be produced by some peculiar acrimonious quality in soot, when incorporated and fermenting with the secretions on the skin on some persons, whose constitutions are disposed to undergo a certain change, or receive a new modification of their inherent properties."[13] In other words, there was something about soot, or something *in* soot, capable of causing cancer, and there was something about certain people, or something *in* certain people, that made them vulnerable to the assault.

In that single groping sentence, Percivall Pott's son-in-law presaged the next two centuries of research into the carcinogenic effects of organic (carbon-based) chemicals. Chemists would spend the next two hundred years sorting out the components of soot and other byproducts of incineration. They would first separate those components into broad groupings like "coal tar" and "soot," then smaller families such as "amines," "phenols," and "polycyclic aromatic hydrocarbons," and finally a galaxy of specific molecules such as anthraquinone, benzidine, and benzo(a)pyrene. Most of the scientists engaged in those tasks would do so in the hopes of creating new manmade hydrocarbons—initially for dyes and then for medicines, plastics, and thousands of other uses. But a few unconventional researchers would choose the harder path, taking up the gauntlet thrown down by Paracelsus and Pott by searching for the "peculiar acrimonious quality" that made some of those manmade compounds so dangerous to the humans who handled them.

In the early years of the chemical plant in Toms River, it had been easier for managers to ignore or at least downplay signs that the factory was polluting groundwater and the river. But as the company expanded, the uncomfortable reality could not be avoided. The waste basins sometimes overflowed, and even when they did not, huge volumes of wastewater percolated down into the aquifer because most

of the holding ponds were unlined. Some of the worst chemicals used
at the plant—including arsenic acid, which was supposed to be kept
out of wastewater at all times, were sometimes poured directly into
sewer drains in incidents the company invariably would later describe
as mistakes.[14] The wastewater system was not meeting the standards
laid out in its permit, though there was still no indication that the
state would enforce those standards.[15]

More ominous for Ciba were the signs that it had fouled its own
water supply after just two years of operation. As early as the summer
of 1954, employees complained that the water in the drinking fountains smelled and tasted awful. A few weeks later, the factory's newly
hired supervisor of waste disposal, Morris Smith, discovered the reason: Chemicals seeping out of an unlined wastewater basin had contaminated one of the wells that supplied the factory with fresh water.[16]
The well was 100 feet deep and 850 feet north of the closest wastewater lagoon—a startling indication of how quickly the liquid waste
was moving through the porous soil. By 1956, Smith had determined
that the chemical plume had spread more than two thousand feet
from the lagoons and was expanding in all directions, including
toward the river.[17] Two years later, a consulting firm brought in by the
company confirmed that all of the wastewater basins were leaking,
even the lined ones; their supposedly impermeable asphalt bottoms
had "become porous because of the corrosive nature of the waste
water."[18] Worst of all, test wells at the eastern edge of the property
showed that the plume of contaminated groundwater had spread beyond company-owned land and was creeping southeast toward the
fast-growing Oak Ridge Estates neighborhood. The Toms River
Water Company was gradually extending its mains into the subdivision, but some Oak Ridge families still got their drinking water from
shallow wells in their backyards.

Just seven years had passed since the chemical plant had opened
for business, but it was already obvious that even two square miles of
land was not enough to contain its colossal waste flow. Something
was going to have to change, especially after Ciba decided to close the
Kimberton and Cincinnati plants in 1958 and shift more production
to Toms River. If the company could not cope with the waste byprod-

ucts of anthraquinone vat dye manufacturing, how could it possibly deal with the additional waste from azo dyes, epoxy resins, and all the other specialty chemicals that would soon be made at the newly re-named Toms River Chemical Corporation? The wastewater from azo dyes would be brightly colored and highly acidic, and the resin opera-tion would add many new toxic compounds to the effluent. It could not be hidden or ignored.

The company could have responded to this challenge by redesign-ing its processes to reduce the amount of waste or by building a mod-ern wastewater plant to provide more than rudimentary treatment. One of Ciba's competitors, American Cyanamid, in 1958 had built an activated sludge treatment plant at its factory in Bound Brook, sixty miles north of Toms River. Ciba had considered building a similar one in Toms River as early as 1955, but building a modern treatment system would have been expensive.[19] Even merely lining all of the ex-isting basins with asphalt would have cost $100,000 and would have required shutting down the entire plant for three weeks.[20]

Toms River Chemical found a much cheaper solution: The com-pany added a second liner to its initial holding basin and then dug a new set of unlined lagoons at the eastern border of its property, right beside the Toms River. Unlike the old settling lagoon, chemicals seep-ing out of the five new lagoons would not threaten the company's re-maining water wells, which were a half-mile away. Instead, waste leaking out of the new lagoons would go straight into the adjacent river.[21] The riverside location was ironic, because the new lagoons were dug on the site of the "fish pond" that had been touted as an environmental showpiece in 1952 but had never been stocked with fish because of doubts that anything could survive. Now the pond was being bulldozed, enlarged, and transformed into yet another dump for barely treated waste. The State Department of Health could have insisted on a more effective treatment system, just as it could have back in 1952. But state officials did no such thing; instead, they rubber-stamped the company's plan, concluding that it was consis-tent with the prevailing industry standard, and issued a new permit.

With the completion of the new lagoons by the river in 1959, virtu-ally all of Toms River Chemical's liquid waste was now ending up in

the beleaguered Toms. By now, the narrow stretch of river by the factory was little more than "a handy receptacle for chemical wastes left
over from our manufacturing processes," as a company newsletter
candidly put it in 1961.[22] The company had found a way to protect its
own water supply, but more chemical pollution than ever was heading
downstream toward the riverside wells on Holly Street that supplied
the town with its public drinking water. It was a nice arrangement for
Toms River Chemical, but not for the citizens of Toms River.

Wastewater was only the most obvious of the company's pollution
problems. Solid waste dumping and smokestack emissions, both relatively minor issues in the factory's early years, suddenly were big concerns starting in 1960, now that the plant was making azo dyes, epoxy
resins, and a slew of arcane specialty chemicals. The types of waste
were familiar: gummy distillation residues, solid chunks of unreacted
chemicals, solvent-soaked filter cake, and chalky earth mixed with
waste. Instead of just a few drums each day, however, the company
was now filling dozens of fifty-five-gallon drums every shift. The azo
dye manufacturing process was not as wasteful as vat dye production,
but it did generate a half-pound of filter cake and two-thirds of a gallon of mother liquor (a concentrated liquid residue) for every pound
of azo.[23] There were two new waste sludges to deal with, too.[24]

All that sludge and all those fifty-five-gallon drums would have to
go somewhere; once again, Toms River Chemical selected the cheapest possible destination: an open pit. This time, company managers
could not claim to be ignorant about the environmental consequences.
By 1960, they knew that the company had wrecked several of its own
water wells by dumping liquid waste into unlined lagoons, and they
also knew that the plume of underground contamination was expanding in all directions. Similar incidents elsewhere had prompted
Dow, Monsanto, and DuPont to experiment with new treatment processes and off-site transfers of highly toxic wastes instead of relying
entirely on open-pit dumping on their own properties.[25] Not in Toms
River, however. Toms River Chemical would not build a modern,
lined landfill until it was finally forced to do so in 1977.

The company did not bother digging a new pit for its newest dump.
Instead, it used the leaky settling lagoon, which was dry now that the

wastewater had been re-routed to the new lagoons by the river. Now azo sludge and a stream of drums from Building 108 would be carted to the shallow basin and dumped. To conserve space, workers used earth moving equipment to mash the drums before burying them. The drum-crushing operation extended the life of the dump, which was used until 1977, but it also ensured that the chemicals inside the drums would leak even faster into the water table—just as liquid waste had leaked during the site's previous incarnation as a settling lagoon.

As laboratory assistants in the 1960s, John and Ray Talty and George Woolley visited the dump frequently, usually lugging five-gallon cans filled with potent solvents and mother liquor residues. Their destination was the southern edge of the basin, where several cylindrical concrete tubes, or standpipes, had been embedded in a newly poured concrete slab. The standpipes (the workers called them the Smudge Pots) were open all the way to the ground, which meant that whatever was tossed inside would go straight down to the sandy soil. At least once a day, someone from the lab would make the long trek to the Smudge Pots. More than 120,000 gallons of undiluted chemicals were dumped this way until the practice finally ended in 1970.

"We would take these cans out there and pour the stuff in," Ray Talty remembered, "and then my responsibility was to try to burn off all of the solvent." Talty would light a piece of paper with a match and toss it inside the standpipe. Often, however, the mixture was too watery and would not ignite. "When that happened," he said, "you'd just leave it alone, because it would go right into the ground anyway."

Elsewhere at the plant, other chemicals were burning on a much larger scale. The newly expanded factory was not just polluting water and soil; it was now fouling the air too. That was a change from the years when vat dyes were the only product. Explosive solvents were such important ingredients in vat dyes that it was too dangerous to try to incinerate leftover waste. Now, however, the factory was making a much wider variety of products, some of which did not use so much solvent. As a result, there was a large smokestack on top of the new Building 108, and a smaller one on top of one of the azo buildings. At a glance, experienced employees could look up at the sky and tell what kinds of chemical reactions were occurring based on the color of the smoke rising from the various stacks. Yellow smoke from the azo complex signaled a bromination reaction. Reddish brown smoke from Building 108 meant that a resin nitration was under way.

The factory's neighbors in Oak Ridge and other nearby neighborhoods could not read the smoke signals, but they could see the smoke—and smell it. Almost as soon as the new production buildings were finished, Toms River Chemical started getting complaints about odors, and the complaints increased whenever a plume of colored smoke was visible. It was the first negative attention the company had received since coming to Toms River, and its managers responded by making the pollution less visible without actually curbing it. Instead of adding expensive pollution-control equipment, Toms River Chemical merely adjusted its round-the-clock production schedule so that, as often as possible, the colored smoke would only emerge at night. Jorge Winkler, the Swiss chemist who came to the plant in 1967 and eventually became its director of production, explained the practice this way: "When it was feasible, it was better to do these reactions in the evenings or at night because of the visual pollution."

Not everything, however, could be camouflaged. By the 1960s, the Toms River Chemical Corporation was generating too much and too many types of chemical waste to escape notice. That was why George Woolley's attempt at a nostalgic river swim in 1962, after a three-year absence from the shady pool that had been his summer hangout as a teenager, had gone so badly. The river water looked different and smelled different because it *was* different. By 1962, the company was discharging twice as much wastewater into the Toms as it had in 1959, and azo dye waste was now a major component. Azo dyes were soluble in water and could color a slow-moving stream almost as easily as they could color a silk blouse or a woolen suit.

The tinted river water was an early and indistinct fingerprint. It was one of the first signs—along with the contaminated water wells, the colored smoke, and the foul smells—that the people of Toms River, and the employees of the Toms River Chemical Corporation, might come to regret their unquestioning support of the fast-growing company that was reshaping their town so quickly and in so many ways, not all of them yet apparent.

CHAPTER FOUR

~~

Secrets

The river ran low during the parched summer of 1965, the driest in Ocean County in a century. It was so low that about one-sixth of the flow under the Route 37 bridge originated not as rain or groundwater but as wastewater discharged by the Toms River Chemical Corporation two miles upstream. Below the factory's outfall pipe, the river was tinted and frothy, and a strange dark-brown fungus clung to its banks. The stench was obnoxious, but that was nothing new. The Toms had stunk for the previous four summers, too. What was new was that the odor no longer came solely from the river and the waste lagoons at Toms River Chemical. When residents of the nearby Oak Ridge neighborhood turned on their kitchen taps, they could smell it in their household water now too—the water that was piped to them by the Toms River Water Company, which operated three shallow wells (two newly dug) beside the river on Holly Street, two and a half miles downstream from the factory.

The first people to make the connection between the malodorous tap water and Toms River Chemical were plant employees, many of whom lived in Oak Ridge. For them, the smell was nauseatingly familiar. They knew it from the factory's own drinking fountains, which drew from company wells that had been tainted with dye wastes since the mid-1950s. Almost no one drank from the fountains at Toms River Chemical—not more than once, anyway—and now the familiar smell was in their water at home, too.

Jim Crane, the company's manager of chemical engineering, lived in the neighborhood and was among the first to notice the odor at home, while taking a shower on a scorching July day in 1965. It must have been a supremely frustrating lavation for Crane; he had been try-ing to cope with an escalating series of pollution problems at the plant ever since 1959, when the production lines for azo dyes and res-ins shifted to Toms River. Crane had no special training in environ-mental matters; he had supervised DDT manufacturing in Cincinnati, and his main job at Toms River was to make the manufacturing pro-cess as efficient as possible. Because Toms River Chemical had no environmental department, however, coping with the manifold conse-quences of the plant's burgeoning waste discharges was an unwel-come part of Crane's portfolio.

The Swiss had hoped that dilution would mask its pollution of the Toms River, just as they had hoped it would in the Rhine and the Ohio. But the natural flow of the Toms averaged just 120 million gal-lons per day (about five hundred times less than the Ohio) and dropped to only about thirty-six million gallons when the weather was hot and dry.[1] That was not nearly enough volume for significant dilution. Toms River Chemical was now not only discharging five million gal-lons of wastewater directly into the river every day, it was also with-drawing at least five million gallons of clean upstream water to cool equipment (among other purposes) before returning it to the Toms. This "cooling water" picked up nitrobenzene and other chemicals during its labyrinthine journey through the plant. In addition, an-other two hundred thousand gallons of chemical-laced groundwater flowed off the property every day and seeped into the river through its sandy banks. The feeble stream was simply overwhelmed by the del-uge of industrial waste.

Signs of distress appeared almost immediately after full-scale pro-duction of azo dyes began in 1960. Unexpectedly high bacteria counts in the factory effluent (probably from sewage) forced the company to start adding chlorine to the wastewater, which increased the "medicine" odor in the Toms River. On one hot summer day in 1960, the chlorinator broke down and bacteria counts spiked in the river, forcing the closure of several bathing beaches miles downriver.[2] The following spring, Crane warned his bosses that "this could be very obnoxious during swimming season"—and it was.[3] That summer, inspectors from the state Division of Fish and Game paid the first of many visits to the factory. After accompanying the inspectors to view the brown fungus below the outfall pipe, a company manager, Al Meier, wrote a candid memo to his superiors. "As much as I personally enjoy swimming in fresh water, I would not swim in the waters below our plant effluent even if I were paid a reasonable sum," he wrote, adding that he expected the state agency to soon launch a formal investigation. "I feel that we are constantly skirting on the thin edge regarding our waste water treatment problems," he concluded.[4]

By the summer of 1963, Toms River Chemical was over that thin edge. After the local papers published reports of dead fish floating in the river, organized opposition to the company appeared in town for the first time, in the form of a new group called the Ocean County Fish & Game Protective Association. The group blamed the factory discharges for killing marine life, including perch and flounder found covered with sores seven miles downstream. The group found allies in one of the local papers, the *Ocean County Sun,* which publicized the fish kills, and in Philip Maimone, a developer who also owned a Cadillac dealership. Back in 1949, Maimone had sold Ciba most of the land for its factory complex. He still owned more than six hundred acres of riverfront property and now protested that the company's pollution was wrecking his investment.

Whether Toms River Chemical was really responsible for fish kills occurring miles downstream was unknowable, but there was no doubt that almost nothing could live in its undiluted effluent—especially during hot summers. Gill-breathing fish cannot survive without dissolved oxygen, but carbon-based liquid waste—including sewage and organic chemicals—contains microorganisms that are voracious consumers of dissolved oxygen. The bad news for local fish, and local fishermen, was that bacteria and other microbes thrived in Toms River Chemical's effluent. By the summer of 1963, microbes downstream from the factory outfall were consuming oxygen at a rate more than ten times higher than in the unpolluted upstream stretches and more than three times higher than state rules allowed. The "BOD" rate, for biochemical oxygen demand, was equivalent to a city of thirty thousand people dumping their raw sewage into the river.

Fish that managed to survive in the polluted river and were then caught by anglers had a strange taste when cooked, many local fishermen insisted. Because of their complaints, the state Fish and Game Division (which by 1963 had finally launched the formal investigation Meier had grimly predicted two years earlier) conducted a "taste test" to see if white perch caught in the Toms had a different taste than those caught in a much cleaner river in South Jersey, the Mullica. The results were ambiguous, perhaps because Toms River Chemical found out about the test ahead of time and arranged to postpone all "unusual or unnecessary losses or discharges" from the factory until the test was over.[5]

Facing something new and deeply disconcerting—public criticism—company managers decided, for the first time, to acknowledge publicly what had been obvious for years to anyone in town with a working sense of smell: Toms River Chemical had fouled the iconic river that lent its name to the town and the company. This time, there were no absurd claims that the factory's effluent had somehow improved the quality of the river water. Instead, the general manager of the factory (by now it was a Swiss national named Robert Sponagel) acknowledged that "at times our wastes add color and medicinal odors" to the river. He said that the company was considering building a "chemical recovery plant" to reduce its pollution but made no promises that it would eliminate odors and discoloration. Reminding reporters of the company's annual payroll of $8 million (about $56 million in today's dollars), Sponagel added, "If the responsible people of this area want a prosperous community with continuing growth, we must realize there will be changes in our natural surroundings. Very few of us are willing to live like the Indians, in spite of our idyllic dreams."[6]

Many in Toms River seemed inclined to agree with him. Only the old-timers remembered when the river was clean and the fishing abundant, but they also remembered when land was all but worthless and the moribund local economy subsisted on egg farming and a few summer beachgoers from Philadelphia. Now land prices were soaring, and the economy was expanding almost as rapidly as Toms River Chemical's workforce, which by 1964 was nearly twelve hundred strong. Certainly the State of New Jersey showed no inclination to spoil the party. State officials not only failed to enforce the terms of their own discharge permits (which banned "detectable odors," among many other ignored provisions), they actively advised Toms River Chemical on how to stymie its critics—encouraging the company to adopt a policy of "complete silence" because its news releases "are only agitating people and supplying them with information."[7] A senior state environmental official, Robert Shaw, gave voice to the state's overall attitude in a frank newspaper interview in 1963: "If anyone believes that New Jersey will remain what it was years ago, whether in regard to population, open spaces or streams or any other environmental factor, he fails to appreciate what's going on in New Jersey."[8]

Ciba, Geigy, and Sandoz understood what was going on in New Jersey. Their investment in Toms River Chemical was paying off, and they saw no reason to tamper with a good thing. America's love affair with bright colors—a passion with cultural antecedents stretching all the way back to Heracles' mythic discovery of Tyrian Purple on the Levantine shore—was in full swoon. Back in 1960, Toms River Chemical sold slightly more than a million dollars' worth of dyes and other products per month. By 1964, monthly sales had topped four million dollars, and production was also at a record level: more than five million pounds per month. Toms River Chemical was now the second-largest producer of epoxy resins in the United States and the fifth-largest dye maker.

As for the consequences of this frenetic growth, Jim Crane described them in a confidential memo to Robert Sponagel and other senior managers in 1965.[9] Effluent from the plant, he wrote, was coloring the river water for eight miles downstream. Azo compounds and nitrobenzene could be measured in the water and in bottom mud all the way to Barnegat Bay. Yet there was no sign that Toms River

Chemical was changing its practices. The "chemical recovery plant" Sponagel had discussed back in 1963 still did not exist; the company was still handling its solid and liquid wastes in essentially the same primitive fashion it had chosen when the plant opened in 1952. The only significant changes were that the wastes were more toxic now due to the addition of azos and resins to the product stream, and the volume had more than tripled.

From a business perspective, the company's unwillingness to take action to curb its pollution made sense. Almost any step Toms River Chemical could have taken to reduce its discharges would have cut directly into its profits. Building an entirely new waste-handling system, one that would cost millions of dollars in capital and yearly operating costs, would have required shutting down the plant for months. To the Swiss, that was unthinkable. Their resistance to any shutdown stiffened after a brief strike at the plant in October of 1964 cost the company hundreds of thousands of dollars in lost production. Even smaller cleanup actions that might have saved the company money in the long run were given low priority out of fear that they would slow the manufacturing process. Jim Crane tried unsuccessfully for years to convince supervisors to recapture and reuse expensive and very hazardous production chemicals like epichlorohydrin and nitrobenzene; instead, thousands of pounds of those chemicals every day were spilled or dumped into factory drains that led to the lagoons and eventually the river.[10]

But Crane did not strike out completely in his pleas for Toms River Chemical to do something, *anything*, to clean up the river. Starting in early 1964, he and others in the company began pushing an audacious idea that promised to eliminate the factory's river pollution problems—the visible ones, anyway—in one fell swoop. If it worked, Toms River Chemical would no longer have to deal with angry fishermen or disgruntled neighbors; their complaints of tinted, smelly river water and dead or strange-tasting fish would go away virtually overnight. Best of all, from the company's point of view, this new idea was a bargain, which is why Robert Sponagel and his bosses in Basel were willing to consider it. In their eyes, it was a much better option than the only realistic alternative, which was to build an activated sludge treatment plant that would have cost at least $2.5 million plus another $250,000 a year in operating costs and—most importantly—would have required closing down the factory for weeks or months.[11]

Their big idea was to bypass the river by building a ten-mile pipeline to carry all of Toms River Chemical's wastewater—about five million gallons per day and growing—to the Atlantic Ocean. The pipe would cost almost $4 million, but the operating costs afterward would be minimal. Most importantly, the company would not have to suspend production or provide any additional treatment to its effluent, which could remain just as toxic, smelly, and tinted as ever. Sponagel was so enthusiastic about this bold idea that he proposed the pipeline be wide enough (at twenty-eight inches) to handle as much as fifteen million gallons per day—leaving plenty of room for Toms River Chemical to ramp up its own discharges or sell excess capacity to other companies in search of a hassle-free dumping ground.

By today's standards, dumping minimally treated waste into the coastal ocean seems inconceivable, but it was not at all an outlandish idea in 1965. Many cities around the world, including a half-dozen in northern New Jersey, had been discharging poorly treated sewage into the ocean for years (a few still are today, though typically only after providing higher-level treatment).[12] Even in the 1960s, however, it was rare in the United States for private companies to use ocean outfalls, and the few that did were usually grandfathered by government approvals granted years earlier. Securing permission to build a new one might not be easy.

Executives from Toms River Chemical did not need to worry about their immediate neighbors. Support in Toms River was as reliable as ever, and town officials loved the idea of shifting the pollution elsewhere. But the reception was very different in the beach communities out on the Barnegat Peninsula, where local leaders were livid at the prospect of an industrial waste dump just a half-mile offshore. They were not at all mollified by tests, undertaken by Toms River Chemical, that concluded that the color and odor would be sufficiently diluted to be invisible to beachgoers.

The ever-complaisant New Jersey State Department of Health, which had always given the Swiss whatever they wanted, green-lighted the ocean pipeline in the fall of 1964, leaving only the U.S. Army Corps of Engineers as the final hurdle. The beach communities appealed to the region's new congressman, James Howard, a Democrat who had little love for the Republican bosses who played golf with Toms River Chemical executives at the company-owned Toms River Country Club. Howard convinced the Army Corps of Engineers to delay issuing the permit while the U.S. Public Health Service studied a long list of environmental concerns, including possible impacts on clams and other marine life. The wrangling held up the project for months and continued into the blistering summer of 1965 with no sign of resolution.

As the river fell and the stench worsened during that summer of high heat and record low rainfall, Toms River Chemical's problems escalated. In July, Philip Maimone, the Cadillac dealer who owned more than six hundred acres downstream from the factory outfall, sued the company for dumping "poisonous and deleterious effluents" into the river.[13] A week later, responding to reports of another fish kill in the river, an inspector from the state Division of Fish and Game visited a site five miles downstream and saw thousands of fish, crabs, and eels "dead, dying or in distress." The inspector, Bruce Pyle, noted that the fish had died from lack of oxygen and that Toms River Chemical is the "principal source of oxygen depleting matter in the river." Ever since its first inspection in 1961, the state agency had been warning Toms River Chemical to correct the problem. Now, Pyle wrote his supervisor, it was time to prosecute the company, finally, for destroying fish habitat. It would be an easy case to prove, he predicted.[14]

After fifteen years of operating with impunity, Toms River Chemical was now besieged on multiple fronts.[15] Yet the situation did not appear to be anything the company could not handle. Relatively few people in town were paying close attention to the issue; a much bigger controversy in the local papers during that long, hot summer was whether flying the United Nations flag at town hall was a gesture of international cooperation or "evidence of communist conspiracy," as one newspaper article put it.[16] Toms River Chemical was still the economic colossus of Ocean County, the engine of the region's headlong growth. The company had powerful friends, a huge payroll, and a multimillion-dollar revenue stream. It seemed more than a match for Philip Maimone and the New Jersey Division of Fish and Game.

And then Jim Crane smelled chemicals in his shower, and suddenly the company had bigger problems than dead fish and an angry Cadillac salesman.

The idea that dangerous compounds in the environment might harm innocent bystanders, and not just industrial workers, was slow to take hold. Paracelsus recognized that nearby residents could be affected by emissions from workshops and mines, and so did Bernardino Ramazzini, but they focused their investigations on work-related illnesses because they reasoned, correctly, that most people are exposed to poisonous substances at much higher concentrations and for much longer periods at their workplaces than at their homes. There were sporadic attempts to investigate disease patterns outside of the workplace, but they were rare. In 1761, fourteen years before Percivall Pott published his observations about scrotal cancer in chimney sweeps, another London physician, John Hill, published his *Cautions Against the Immoderate Use of Snuff,* in which he described two patients with nasal cancer and speculated that users of tobacco snuff were vulnerable to the disease.

But how could Hill know what had really caused those nasal tumors? How could Pott know? How could *anyone* know? Answering those questions in a workplace setting was extremely difficult; in a residential setting, it was close to impossible. In a neighborhood, where exposures to harmful substances tended to be much lower, how could any investigator ever hope to credibly identify a specific cause, especially for a disease like cancer that took years to develop? Trying to determine the environmental trigger of a slow-developing disease was like trying to identify a criminal based on a smudged fingerprint left at the scene of a crime: It required a subjective interpretation of an indistinct impression left behind long after the perpetrator had fled. The neurological ailments Ramazzini observed in glass workers in Venice, for instance, often took many years to develop and could have been caused by exposure to lead, arsenic, antimony, or mercury (the artisans worked with all four metals) or something else entirely.

This was the central dilemma of epidemiology, a term coined in the mid-nineteenth century for the study of factors influencing health and disease across populations.[17] Identifying an exposure that appeared to increase the risk of disease in a particular population—whether neuropathy among Venetian artisans or bladder cancer among German dye workers—was interesting and perhaps useful, but what did it prove? It did not prove that the chemical caused any particular case of the disease, since there were probably other potential causes, too. It did not even prove that the apparent link between chemical and disease was important and not a coincidental distraction from the still-hidden true cause. It did not *prove* anything at all. This inherent uncertainty would take on extra significance as the age of industrial chemistry dawned. With the rise of large-scale manufacturing, the outcomes of environmental health debates could affect the economies of entire nations.

As factories sprouted across England in the early nineteenth century, the successors to Ramazzini and Pott continued to look for connections between cancer and pollutants, focusing almost always on the workplace. One of the most notable of these medical detectives was a London physician named John Ayrton Paris, a skilled mineralogist and prolific writer whose eclectic interests included applying medical evidence to legal disputes and using toys (everything from shuttlecocks to soap bubbles) to teach children the principles of science. From 1813 to 1817, soon after completing his medical training, Paris moved from London to the Cornish harbor town of Penzance, where his patients included laborers in the small copper- and tin-smelting workshops of Cornwall. He took the time to inspect those small factories in person, as Ramazzini had advised more than a century earlier, and noticed that on nearby farms "horses and cows commonly lose their hoofs, and the latter are often to be seen in the neighboring pastures crawling on their knees and not unfrequently suffering from a cancerous affection in their rumps." Paris blamed the "pernicious influence of arsenical fumes" from the factories, and thought there were human casualties as well. "It deserves notice that the smelters are occasionally affected with a cancerous disease in the scrotum, similar to that which infests chimney-sweepers," he wrote in 1822.[18]

But Paris, as an expert on court proceedings, well understood the weakness of an argument that relied on vague characterizations like "occasionally" and "not unfrequently" in an era in which industrial production was beginning to generate great wealth. What was needed was a much stronger standard of proof, one rooted in the solid ground of mathematics. Two Frenchmen played key roles in providing it.[19] The first was Pierre Louis, who published a study in 1835 that proved what he and many other physicians had long suspected: Bloodletting, a mainstay of humoral medicine since ancient Greece, did not work. He reached this revolutionary conclusion by going beyond the anecdotal evidence of individual cases and instead applying what he called "numerical medicine." Louis analyzed 174 patients with pneumonia or related conditions and discovered that no matter when leeches were employed—early in the progression of the disease or late—they had no impact on whether or when a patient recovered. Bloodletting may have *seemed* to be the critical factor, but in fact it was irrelevant or harmful, as Louis showed.

Louis's work not only helped hasten the long-overdue demise of humorism—by 1837, just seven thousand leeches were imported into the city of Paris, down from thirty-three million ten years earlier—it also helped lay the groundwork for modern observational epidemiology. In order to find out whether a treatment was really healing an illness, or if a pollutant was really causing it, an investigator would have to come up with an experimental design that could eliminate alternative explanations. Anecdotal observations, like those of Percivall Pott or John Ayrton Paris, were not enough.

Another Frenchman who made a key contribution was Siméon Denis Poisson, a mathematician whose 1837 study of Parisian jury verdicts became a cornerstone of modern statistical analysis. His ideas were an extension of the "law of large numbers"—namely, that the outcome of any particular random event cannot be predicted with confidence, but if you repeat the event enough times under the same conditions, and if there are only a certain number of possible outcomes, then the aggregate results can be predicted very accurately. Predicting the outcome of a single flip of a fair coin, for example, requires a guess that will be wrong half the time. But if you predict that "heads" will turn up 50 percent of the time and then flip a coin one hundred times, your prediction is very likely to be accurate, plus or minus a few percent. If you make two hundred flips, the results will be even closer to 50 percent. In fact, the more times you flip the coin, the more accurate your initial prediction of 50 percent will be.

Siméon Poisson extended this very simple concept to circumstances in which an event occurs rarely despite many opportunities. (Traffic accidents, for example: The likelihood that someone will be in an accident on any particular day is very low, yet many accidents occur every day because so many people drive cars.) Poisson discovered that if the overall number of events is sufficiently large, it is possible to predict the "normal" or random distribution not just of coin flips but also of unusual events such as carriage accidents, deadlocked juries, or rare diseases. This "Poisson distribution" would eventually become a huge breakthrough for epidemiology, although its value would not be widely recognized until much later. Using his formulas and those of his successors, a statistician could analyze what appeared to be an unusual number of cases of a disease in a particular place and time—say, bladder tumors in workers at an aniline dye plant—and determine the likelihood that the apparent cluster was not a mere chance occurrence but instead was a true cluster for which there might be a specific environmental cause.

There was a catch to Poisson's insight, however: A disease cluster could be confidently declared to be nonrandom only if there were a sufficiently large number of cases available to establish the "normal" random distribution. That was a formidable hurdle when rare cancers were at issue because predictions based on the law of large numbers were more like guesses when only a few cases could be included in the analysis. There was also the confounding issue of latency: As Percivall Pott recognized in his study of chimney sweeps, many years could pass between a triggering event and the appearance of a tumor large enough to be diagnosed. How would it ever be possible to identify, with high confidence, a culprit that did its nefarious work years earlier? As Pierre Louis had pointed out, a causal agent could be confirmed only after the elimination of reasonable alternative explanations. Which specific component of the "arsenical vapours" identified by John Ayrton Paris had sickened those factory workers in Cornwall? Or did something else entirely make them sick? Or was the apparent disease cluster just a coincidence, as Poisson's random distribution might suggest, were there enough cases to apply it?

To identify definitively the cause of a disease based on an observed pattern of cases, practitioners of the new "numerical medicine" of Louis and Poisson would not only have to undertake an analysis that eliminated alternative causes, they would also have demonstrate that the apparent pattern was not a coincidence. Those were formidable hurdles, especially if the disease at issue was rare and progressed slowly, as with most cancers. For all its advances in mathematical technique, the rise of statistical epidemiology had the unintended effect of making it harder than ever to investigate patterns of cancer cases in places like Toms River. It also gave medical investigators powerful new reasons to turn their attention away from cancer and to focus instead on fast-moving infectious diseases. Soon they would do so, with spectacular results.

Jim Crane's discovery that hazardous waste from Toms River Chemical had contaminated the town's public water supply set off a crisis—but one that was kept secret from the people who drank that water.[20] By mid-August of 1965 the Toms River Water Company knew that azo dye wastes had contaminated three shallow wells (they were about seventy feet deep) that supplied drinking water to almost all of the seven thousand homes and businesses in town that were water company customers. Yet there were no public warnings, no newspaper articles. Life in Toms River continued as usual except for the unfortunate fact that the weather was so uncomfortably hot and dry that residents drank even more water than usual, even if it sometimes smelled. Just two entities knew what was really happening, and they were old friends used to working together closely and sometimes secretly: the Toms River Water Company and the Toms River Chemical Corporation.

For years, chemists at Toms River Chemical had assisted the water company in analyzing the quality of the town's drinking water supply, since the expertise of the Swiss-trained chemists far outstripped that of anyone at the little water company, which was struggling to meet the soaring demand. In ten years, its customer base had quadrupled. On hot days, Toms River Water was pumping more than two million gallons from its four wells, three of which were on Holly Street near the river.

Even before Jim Crane smelled chemicals in the shower, water company officials knew something was wrong with at least one of the three Holly wells. According to an internal Toms River Water report dated March 23, 1965, water in one well had such a strong odor and was so visibly contaminated with what the report described as "trade wastes (dye)" that the water company had started adding very large doses of chlorine (so large as to be unsafe by modern standards) to reduce the color.[21] The dry summer made the situation even worse. Pumping rates rose with the increased summertime demand, as usual, but there was no rain to recharge the underground aquifer. Instead, the Holly wells sucked up more of the polluted river water, which is why, by July, Crane could smell dye wastes in the tap at his Oak Ridge home. He and his staff responded quickly, collecting and testing samples from several water lines in the neighborhood and then, with the water company's permission, testing the Holly wells.

Two weeks later, in mid-August, Toms River Chemical was ready to tell the water company what its chemists had found: dye chemicals, almost everywhere they looked.[22] Analytical methods were primitive in 1965, but Ciba had been working with aniline dyes for almost a century; its chemists knew how to spot aniline-like molecules in water, even if they could not always distinguish between similar molecules. In addition to aniline, whose toxicity had been obvious since the industrial poisonings of the 1870s, these similar-looking molecules included benzidine, which had already been implicated as a cause of bladder cancer, and nitrobenzene, the highly toxic "bloodstream" of the vat dye operation. As they tried to determine dye concentrations in the town's drinking water supply, company chemists devised an overall measurement for what they called "diazotizable amines" because they were components of azo production.

The contamination, the chemists discovered, was severe. Levels of diazotizable amines in the town's wells and water pipes ranged from five to seventy parts per billion, and later as high as 160 parts per billion. In the 1960s, there were no specific limits for any of those chemicals in drinking water; they were too obscure and difficult to analyze. But by today's standards, the concentrations were alarming. For benzidine, for example, current Environmental Protection Agency guidelines call for no more than one part per *trillion* in drinking water—five thousand times less than the lowest concentration of aromatic amines found in Toms River water in 1965.[23]

There was only one way dye wastes could have reached the water supply. The Toms River Chemical Corporation must have discharged them into the river, and the three shallow riverside wells operated by the Toms River Water Company on Holly Street, more than two miles downstream, must have then sucked in the contaminated water, pulling it through the sandy riverbank and into the wells and eventually pumping it to the kitchen taps of seven thousand unsuspecting customers all over town.

Within a couple of weeks, the chemists from Toms River Chemical had figured out what had happened and explained it to the water company—in secret, of course. Toms River Water could have responded by shutting down the wells, but did not. August was a time of intense water demand, thanks to the influx of summer tourists, and that August was the driest on record. Toms River was growing frenetically, and having a verdant lawn was part of the new ethos of a town in which appearance and property values meant everything.[24] There was no way to shut down the Holly Street wells during the summer without having to explain exactly why customers would suddenly have to make do with less than half the water supply they were used to. The water company and the chemical company would also have to face uncomfortable questions about how long the people of Toms River had been sipping toxic chemicals in their morning tea, and why the companies had done nothing to stop it earlier.

The wells stayed open, continuing to operate at their maximum capacity, and the people of Toms River kept unknowingly drinking tainted water. The water company did add filters to its Holly Street pumping station, and Toms River Chemical took the unprecedented step of reducing production over the Labor Day weekend, when demand for drinking water was high and the river was full of fishermen. Both stopgap measures were soon abandoned, however. The factory went back to full-scale production, and Toms River Water stopped using the filters after tests at Toms River Chemical showed that they weren't working: concentrations of diazotizable amines were as high as ever. In October, the water company finally, and quietly, shut down two of the Holly Street wells. By then, the weather had cooled and demand had declined enough that no one would notice the drop in water pressure. Even so, Toms River Water had to keep one of the Holly wells open—the one in which the amine concentrations were lowest. Again, water users were told nothing.

Toms River Chemical and Toms River Water had managed to keep their secret in 1965, but the summer of 1966 would pose a bigger challenge. They would have to worry about keeping odors down and water flowing for an entire summer, not just in August. As the two companies worked in secret to devise a solution, a long-awaited gift arrived from Washington, D.C.: A construction permit from the U.S. Army Corps of Engineers for Toms River Chemical's pipeline to the sea. The permit came with several conditions, including a requirement that the inside and outside of the twenty eight-inch steel pipeline be lined with a thick enamel made from coal tar—an ironic choice, considering that most of the contaminants the pipe would carry were also derived from coal tar. (The liner was supposed to be leak-proof, though subsequent events would demonstrate the folly of that assumption.)

With the permit in hand, Toms River Chemical moved quickly. By January of 1966, it had beaten back a legal challenge from the shore communities, selected a contractor, and bought ten miles of piping for the project. By March, construction was under way. Meanwhile, lawyers and managers for Toms River Chemical and Toms River Water were secretly negotiating a plan to get through the summer without water customers finding out that they had been drinking dye wastes.[25] Here, they were fortunate again. Tests at Toms River Chemical showed that adding chlorine dioxide to the drinking water improved its smell and taste and reduced the diazotizable amines from 100 parts per billion to about 20 parts per billion—still far higher than would be permitted under a modern standard, but a major improvement. Adding lots of chlorine to the water also made it harder for anyone to smell the contamination. Plus, chlorination was cheap: just $2,000 to set it up, plus $50 a day for salt. Toms River Chemical readily agreed to cover the expense as long as it could reimburse the water company in secret and without admitting fault.[26]

Their luck held. The summer of 1966 was cooler and wetter than the previous years, and the chlorination treatment worked well enough that there was no outcry over smelly drinking water. On July 11, the ocean pipeline began operating, having overcome a series of glitches, including an incident in May in which a large section of pipe came loose from its moorings and bobbed to the surface of Barnegat Bay.

Even after the diversion to the Atlantic, dye waste from Toms River Chemical was still reaching the river, of course. The company was still withdrawing up to ten million gallons of cooling water every day from the Toms and returning it there in less-than-pristine condition. In addition, so much waste had already been dumped on the factory property's sandy soil, and would continue to be dumped there for another two decades, that about two hundred thousand gallons of chemical-laced wastewater was still seeping through the soil and into the river every day.[27] Nevertheless, the most visible sign of the company's despoliation of the river—its direct discharge of liquid wastes—had finally ended after fourteen years and more than ten *billion* gallons of wastewater. The state Division of Fish and Game dropped plans to prosecute the company; its jurisdiction did not extend to the ocean.

Their shared secret was safe, so Toms River Chemical and Toms River Water formalized their accord in a confidential legal agreement in February of 1967. Toms River Chemical agreed to reimburse the water company for the $45,000 cost of drilling a new well in another part of town plus the small expense (just $50 a day) of chlorine treatment of the Holly wells. In return, Toms River Water gave the chemical company something far more valuable: a liability waiver. The final clause of the confidential agreement stated: "Toms River Water Company hereby agrees to release, discharge, indemnify and save harmless the Toms River Chemical Corporation from any and all claims or causes of action which it may or might have or which may or might hereafter arise from the use of the Holly Street wells."[28]

There were a few more loose ends to tie up, and Toms River Chemical took care of them with dispatch. Later in 1967, the company cut a deal with its longtime adversary Philip Maimone, the Cadillac dealer who owned developable land along the river southeast of the factory. Maimone dropped his lawsuit in return for a land swap and a $52,000 payment from Toms River Chemical, which insisted that the settlement also include a statement by Maimone that the factory was no longer polluting the river.[29] The company had already bought the adjacent Luker Farm for just $35,100. Once owned by a descendant of the legendary Tom Luker (arguably the "Tom" of Toms River), the farm was a wreck. Toms River Chemical was able to buy it at a sheriff's sale just before Christmas in 1965, and to tear down what little remained. With the purchase of the old farm and the land swap with Maimone, the company had an even bigger buffer zone, plus newly won protection from liability for its years of pollution. Not only had Toms River Chemical weathered its first storm, it had set itself up for another generation of unfettered chemical manufacture, and waste dumping, deep in its forest redoubt.

As quickly as it had emerged, the controversy over the river contamination faded from view. The river looked cleaner now, and the drinking water smelled better. The factory's smokestacks belched colored smoke only at night, and its wastewater gushed into the ocean out of sight, forty-five feet below the surface and a half-mile offshore. People in Toms River stopped thinking about Toms River Chemical. When the company was mentioned in the local papers, it was usually because of a charitable project, not an environmental controversy. Would anyone else in town ever find out about the contamination of the town's drinking water? That seemed unlikely, and it seemed even less likely that Toms River Chemical would be blamed if it did. Both assumptions would eventually be proved wrong, but by the time the truth emerged, few would care about the long-ago contamination of the Holly Street wells. By then, the town would be in midst of an uproar that would make the events of 1965 and 1966 seem quaint by comparison.

CHAPTER FIVE

~~

Sharkey and Columbo
at the Rustic Acres

The Fernicola brothers, Nick and Frank, grew up in the dirt cowboy subculture of the New Jersey waste industry. Their father, also named Nicholas (his son always went by "Nick" to distinguish them), operated a drum reconditioning business starting in the 1940s on Avenue L in the Ironbound section of Newark, across the street from a slaughterhouse. Even by the standards of that heavily industrialized neighborhood, it was an extraordinarily filthy way to make a living. Nicholas Fernicola specialized in cleaning, repainting, and reselling the fifty-five-gallon steel drums that carried the foulest dregs North Jersey manufacturers could produce. There was no better place than Newark to be in that line of work. It was the "drum capital of the world," as Frank Fernicola would wistfully describe it years later.[1]

The big money, though, came not from refurbishing waste drums but from making them disappear. Back in the 1960s, when the chemical industry was roaring in North Jersey, the forests and farms in the central part of the state were the equivalent of Sutter's Mill in 1849 California. The rush was not to pull gold out of the ground but to dump chemical waste into it. Up in Newark, the landfills were expensive and so crowded that the lines of trucks waiting to unload their drums would stretch for blocks. A generation earlier, Ciba had found space and privacy in the deep woods of Ocean County to manufacture dyes and plastics on a massive scale. Now the chemical-waste haulers from Newark, Elizabeth, and Perth Amboy started to follow suit, pointing their big rigs south on Route 9 in search of cheaper dumping grounds. (The Garden State Parkway would have been faster, but trucks were banned north of Toms River.) They found plenty of willing partners among the farmers of Monmouth, Burlington, and Ocean counties. The real estate boom had not yet reached into the rural inland areas of the state, and chickens could not compete with hazardous waste as a cash crop, since farmers typically were paid anywhere from $20 to $50 per drum of waste dumped on their land.

One of the more enterprising landowners was a man named Edward Wilson, who worked for Morton International in the 1950s and 1960s, when the salt maker was broadening its business to encompass chemical manufacture. Wilson offered his family farm in Ocean County's Plumsted Township as a dumpsite for Morton's toxic wastes, which included halogenated solvents, chlorinated compounds, volatile organics, and heavy metals. His neighbor, Dayton Hopkins, was even more eager: He let Morton dump on *three* of his farms, including his family's own fifty-seven-acre homestead. At the worst of those sites, called Goose Farm, drums were tossed into a pit that was three hundred feet long and one hundred feet wide.[2]

Most of the rural townships also operated small municipal landfills—back then, people called them dumps—which were also popular destinations for industrial waste from out of town because they charged less and had much shorter lines than their counterparts to the north. The town governments were just as eager as individual farmers to open their gates to the eighteen-wheelers from Newark. If the dumps were not burying waste, they were burning it; on Fridays, the usual burn day, the trails of black smoke would billow for miles. Taking in industrial waste from out of town served everyone's short-term interest. Farmers and rural townships got much-needed cash, while haulers and chemical companies got cheap and secluded dumping grounds where no one asked too many questions about what was in the drums. If anyone was worried about the longer-term environmental consequences, there is no evidence of it. A generation later, when investigators finally assessed the damage, they identified two dozen major hazardous waste sites in Ocean County alone, including seven farms and three town dumps. Cleaning them all up would cost hundreds of millions of dollars.[3]

As easy as it was to dump legally in the hinterlands of central Jersey, many haulers wanted even sweeter deals. They preferred to do their dumping for free, deep in the pinelands, without anyone's permission or knowledge. No one stood in their way. In those days, the closest thing New Jersey had to an anti-dumping law was a misdemeanor public nuisance statute; it was invoked very rarely and only in the most blatant cases. The Ocean County Prosecutor's Office did not begin going after dumpers until 1980, when an investigator named Dane Wells started trying to track them down. It was harder than it looked. According to Wells, the "midnight dumpers" (though they sometimes operated in broad daylight) would use "horseshoe roads" in the woods, narrow dirt paths with only one entrance and exit. "That way, they could just bring the eighteen-wheelers in and not have to turn them around after they dumped the drums. Or, if it was liquid, they would just open the spigot and let it run out," he explained years later. The more advanced operations, the ones with links to the organized crime families influential in the North Jersey hauling industry, would post lookouts with walkie-talkies.

Frank Fernicola followed his father into the drum business, working first in Newark and then Toms River, where his clients included Toms River Chemical. He also did some waste trucking but had to give up his hauling permit after he was convicted in the late 1960s of illegally dumping chemical drums at the old Manchester Township landfill, about ten miles west of Toms River.[4] Frank even made a brief foray into the incineration business: He hauled sodium waste from a North Jersey chemical plant to Beachwood, just south of Toms River, and burned it in an open pit. That escapade ended with a bang: An explosion summoned a town fire truck, and Frank lost that permit, too.

His older brother Nick, on the other hand, had stayed away from the waste business for years—a high school summer spent working for his dad was more than enough. He was crazy about motorcycles and worked as an auto mechanic before opening a used car salvage and sales business in Newark. In 1966, when he was thirty-one, he followed his brother south to Toms River, running a gas station for a few years and then moving briefly to Idaho. When he returned to Toms River, Nick Fernicola bounced through a series of construction jobs: At various times he drove a front-end loader, laid pipe, and set explosives. All the while, he kept an eye out for a more lucrative opportunity, perhaps even one in the family business. He found it one January night in 1971 at the Rustic Acres, thanks to two guys named Sharkey and Columbo.

The Rustic Acres was a blue-collar landmark in Ocean County until it was finally torn down in the late 1970s. The bar was out on Route 37, midway between the two largest employers in the county: Toms River Chemical and the Lakehurst Naval Air Station. Its wooden tables and stools would fill up at the four o'clock shift change, and again at midnight. Frank Fernicola shot pool at the Rustic Acres five nights a week, sometimes accompanied by his brother Nick. Frank had hauled drums for Union Carbide and had friends who worked at the company's huge chemical plant on the Raritan River, about sixty miles north in the town of Bound Brook. At the time, it was one of the country's largest plastics factories, generating hundreds of thousands of pounds of toxic waste. Frank knew that Union Carbide was looking for a hauler to get rid of thousands of deteriorating drums of waste accumulating on the Bound Brook property, and soon his brother knew it, too. To Nick, it sounded like an attractive business opportunity. It sure as hell beat punching a clock as a construction laborer.

To turn a profit, though, Nick Fernicola would need a cheap place to dump all that waste, a place where no one would ask any uncomfortable questions about what was inside the rusty fifty-five-gallon steel drums. Enter Sharkey and Columbo. The two men (Fernicola would later claim to investigators that he never knew their real names) worked at the Dover Township Municipal Landfill, the town dump for Toms River. They were also regulars at the Rustic Acres, and one night Nick Fernicola told them about all those Union Carbide drums

that needed a final resting place. The two men agreed to introduce
him to the town's superintendent of public works, who ran the landfill. Before long they had a deal: For a bargain-basement price of $10
per truckload, Fernicola could dump the drums in the town landfill
on Church Road—but he could dump only at lunchtime, when Sharkey was running the shift, and he had to pay Sharkey directly. Many
years later, in a court deposition, Fernicola acknowledged that the
payment was not an official fee. When a lawyer asked whether it was
a bribe, Fernicola responded, "Could be."[5]

Fernicola drove up to Bound Brook to give Union Carbide the
good news: He had found a way to get rid of their drums. In fact, he
claimed—falsely, as it turned out—that he had lined up *three* "authorized land fill areas," according to a document he signed and gave to
Union Carbide on February 16, 1971.[6] "My company will handle all
removal and disposal of said drums at a rate of $3.50 per drum and
will assume all risks and problems arising from such removal and disposal of said drums," he wrote. Fernicola was eager to begin. He
would be clearing a hefty profit, since he could dump an entire load of
about thirty drums for just $10. He rented three trucks and hired two
drivers to help him with the hauling. When Fernicola carried his first
load to Toms River on April 1, a low-level manager at Union Carbide
followed him to confirm that he was dumping at the town landfill. It
would be the last time for eight months that Union Carbide would
take an interest in where thousands of its waste-filled drums were
going.[7]

A strike at the Bound Brook plant initially slowed down the operation, but soon Fernicola and his drivers were making two trips a
day—or sometimes just one, on days when they parked their rigs at
Freehold Raceway and spent the afternoon betting on the horses. It
was not all fun and games, though. Sharkey, belatedly recognizing
how much profit Fernicola was clearing, began to squeeze him, up-
ping his price to $20 per truckload and threatening further hikes, ac-
cording to Fernicola. He responded by convincing Union Carbide to
agree to pay him an extra dollar per drum. Fernicola also tried to
generate more income with a scheme that hearkened back to his father's trade in Newark. Instead of rolling the drums into the pit, he

started tipping them and letting the waste run out and splash directly onto the earthen floor of the unlined landfill. Then he would put the empty drums back on his truck and try to resell them. Unfortunately for Fernicola—and for the families of Toms River, as things turned out—most of the drums were already beat-up and leaking and thus valueless for resale.

Whatever its rewards, Fernicola's hauling operation was not without its risks. One rainy day in early September, some drums freshly deposited in the town landfill exploded after coming in contact with water. One of them blew fifty feet into the air. The landfill manager blamed Fernicola, who insisted he had nothing to do with it. By then, though, his arrangement was already unraveling. According to Fernicola, Sharkey had upped his demand to $50 per load—and Fernicola had turned him down flat. He was already working on a new plan.

In 1837, two years after the pioneering statistical analyses of the French physician Pierre Louis had laid the groundwork for investigations of disease clusters, tuberculosis claimed a woman known to history only as "Miss Langford." There is no record of her first name, only that she was a farmer's daughter from Shropshire, in the West of England, and that she had been married for three years to a struggling young London physician named William Farr, an admirer of Louis' statistics-based "numerical medicine." The bereft widower consoled himself by studying the death records of the City of London, looking for patterns in the waves of communicable disease that regularly washed across the metropolis, killing thousands in their wakes, including his own Miss Langford. Perhaps, he thought, the same kind of rigorous statistical analysis that Pierre Louis had used to such great effect to discredit bloodletting could also be used to understand and even master the much greater terrors of tuberculosis and cholera.

He was not the first to try. Investigators had been trying to discern and interpret patterns of disease outbreaks ever since Hippocrates, who wrongly thought that swamp water caused malaria because the disease was more prevalent in swampy areas (he knew nothing about mosquito-borne parasites). The Greeks were bested by Avicenna, the Persian physician whose fourteen-volume *Canon of Medicine,* completed in 1025, was hurled into a bonfire by Paracelsus five centuries later in Basel. The torching was undeserved. Despite Avicenna's wrongheaded devotion to humoral theory, his many insights included the identification of tuberculosis as an airborne communicable disease. He proposed quarantines and suggested that contagious diseases could be transmitted through water and soil as well as "pestilential" air.[8] The spread of his ideas to Western Europe led to the founding of hundreds of quarantine hospitals for lepers. When the bubonic plague arrived in 1348, many cities—including Basel—adopted laws requiring improved sanitation and, later, the isolation of anyone afflicted with a disease considered contagious. Isolating its victims did little to curb the Black Death, which was caused by a bacterium carried by fleas and rats, but cleaner streets helped marginally.

Fear of communicable disease also led to another crucial innovation: the collection of statistics on disease rates, which began in London in the 1530s with the issuance of weekly bills of mortality. Originally devised as an early warning system for outbreaks, the weekly reports were scrutinized by anxious English nobility who would flee to their country estates at the first sign of the Black Death's return to the teeming, filthy city.[9] In 1837, Parliament created the General Register Office to centralize and expand collection of vital statistics, including the successors to the old bills of mortality. The young widower William Farr, meanwhile, was busy publishing articles based on his research into disease statistics, including one influential book chapter in which he argued that there was a mathematical symmetry, and thus predictability, to the rise and fall of disease outbreaks.[10] His work caught the attention of some prominent Londoners, and in 1839 Farr was appointed "compiler of abstracts" in the new General Register Office.

Communicable diseases turned out to be ideal proving grounds for the new "numerical medicine" Farr had embraced. Certainly they were more suitable than cancer. Copper smelting might cause scrotal cancer, as John Ayrton Paris suspected, but that was mere speculation unless other possible causes—including mere chance—could be identified, tested, and excluded. Highly transmissible diseases like cholera or tuberculosis, on the other hand, were much easier to study as they sliced through a city, afflicting thousands of people at a time. These diseases could be diagnosed just days or even hours after infection, and their transmission routes were often apparent—or soon would be.

In fact, the causes of these outbreaks seemed so obvious to Farr that he did not wait for conclusive evidence before campaigning for preventative measures. Like most of his peers, he thought that tuberculosis and cholera were caused by "miasma"—air poisoned by vapors from garbage and feces—so he championed legislation to improve sanitation. As he wrote in 1847, "This disease-mist, arising from the breaths of two millions of people, from open sewers and cess-pools, graves and slaughterhouses, is continually kept up and undergoing changes. In one season it is pervaded by cholera; in another, by influenza; at one time, it bears smallpox, scarlatina and whooping-cough among your children, at another it carries fever on its wings. Like an angel of death it has hovered for centuries over London. But it may be driven away by legislation."[11]

And it was. Cholera was the turning point. The Victorians did not know its cause (they were wrong about "miasma") and had no effective treatment for it, but Farr and his colleagues managed to devise a robust response that more or less held cholera outbreaks in check. In the process, they pointed the way toward a new kind of medical science that focused on patterns of illness within and across communities. They created modern epidemiology. In fact, just about all of the key ideas that would eventually be utilized to try to understand pollution-induced cancer in Toms River would be developed not through investigations of lumbering maladies like cancer but through much simpler investigations of fast-moving communicable illnesses.

The first important test of Farr's ideas was the cholera that swept through London in 1848 and 1849. Farr assumed that the miasma, or fetid air, was at its worst near the filthy Thames River, so he checked the General Register Office records to see whether cases were concentrated near the river. As his successors would do in Toms River exactly 150 years later, Farr also looked to see where the affected neighborhoods got their water. In both cases, his analysis implicated the

Thames: Cholera rates appeared to be high among those who drank contaminated river water. When cholera returned in 1853, Farr again counted cases in various parts of the city, noticing that there were slightly fewer in an area where the water supply had recently been improved. As he pressed the city's water companies to find cleaner sources of drinking water, Farr speculated that a much larger epidemic would strike the following year. He made little headway with the companies, but a quirky forty-year-old physician named John Snow began paying close attention to Farr's reports and started compiling his own data, too.

Farr's prediction was correct. Cholera returned in the summer of 1854, killing tens of thousands in a much larger outbreak. John Snow regarded the epidemic as a preventable tragedy. A bachelor who did not smoke, drink alcohol, or eat meat, Snow was a pioneer in the use of anesthesia (earlier that year, he had given chloroform to Queen Victoria during the birth of her son Leopold). He took pride in being unencumbered by orthodoxy, including the prevailing miasmic theory of disease. Since cholera's symptoms were centered in the gut, Snow thought it spread via bad water, not air. He practiced what he preached by refusing to drink water unless it had been boiled first.

Like Farr, Snow thought that cholera's secrets could be unmasked by analyzing the geographic distribution of cases around London— and he had a more sophisticated idea of how to do it. Snow knew that after the cholera outbreak of 1849 devastated South London, one of the district's two major water suppliers, the Lambeth Company, had responded to Farr's entreaties by moving its intake pipes to a cleaner, upriver portion of the Thames. Meanwhile, its main competitor in South London, the Southwark and Vauxhall Company, continued to draw from a highly contaminated stretch of the river in the central city. That set up the conditions for what Snow later called a "grand experiment" in which his theory that cholera was a waterborne contagion could be tested in neighborhoods supplied by two different water sources. When the major epidemic struck in 1854, Snow used Farr's data tables to test his theory. He found that the death rate in houses supplied with contaminated water from Southwark and Vauxhall was almost nine times higher than in nearby houses that used the much cleaner Lambeth water, and more than five times higher than the death rate in the rest of London.

It was a penetrating observation, but Snow's iconic status as the pivotal figure in the history of public health rests on what he did next. In the late summer of 1854, as what he later called "the most terrible outbreak of cholera which ever occurred in this kingdom" raced through the Soho district, near his home, Snow decided to attempt a more direct investigative approach—one that would be emulated in Toms River.[12] With Farr's help, Snow got the names and addresses of victims in three of the hardest-hit districts in Soho and resolved to determine the circumstances of each of the eighty-three deaths that had occurred there between August 31 and September 2, 1854. With assistance from a local minister, Henry Whitehead, Snow spent the next four days questioning families or friends of each victim, asking the same questions each time to avoid bias. He quickly discovered that all but a handful of those who had died had gotten their water from the same well, on Broad Street. Intent on showing exactly when and where the victims were infected, Snow drew maps showing where they lived and also how the epidemic had waxed and waned over time. Both kinds of mapping—geographic and temporal—would become standard techniques in Toms River and elsewhere.[13]

There was still no direct evidence that the Broad Street well was the source of the epidemic in Soho—microscopic examination of the water was inconclusive—but John Snow refused to wait for more deaths. When he met on September 7 with the board that served as the parish's local government, Snow presented his circumstantial evidence so forcefully that the board agreed, as an experiment, to disable the well by removing its pump handle. The outbreak, which was already waning as residents fled the area, quickly ended. Still, city authorities were reluctant to accept his theory that cholera was a waterborne contagion, even after Snow identified an old cesspool that was three feet from the well as the likely source of the outbreak.[14] To Snow's dismay, the pump handle was replaced soon after the outbreak ended because residents wanted the convenience of a nearby well. As in Toms River, comfort trumped precaution; memories were short. Cholera would strike London again in 1866, though this time Farr was able to prevent another outbreak by identifying a particular water company as the contamination source. By then, the frustrated Snow was dead, felled by a stroke at forty-five, despite his abstemious lifestyle. His ideas would not be fully vindicated until 1885, when Robert Koch identified a rod-shaped bacterium that thrived in sewage-contaminated water, *Vibrio cholerae,* as the cause of cholera.[15]

The creativity, doggedness, and ultimate frustration of John Snow's investigation of the Broad Street outbreak presaged the experiences of generations of public health researchers who would follow him. They would use many of the techniques pioneered by Snow and Farr, including close analysis of vital statistics, standardized interviews, and meticulous geographic and temporal mapping of cases and suspected environmental causes. And in some of those cluster studies, including the one in Toms River, they would take the disease-tracking tools Snow and Farr had developed for cholera and try to adapt them to investigate cancers and other chronic illnesses that took many years to develop. Compared to cancer, they would discover, tracking a fast-moving infectious disease like cholera was a cakewalk.

The conquest of communicable disease that began in earnest in 1837 with William Farr's scrutiny of the London mortality lists not only gave birth to modern epidemiology, it also clinched the case that waste in all forms, including sewage, was too important a problem to be left solely to private enterprise. Yet the necessity of regulating waste also created unexpected opportunities for canny entrepreneurs like Nick Fernicola, whose actions would pose health risks that Farr and John Snow could never have imagined.

A simple pecking order has always characterized mankind's relationship to waste: The wealthy throw out what they do not want, the poor scavenge what they can, and whatever remains is left to rot. A few ancient civilizations were adept at reusing excrement, ash, and other waste as fertilizer. The more common waste disposal technique, however, was simple dumping, which is why heaps of midden—shells, feces, bones, pottery shards, and other refuse—are conspicuous features of archaeological sites.

As cities emerged from nomadic and agrarian societies, however, vermin and odors made open-pit dumping a problem. Rulers responded with the first waste-management laws, ordering the burning or burial of waste—preferably out of town. The oldest known landfill, a series of earth-covered pits in the Cretan capital of Knossos, is about five thousand years old. Ancient Jerusalem's dumpsite was beyond the walls of the Old City in the Valley of Hinnom. In the days of the Judaean kings, according to the Bible, cults would go to the valley to sacrifice children to the pagan god Moloch. By Jesus' time, Hinnom was a foul dump full of rotting garbage, animal carcasses, and smoky, acrid fires. It was, in a word, hellish, which is why the valley's other name, Gehenna, came to stand for the place where sinners were tortured in the fires of eternal damnation.

In the centuries that followed, Gehenna's successors arose at the edges of many major cities, with little oversight or interference from governments. They were lawless places where only criminals and the poorest of the poor would go. The culture of scavenging that arose there, and also in the back alleys and fetid waterways of the great cities, was both intricate and dangerous; its desperate practitioners faced occupational risks that rivaled those of Percivall Pott's chimney sweeps. Variants survive today in slum-ridden megacities like Cairo, Mumbai, and Buenos Aires, but the epitome was early nineteenth-century London, where a scavenger army of tens of thousands of impoverished men, women, and children, each with a defined specialty, scavenged the dregs of the metropolis. There were toshers in the sewers and mudlarks on the riverbanks, rag-pickers atop rubbish heaps and bone-pickers behind kitchens. "Pure-finders" scooped up dog manure for tanneries, dustmen collected ash, and night-soil men emptied cesspools. Even coarse dust was sieved at the dumps and reclaimed for the manufacture of bricks. Teeming cities like London and Paris could not have functioned without the ad hoc scavenging system, but the cost was very high. The scavengers worked in filth, and as the investigations of William Farr and John Snow demonstrated, filthy conditions were crucial in the spread of communicable disease.

The solution most cities chose was to transform waste collection into a public function overseen by government officials instead of an unregulated private enterprise that depended on legions of poverty-stricken scroungers.[16] Cities built their own waste-collection systems, financed by taxes and fees. Waste handling became a mechanical task, instead of one performed by hand by those at the bottom of the economic pyramid. There were unforeseen consequences, however. Under the old scavenger system, waste was spread throughout the metropolis, but the new publicly owned works concentrated an entire city's sewage and garbage in just a few places where conditions were especially awful. Governments became the largest polluters in their own jurisdictions, having built colossal landfills, incinerators, and sewage treatment plants to replace the thousands of refuse piles and drainage pipes that had speckled major cities. Dependent on unpopular taxes and fees, these waste behemoths were expensive to build and maintain and tended to operate far beyond their design capacities as city populations boomed.[17]

On top of those dilemmas came the explosive growth of industries that generated tremendous volumes of liquid and solid waste. The Great Depression and shortages of raw materials during the First and Second World Wars temporarily masked the problem by stimulating reuse of discarded material, but the postwar economic boom ended that. The City of Cincinnati's experience with the Cincinnati Chemical Works, in which a resentful city government eventually imposed large fees on the company for dumping hazardous waste into the municipal sewage system, was repeated throughout the industrialized world in the 1950s and 1960s. Some cities went further and closed their gates entirely to factory waste. Suddenly, manufacturers needed to find new places to dump and new people to do the dumping. They needed Nick Fernicola.

The Toms River Chemical Corporation, of course, did not have to worry about finding a dumping ground. By the late 1960s, the factory was generating enough hazardous waste to fill several thousand drums every year, but the company had no need for someone to haul it away. The factory property was pockmarked with pits, ditches, and lagoons after a generation of dumping, but there was still plenty of additional space available. Just as the Swiss had envisioned way back in 1949 when they had selected the huge and secluded site, the sponge-like sandy soil seemed to have an infinite capacity to absorb whatever was placed on it. In 1966, for example, six open-pit dumps were operating at the factory, plus a primitive incinerator that accepted twenty-five truckloads of waste every day. There was also an open area that held several thousand drums filled with toxic waste. For liquid waste, meanwhile, the ocean pipeline was turning out to be a boon, just as company managers had hoped. There was enough unused capacity in the pipe for the company to sell space to other manufacturers, which paid Toms River Chemical to take their waste and pump it into the Atlantic. The company even began accepting solid waste drums from other Ciba facilities around the country and burying them on the factory grounds. More than ever, Toms River Chemical was not just in the chemical manufacturing business; it was very much in the hazardous waste disposal business, too.

After the river pollution controversies of the early 1960s, company managers were careful not to talk about all the out-of-town waste they were accepting, just as they never disclosed that dye chemicals had contaminated the town's drinking water in 1965. Even after the ocean pipeline opened in 1966, Toms River Chemical still faced a "staggering number of problems associated with our effluents and/or water supply," as one company manager put it in a memo.[18] Groundwater beneath the plant was so contaminated that it was harder than ever for the company to find drinking water clean enough to use at the factory. The river smelled better now that direct dumping had ended, but toxic chemicals were still seeping into the Toms through the sandy soil, and the factory's neighbors still sometimes complained about the "terrible stench" from the ten-acre holding pond where wastewater was stored before it was pumped to the ocean.[19] Worried about public reaction, company officials made sure that the plant's tepee-style incinerator burned its smokiest materials only at night and told outside contractors not to talk about any of the work they were doing on the factory's waste-handling systems.[20]

Toms River Chemical was so worried that its mistreatment of the river had damaged its reputation that in 1967 the company commissioned a poll of town residents. The results were a relief. Eighty-six percent said Toms River Chemical was a community asset, while just 6 percent called it a liability (the other 8 percent said it was both). Thirty-four percent said the factory was the major cause of water pollution in the region, but only 4 percent would object if it expanded.[21] "You came out smelling more like a rose than a chemical plant," the pollster said in the company newsletter.[22]

There was only one important note of caution: The town was full of newcomers who did not regard Toms River Chemical as the irreplaceable heart of the community. Sixty percent of residents questioned in the poll had lived in the area less than ten years, which meant that very few people remembered how weak the local economy had been before the Swiss arrived in 1952. Potentially more troubling for the company, only 10 percent of those polled worked at the plant or had a close relative who did, a significant change from the late 1950s when, by some accounts, one household in three had a direct connection to Toms River Chemical. Toms River was not quite the company town it had been in the 1950s. Toms River Chemical was still liked, but it was no longer venerated; mostly, it was ignored. If its environmental practices ever again became a focus of public interest, the company would have relatively few friends it could count on.

There was certainly plenty of action to distract residents from paying attention to whatever was going on behind the thick curtain of oaks and pines at the factory site. It seemed inconceivable that Toms River and the surrounding communities in Ocean County could grow more rapidly than they had in the late 1950s and early 1960s, yet they did. The deadly summer riots of 1967 in the predominantly black slums of Newark were both cataclysmic and catalytic. White flight from the urbanized northeastern corner of the state accelerated and then shifted farther south in 1970 after riots in Asbury Park, where there was a significant minority population just twenty miles north of Toms River. "After the riots, people were running down here in droves," remembered John Paul Doyle, a Toms River attorney who served in the state assembly for eighteen years starting in 1974. "The expression was: 'Riots and race gave them a reason, and the parkway gave them a route.'" Between 1960 and 1977, Ocean County's population tripled to 320,000; the 1970 census listed it as the sixth-fastest-growing county in the entire United States. Ocean had been solidly Republican for decades, but the latest migration made it whiter and more conservative. (By 1980, Ocean was 97 percent white.)

Among the flood tide of new arrivals was a young couple: Rusty and Linda Gillick. They settled into a comfortable home in the Brookside Heights section, near what would soon be the site of the town's third high school. Rusty worked at a local bank; Linda was a schoolteacher who stopped working full-time after their oldest son, Kevin, was born in 1971.

Like a fast-growing toddler whose shoes always seemed to be one size too small, Toms River was perpetually short of public services. Its roads and classrooms were overcrowded, and its water and sewer systems were operating at the edge of their capacities—and often beyond them. At first, the municipal sewage plants dumped their wastewater into the river, just as Toms River Chemical had. Later, with the assistance of engineers from the chemical factory, the sewer districts again followed the company's lead by building their own discharge pipes and shifting to ocean dumping. The school board, meanwhile, was under constant pressure to build more schools. In 1964, the board had jumped at the chance to build an elementary school in Oak Ridge even though the playground backed right up to the edge of Toms River Chemical's property. One of the board members, a physician, had objected—"We don't know what's back there," he had warned—but was outvoted, according to Don Bennett, a longtime local newspaper reporter who heard the story from his father, John, the superintendent of the Toms River schools at the time. By the late 1970s, Toms River would be the largest suburban school district in the entire state.

The soaring population growth diluted the importance of Toms River Chemical to the local economy, but the company's connections to the local power structure were still impeccable. Its Swiss executives mainly socialized among themselves but compensated by throwing themselves into the work of community groups trying to cope with the region's headlong growth. Company accountants even helped develop a formula that steered more state aid to local schools. The employee union was active in county politics, and several spouses or siblings of employees held local office, including mayor.

Chemists and engineers at Toms River Chemical also kept up their longstanding practice of assisting the perpetually overwhelmed Toms River Water Company, which needed all the help it could get. Almost every summer, the water company struggled to meet peak demand, which soared from one million gallons a day in 1960 to five million in 1970. (By 1977, peak daily demand would top ten million gallons.) The shallow wells on Holly Street, despite their proven vulnerability to whatever chemicals were still seeping into the river from the factory, remained the source of more than half of the drinking water Toms River Water supplied to its customers. The water company needed more wells—in fact, it needed enough open land to drill *many* more wells. In 1970, it found what appeared to be a suitable site, a parcel just east of the Garden State Parkway. By the summer of 1971, Toms River Water was operating four wells at its new Parkway well field. It added two more wells there the following summer. By 1975, there were eight Parkway wells—six of them shallow—supplying almost half of the town's water during the summertime peak.

The town had gone from depending on one set of wells beside the river to depending on another set beside the parkway, but the exchange seemed to be a good one. The new wells were in the most rural part of town, safely uphill from Toms River Chemical and away from the river. Except for the parkway, there was nothing near the new well field except some houses and churches—and a few run-down egg farms.

Samuel and Bertha Reich were among the wave of Holocaust survivors who settled in Ocean County and took up egg farming after World War II. "It was a hard business, but it was the easiest business to do if you didn't know any English," Bertha Reich remembered. The small farm they bought in Toms River in 1952 was one of many that lined Route 9 in those days, in the neighborhood known as Pleasant Plains. Poultry farming was never a lucrative line of work, and the widespread use of refrigerated trucks starting in the 1960s devastated the local industry by making it possible for huge meatpacking operations in the Midwest to long-haul meat and eggs all the way to New York and Philadelphia. Many of the farms on Route 9 shut down, including the Reichs' farm in 1965. Now that he was out of the egg business, Sam Reich found work as a construction manager in North Jersey, but the job did not pay particularly well, and the taxes on their property kept rising as the town grew. The Reichs' biggest asset was their land, and they began looking for ways to generate income from it. "We wanted some renters," Bertha Reich recalled.

Reich Farm was not far from the town dump, so Nick Fernicola had driven by it for years, including on his evening forays to the Rustic Acres to shoot pool. Sam Reich was a friend of a friend, so when Fernicola heard that Reich might be amenable to a business proposition, he approached Reich, in August of 1971. Fernicola suggested that Reich lease him the former egg farm's back two acres—conveniently out of sight from the main road—as a place to store the empty drums he was accumulating after emptying them in the town landfill.[23] The two men quickly agreed on a rental price that was dirt cheap: just $40 per month, starting August 15. Sam and Bertha Reich would later insist that they had never consented to turning their land into a toxic dump and had agreed only to allow temporary drum storage while Fernicola looked for a permanent site. Fernicola, meanwhile, would claim that any chemicals that ended up on the ground at Reich Farm got there via incidental spillage, not deliberate dumping—despite overwhelming evidence to the contrary. In any case, by September Fernicola's three trucks were making two or three runs a day between the Union Carbide factory in Bound Brook and Reich Farm.

Fernicola had arranged with Sam Reich to rent a building at the farm to store his fifty-five-gallon drums but soon outgrew that space. The drums started piling up outside—first by the dozen and then by the hundred. At first, Fernicola tried to find other places to empty the drums before taking them to the farm. The Dover landfill was no longer a friendly destination, so he briefly dumped waste at the nearby Whiting and Manchester town dumps. That arrangement ended in early November when a laborer in the Manchester landfill struck one of Fernicola's drums with the blade of his bulldozer, setting off yet another explosion and injuring the driver. By then, Fernicola had hit on a novel, if illegal, idea: He would make his own mini-landfill in the back of Sam Reich's farm. He used a front-end loader to scrape out a series of shallow trenches into which he and his haulers could dump drums that were too damaged to be recovered.[24] Instead of using a ramp, Fernicola often would just roll the drums off the backs of the trucks and let them drop into the trench, where they often broke apart on impact. Thousands of gallons of toxic chemicals splashed directly onto the farm's sandy soil, with no barrier to prevent them from seeping down through the sand and into the groundwater that flowed beneath.

By the end of November, Fernicola had dumped five thousand drums at Reich Farm—in addition to the fifteen hundred drums he had dumped that summer at the town landfill. Business was booming, and Fernicola's spurned partners at the landfill resented it. Things came to a head one November night at the Rustic Acres, when Sharkey and Columbo confronted him. What happened next, according to Fernicola, was a scene that combined elements of *The Godfather* and *The Three Stooges.* In a court deposition years later, Fernicola claimed it happened this way: "They said I am taking bread out of their mouths. . . . They threatened me: 'A match would cure all your problems. We know where you're dumping.' So I knocked both of them out cold. . . . I grabbed them in the back of the head and whacked the heads together and they went out, and I left."[25]

There is no way to know how much Fernicola embellished his description of the head-knocking incident, but what is incontestable is that a few days after the fight, the authorities started taking at least a vague interest in Fernicola's operation at Reich Farm. Someone from the town government called the New Jersey Bureau of Solid Waste Management, an arm of the attorney general's office, but that office—whose small staff was responsible for investigating dumping throughout the entire state—declared that it lacked jurisdiction, based on assurances from Fernicola that he was only storing drums on private land, not dumping them.[26] There is no indication that anyone from the state bothered to take a look at the site to see if Fernicola was telling the truth.

Finally, in early December, Sam Reich awoke to the magnitude of what was happening on the rear two acres of his property. He got into an argument with Fernicola about the dumping, which by now was generating a powerful odor. A few minutes later, summoned by Reich, the town police showed up and arrested Fernicola, charging him with dumping without a permit and violating zoning laws. Union Carbide officials, also summoned by Reich, paid their first visit to the farm on December 15. The following day, the company finally barred Fernicola from taking any more of its drums.

The question now was, what to do about the toxic waste? More than five thousand drums were strewn across the back of the old farm, and Fernicola was in no position to remove them. When Reich called the newly created state Department of Environmental Protection to ask for help, he was told that the state had no jurisdiction on private land and that his lawyers would have to take it up with Fernicola and Union Carbide. For their part, town officials were in no rush to see exactly what kind of a mess they had on their hands; they waited until the end of the following month—January 1972—to send a team over to Reich Farm for a detailed look. When the officials finally arrived, they found a foul-smelling, snow-covered wasteland, with thousands of drums lying in open, muddy trenches near some abandoned trucks. Jack Farrell, the town's code enforcement officer, said in a legal filing that his clothes stank for days after his brief inspection and that his rubber galoshes disintegrated after he walked in the muck. The inspectors found drums marked with labels like "Polymer Solution," "Toluene," and "Styrene" as well as large-print warnings stating the contents were flammable and explosive. Many carried this label: "Caution: Leaking Package Must Be Removed to a Safe Place. Do Not Drop. Chemical Waste."

Don Bennett, the newspaper reporter, remembers getting a call from Jack Farrell, who said, "You're not going to believe what I found on this old chicken farm." Bennett hustled over and was shocked by the scene. "The drums were just all over the place, everywhere you looked," he recalled. Farrell and the other town health officials knew so little about toxic waste that the only way they could think of to assess the hazard was to haul four of the Union Carbide drums to the town landfill and attempt an extremely primitive form of chemical analysis—using live ammunition. Under Farrell's direction, a town police officer named Michael Carlino pulled an army rifle from the trunk of his Ford Thunderbird and took aim at the drums, which were set up on an earthen berm one hundred yards away. After emptying two clips from his M-14, Carlino finally scored a direct hit. Bennett, who witnessed the spectacle, said that Farrell turned to him and said, "Well, I guess the stuff can't be that bad. It didn't explode."

While Carlino was target shooting at the landfill, Nick Fernicola was busy haggling with the town and with Union Carbide over what to do with the drums he had dumped at the farm. The company at first claimed that it bore no responsibility for removing them, citing Fernicola's promise a year earlier to assume all liability for disposal. Fernicola, for his part, was engaged in some fancy footwork, telling the town and Union Carbide that he was deep into final negotiations for yet another dumpsite, this one in Berkeley Township, on the appropriately named Double Trouble Road. By February, however, Fernicola's shenanigans were public knowledge in Toms River. The local papers were carrying stories about the illegal dump at the back of a Pleasant Plains egg farm. Meanwhile, Fernicola's "negotiations" with Berkeley were going nowhere. The Reichs had filed a lawsuit against Union Carbide and Fernicola and so had the town, which argued that the site was a public nuisance and a health threat. The state Department of Environmental Protection was preparing to bring charges, too.

It was obvious that Fernicola and Union Carbide were out of options and ready to cut deals. They found willing partners in government officials who were just as eager to close out an embarrassing incident in which they appeared to be either negligent or corrupt, or both. Union Carbide switched tactics and agreed to take back the drums, pay the Reichs $10,000 in damages, and reimburse them for the cost of digging a new water well for the property, since the old well had been poisoned. Crews from Union Carbide and a new waste contractor hauled the leaky drums back up Route 9 to their starting point in Bound Brook and eventually to landfills and incinerators around the region. At first, the company took just thirty-five hundred drums off the property. Only after the Reichs threatened to reinstate their lawsuit did Union Carbide return to haul away another fifteen hundred drums. State and town officials, meanwhile, were similarly focused on making the problem go away as quickly and quietly as possible. No one was prosecuted for corruption at the landfill, which would close nine years later, but Sharkey, Columbo, and the landfill manager were transferred to jobs elsewhere in town government.

By July 1972, just seven months after Fernicola ended his dumping spree, the whole incident had been buried—literally. Having removed all of the drums that were in plain sight, Union Carbide bulldozed Fernicola's trenches, covering up the solvent-soaked soil and some stray drums too. The thousands of gallons of hazardous chemicals that had spilled, or had been poured, into the unlined trenches and onto the sandy soil of Reich Farm were now out of sight and out of mind, and so was all the Union Carbide waste that earlier had been dumped and buried in the town landfill. No one seemed the least bit worried about where all that buried toxic waste might go—in particular, whether it would trickle down through the sand and reach the water-saturated layer of soil, the aquifer, that the entire town depended upon for its drinking water. No one tested the backyard wells of the Reichs' neighbors to see if they were tainted, even though officials already knew that the Reichs' own well had been poisoned. Could the plume of contaminated groundwater spread even farther? By the summer of 1972, the Toms River Water Company was using six newly drilled wells at the Parkway well field, about a mile south of Reich Farm. Might those wells be affected by Fernicola's folly? No one tried to find out, even though the water company was pumping more groundwater every year to try to keep up with ever-increasing demand and even though the natural direction of groundwater flow—southward—ensured that the chemical plume from Reich Farm would head toward the six new wells.

The Reichs stuck it out for another four years before moving away, but their efforts to sell the old farm were hopeless. No one wanted to own a leaking toxic dump. Instead, the Reichs watched in frustration as their neighbors made big profits selling to developers. Forty years after Nick Fernicola's 1971 misadventure, their property was still in limbo and Bertha and Sam Reich—at ages eighty-four and ninety, respectively—were still bitter. They blamed Union Carbide, government regulators, and especially Fernicola, and they defended their original decision to rent out the back two acres for $40 a month—money Fernicola never bothered to pay them—in exchange for four calamitous months of dumping. "All we did was rent out to a guy who was supposedly gathering the empty drums and selling them," said Bertha Reich. "We didn't know he was a crook."

As for Fernicola, he exited the waste hauling business; he had little choice, since the town had seized his trucks. (Four years later, his permanent departure from hauling would be formalized in an agreement with the state in which Fernicola was also required to pay a $100 "settlement.") He moved on to a new venture that fit his talent for prevarication: fixing and selling used cars. Fernicola stayed in Ocean County until his death in 2006 but rarely spoke about his year as a waste-hauling entrepreneur, doing so only when compelled by legal subpoena. When he did talk about it, he did so with rueful self-pity instead of remorse and with characteristic dissembling. It was unfair to single him out, he would complain, because other haulers paid similar bribes to dump drums at the town landfill for much longer time periods, while he did it for less than a year.[27] "My whole thing in the drum business lasted about five months," he told lawyers in a 1993 deposition. "From what I heard, everybody made money but me. I am the only one that lost money and I started it. I lost the trucks. I lost everything."[28]

Others would lose more.

CHAPTER SIX

≈

Cells

The labor was painful, and he took a few frightening extra seconds to draw his first breath, but nothing else about the birth of Michael Thomas Gillick at Point Pleasant Hospital on February 1, 1979, suggested that anything was amiss. A few days later, his parents, Linda and Raymond "Rusty" Gillick, took Michael home to nearby Toms River, where his eight-year-old brother, Kevin, was waiting. Michael immediately became the center of adoration. He was an unusually attractive baby, with blue eyes, long lashes, and a sunny disposition to match his golden hair. He looked like a cherub in a Renaissance fresco, pink and lively. Linda Gillick, who had taught first-graders for years, considered herself a good judge of children. This child, she decided, was perfect. She never changed her mind about that, despite all the horrors that followed. In his mother's eyes, Michael would always be beautiful.[1]

In May, shortly after he turned three months old, Michael vomited after his morning feeding, which was out of character for him. Later that same day, when his father went to Michael's crib to pick him up after his afternoon nap, the infant's eyes were darting back and forth, as if he was staring at a swinging pendulum. By the time his terrified parents got him to the pediatrician's office, Michael's eyes had stopped their bizarre movements, but the pediatrician recommended a visit to a neurologist, who gave Michael an electroencephalogram, a simple test of electrical activity in the brain. It came back normal, but Michael did not stop vomiting. The formerly placid infant also began to twist and turn in his sleep, as if he could not get comfortable. Five days after Michael's symptoms started, his father was changing Michael's diaper when he saw a lump beneath his son's belly button. In a few more days, there were similar lumps on the small of his back and his right leg. Whatever they were, they were spreading with ferocious speed.

Mystery and dread have always been the close consorts of cancer. The disease's causes and progression have never been well understood, and its prognosis (at least until recently) has typically been dire. Cancer is older than humanity: In a Kenyan lakebed in 1932, anthropologist Louis Leakey found a fossilized lower jaw of a hominid ancestor, possibly *Homo erectus* or *Australopithecus,* that included a malignant bone tumor. A tumor has even been discovered in a dinosaur bone at least a hundred and fifty million years old.[2] The oldest surviving description of cancer is a papyrus from approximately 1700 B.C. but is probably a copy of a document at least a thousand years older. The fifteen-foot scroll, stolen from Thebes by tomb raiders in 1862 and sold to an American adventurer named Edwin Smith, includes a description of an attempt to treat tumors of the breast with a cauterizing "fire drill." For "bulging tumors," the scroll's anonymous author notes, "there is no treatment."[3]

The Greeks fared no better. While surgical removal of tumors was sometimes tried in ancient Egypt and India, Hippocrates preferred diets that were supposed to reduce "black bile" and restore the body's "humoral balance." But after watching his diets and more aggressive treatments fail, Hippocrates wrote in *Aphorisms,* "It is better to give no treatment to cases of hidden [internal] cancer; treatment causes speedy death, but to omit treatment is to prolong life."[4] Hippocrates did make one lasting contribution by observing that bulbous tumors, especially when encircled by veins, looked like crabs. He used the Greek words *carcinos* and *carcinoma,* derived from the word *karki-*

nos, or crab, to describe the condition. Celsus, the venerated Roman physician whom Paracelsus boasted of surpassing, translated that to *cancer,* the Latin word for crab.

The description fit, not only because of the shape of the solid tumors but also because of seemingly inexorable nature of cancer's progression. "Cancer the Crab lies so still that you might think he was asleep if you did not see the ceaseless play and winnowing motion of the feathery branches round his mouth," Rudyard Kipling wrote in an 1891 story, "The Children of the Zodiac." "That movement never ceases. It is like the eating of a smothered fire into rotten timber in that it is noiseless and without haste."

The ancient healers were flummoxed by cancer for the same reasons that stymie present-day researchers. Tumors are as diverse as the sixty bodily organs in which they can arise. Some grow slowly, while others spread with stunning rapidity, often by hitching a ride in the bloodstream to other parts of the body, the process known as metastasis. Some are hardened masses that distend the skin, while others are buried deep in the body cavity. A few types of cancer, including leukemias, rarely form tumors at all unless they have metastasized to other organs. Most frustrating of all, various cancers respond very differently to treatments. Despite Hippocrates' attempt at a unifying taxonomy, it turns out that cancer is not one disease but many—more than 150, by most definitions. Their only common characteristic is supercharged cell division, growth run amok.

The first person to grasp the essential nature of cancer was another irascible, opinionated man of science: Rudolf Ludwig Karl Virchow. A diminutive, humorless, and hyperkinetic German born in 1821, Virchow was a physician who also made important contributions in an astonishing number of other fields, including anthropology, paleontology, and the biology of parasitic worms. In his spare time, he designed the Berlin sewer system and helped to excavate ancient Troy.[5] His chief fault was Paracelsian self-confidence, which made him reluctant to accept ideas he did not originate, including the two most important of his lifetime: Louis Pasteur's germ theory and Charles Darwin's theory of universal common descent via natural selection. A fiery liberal who manned the Berlin barricades during the failed revolution of 1848 and was later a leading reformer in the German parliament, Virchow believed that social progress came through vigorous observation and testing, not abstract theory. "Medicine is a social science," he declared in *Die Medizinische Reform* (The Medical Reformation), a radical newspaper he published during the tumult of 1848, "and politics is nothing else but medicine on a large scale."[6]

A prodigy who could read Latin and Greek by age twelve, Virchow became fascinated with microscopes soon after choosing medicine for a career. While still in medical school, he began conducting microscopic examinations of diseased tissue, something almost no one else was doing at the time. His medical approach was as radical as his politics because, even as late as the mid-nineteenth century, the humoral theories of Hippocrates still held sway and illnesses were regarded as mere indicators of the same underlying problem: a bodily imbalance of blood, yellow bile, black bile, and phlegm. That is not what Virchow saw in his microscope. He saw groupings of diseased cells in bodies that were otherwise healthy. Diseases, he reasoned, were not signs of an organism out of balance. Instead, they were distinct, specific processes that could be monitored through close observation of aberrant cells under the microscope. Thus was born the science of microscopic pathology, the essential discipline of modern medicine, which Virchow championed for the rest of his long and extraordinarily productive life.

Others had previously suggested that the cell was the basic unit of life, but Virchow was the first to propose that it was also the basic unit of disease.[7] An epigram he popularized but did not originate, *omnis cellula e cellula* (all cells arise from cells), therefore meant that cell division must be the means by which illnesses spread inside the body. Pathological processes begin, he reasoned, when a healthy cell malfunctions under the influence of some outside force. Virchow erred in rejecting Pasteur's idea that the outside disruptor could be a living microorganism, a germ. Actually, both men were partly correct: Some diseases were microbial in origin and others were not, but almost all involved the disruption of cells in specific parts of an otherwise healthy body. With their separate insights, Virchow and his rival Pasteur drove a stake through the heart of classical humorism, finishing what Paracelsus had started in Basel more than three hundred years earlier.

Cancer fit neatly into Virchow's ideas about the cellular nature of disease, and his close observations of malignant cells helped him form his theories. In 1845, two years out of medical school, he was the first to observe that some sick patients had far too many white cells in their blood and too few platelets and red blood cells. He coined the word *leukemia* to describe their condition because it meant "white blood" in Greek. Later, he noticed that a swollen lymph node above the left collarbone—now known as Virchow's node—was often an early sign of cancer, an indicator still used by physicians today. Virchow poured all of his ideas and observations about cancer into an eighteen-hundred-page, three-volume work entitled *Die Krankhaften Geschwülste* (The Malignant Neoplasms), published in 1863. Many of his core beliefs have been vindicated, including the cellular nature of cancer, the central role of rapid cell division in its development, and the importance of an initiating event, an "irritation," to begin the process by disrupting a previously healthy cell. As Virchow aged, his irritation theory fell out of favor and researchers drifted toward competing theories of carcinogenesis. He died in 1902, at age eighty—too soon to see his ideas return to vogue, and in variations that even someone with Virchow's remarkable foresight could not have anticipated.

One of the many reasons the discovery of a malignant tumor in a child is so emotionally wrenching is that it is so surprising. Cancer is, in the main, an affliction of the elderly. In any given year, a person over age sixty-five in the United States is almost ten times more likely to be diagnosed with cancer than someone younger.[8] In fact, between ages five and sixty-nine, the likelihood of getting cancer in any particular year rises with each year of life, and it does so in increasingly large intervals: from about one in nine thousand in the fifth year of life to about one in fifty-seven in the sixty-ninth year. There seem to be many reasons for this: complex molecular changes that occur in the cells of older organisms, immune deficiencies associated with aging, and the many years and intricate biochemical steps required before some types of tumors (prostate lesions, for example) can begin growing.[9]

And yet, about thirty-eight times a day in the United States, and perhaps eight hundred times elsewhere on Earth, a child is diagnosed with cancer. That is still a rare occurrence, with fewer than one in six thousand American children diagnosed per year, but not as rare as it used to be. Childhood cancer incidence jumped by more than one-third between 1975 and 2005—more than twice as much as overall cancer incidence.[10] Improved case reporting and earlier diagnosis explain some portion of the overall increase but do not explain why cancer cases among children are rising at a much faster pace than cancers in adults. Nor has science credibly explained why a child would get an old person's disease in the first place.

For Michael Gillick's parents, the diagnosis of neuroblastoma, a cancer of the nervous system, was a thunderbolt from a cloudless sky, a sucker punch from the blind side. They were on the floor before they even knew what hit them. A week after his father discovered the first lump, three-month-old Michael was strapped down on an operating table at New York Hospital in Manhattan. A surgeon made an incision from his neck to his groin and tried to remove a softball-sized tumor that had enveloped Michael's left kidney and adrenal gland and the blood vessels near his heart. When the discouraged surgeon realized that the tumor was far too large to be safely cut out, especially in light of Michael's sky-high blood pressure (170 over 100, instead of the normal 70 over 50 for an infant), the doctor sewed him back up, noting on his way out that the cancer had spread to the liver and lymph nodes. Michael, the surgeon told his stunned parents, had a 50 percent chance of reaching his first birthday. A cure was unlikely (at the time, the long-term survival rate for metastatic neuroblastoma was only about 5 percent), but aggressive chemotherapy might buy him some time.[11]

Michael's cancer, like most others, was named for the cells where the malignant transformation began. Neuroblasts are one of the everyday miracles of fetal development, primitive stem cells that form during the first days of an embryo's existence and gradually transform into neurons, or nerve cells. In most of us, neuroblasts do their job and then quietly retire, occasionally stirring again to supply additional neurons when needed. But Michael's neuroblasts never stopped running at top speed. While he was still in the womb, they started to divide frenetically, clumping together in fast-growing tumors that first formed in his adrenal glands but could appear in any organ with nerve cells, including his eyes, spine, and brain. The cancer was named in 1910 after its origin in neuroblasts was confirmed, but Rudolf Virchow was the first to describe it, in 1864, based on a case description of a child with a large abdominal tumor. Michael's cancer probably began in the first weeks of his mother's pregnancy. A year later, the tumors were large enough to compress many of his vital organs, including his heart, lungs, spine, bowels, eyes, and brain. They also distorted Michael's facial features, transforming his countenance from Raphaelite cherub to something much more abstract.

By themselves, the symptoms of neuroblastoma were more than sufficient to kill Michael, but now he also had to face the horrific side effects of chemotherapy and high-dose steroid treatments, including chronic diarrhea and vomiting, severely stunted growth, a hypersensitive stomach, pounding headaches, brittle bones, and collapsed veins. It was all too much for the Gillicks, and after weeks of chemotherapy failed to slow the growth of the main abdominal tumor, they decided to take Michael home. When he turned six months old, his family celebrated Michael's first birthday, assuming that it would be their only opportunity to do so. Linda Gillick called funeral parlors and even purchased an infant-sized coffin so that she would not have to go through the agony of buying one when the time came. Her son had been given the medical equivalent of a death sentence; he was trapped in a prison from which there was no realistic chance of escape.

Michael Gillick did not escape his cell, but he did not die either. He made it to Christmas, then to his real first birthday and then to another Christmas, and another. His tumors did not spontaneously disappear, as sometimes happens with neuroblastoma, but their growth slowed. Contrary to the doctors' predictions, the malignancies did not crush any of his vital organs, though there were many terrifying emergency hospitalizations. Radiation therapy cost Michael his golden locks without eliminating the facial tumor that damaged his hearing, blurred his vision, and several times even temporarily blinded him—an especially terrifying experience for a child.

Three types of chemotherapy similarly failed. At Michael's insistence, the grueling treatments stopped when he was eight. Other than morphine, one of the few helpful drugs his ravaged body could tolerate was a blood pressure medication called Regitine. But that drug was no longer being produced in pill form by its manufacturer, a company the Gillicks had heard of because *everyone* in Toms River had heard of it: Ciba-Geigy, the Swiss corporation that owned the local chemical plant. Michael was slowly working through his stockpile of the pill.

Several of Linda Gillick's neighbors worked at the chemical plant, but she was far too preoccupied with Michael to pay attention to the occasional articles in the local papers about pollution there and at the nearby egg-farm-turned-dumpsite known as Reich Farm. Like most residents of Toms River, the Gillicks lived east of the chemical plant—downwind, since the prevailing winds came from the west. And like many other families in town, their water came primarily from two well fields operated by the Toms River Water Company: the Holly wells, just downstream from the chemical plant, and the Parkway wells, a mile south of Reich Farm. In the early 1980s, those geographic proximities had no particular resonance to the Gillicks or anyone else in Toms River. The air smelled all right, and the tap water looked fairly clear—even if the water pressure was annoyingly low during the summer. So many people were moving into town that the water company always seemed to be struggling to keep up.

It was not surprising that the condition of the air and water in Toms River had escaped the attention of even alert citizens like the Gillicks. Some of the contamination problems were invisible and others were masked, such as the nighttime smokestack emissions. Behind its thick curtain of trees, Toms River Chemical was spending as little money on pollution control as it could. The company's executives had been explicit about this in 1968, according to an internal memo describing their reaction to a report on various waste treatment options: "After review of this report by top management, the policy of making these expenditures only as fast as absolutely necessary was stated."[12]

Environmental upgrades were "all money losers," as one manager put it, and in the early 1970s Toms River Chemical was making more money than ever. In 1973, the company produced a record 131 million pounds of dyes, resins, and other chemicals—up from 78 million in 1963. Fourteen hundred employees, also a record, toiled in thirty buildings on the huge factory complex, which was now owned by a new corporate entity: the newly merged Ciba-Geigy Corporation.[13] For all the growth and change, however, the company's waste-handling practices had changed very little, except that most of its wastewater now went into the ocean instead of the river. Toms River Chemical still sent liquid waste to unlined lagoons and dumped almost nine thousand drums a year "over the edge of a cliff" and into an unlined pit, spilling their toxic contents.[14] Until the state finally put a stop to it in 1970, workers still ignited solvents in "smudge pot" standpipes and burned an "almost unbelievable array" of waste chemicals in an archaic tepee-style incinerator.[15] The company actually had two te-pees, but one blew its top, volcano-style, when a drum inside it exploded. "Now they have a topless tepee," observed a bemused contractor.[16] There were other accidents, too, including three fires at the drum pit in 1968 alone.

In a brutally frank presentation to company managers in 1969, a newly arrived engineer named William Bobsein detailed all of the pollution-related messes at Toms River Chemical. Sixteen years later, Bobsein's career would take a turn for the worse, but back in 1969 he had inherited James Crane's role as in-house doomsayer, a teller of hard truths who lacked the power to fix the problems he saw. Tougher state regulations were coming soon, but the company was unprepared for them, Bobsein told the managers, punctuating his toxic tour of the plant with what he called "dirty pictures"—color slides of "drum mountains" and "drum cemeteries . . . despoiling the countryside" plus rooftop photos of "intense smoke" emitted by the factory's many stacks.[17]

Few residents of Toms River knew or cared about a litany of environmental ills they could not see or even smell most of the time. But in the world outside, attitudes were changing. The first Earth Day, April 22, 1970, was an epochal event, attracting twenty million Americans to hundreds of rallies and rocketing environmental protection from obscurity to a top-tier national concern. In Toms River, the events were benign—schoolchildren went on hikes and picked up litter—but the New Jersey legislature marked the date by creating a Department of Environmental Protection, which assumed the pollution-control duties formerly—and indifferently—performed by the state health department. An even more important change was that Congress and President Richard Nixon, mindful of public opinion, suddenly became very aggressive about asserting a federal role in combating pollution, a task that had almost always been left up to the states and cities.

It was a drastic shift. Major corporations like Ciba were accustomed to running over and around local regulators, who could not hope to match the companies' legal and technical expertise or their financial resources. Before public opinion began shifting in the 1970s, mayors and governors were much more likely to side with corporations than with their own regulatory agencies in disputes over pollution. Even in a state like New Jersey, where antipollution laws that included potential criminal penalties had been on the books for decades, environmental violations had always been considered to be a matter for civil negotiation, not criminal prosecution. This was the milieu in which Toms River Chemical had thrived for so long.

The brand new U.S. Environmental Protection Agency, however, would be much more difficult to stiff-arm. Its newly hired lawyers and engineers were specialists with advanced technical training. Many were also idealists who were less subject to political pressure than their state and local counterparts. They were from out of town, and their priority was to establish the primacy of federal environmental law, not to protect local jobs. Toms River Chemical began gearing up for conflict, creating a nineteen-member department to oversee pollution issues and conducting "teach-ins" on environmental compliance for more than a thousand employees.[18]

EPA officials paid their first visit to the Toms River factory in July of 1971; they brought a lawyer to show that they meant business. Congress had not yet given the new agency authority to regulate water discharges, but Nixon had already announced that the EPA would begin enforcing a legislative relic known as the Refuse Act of 1899.[19]

This dusty law had been mostly ignored for seventy-two years; its breathtakingly broad language forbade the discharge of "any refuse matter of any kind or description whatever" into navigable waters or their tributaries without the permission of the U.S. Army and its Corps of Engineers.[20] When company lawyers trotted out their shopworn claims that Toms River Chemical was already doing all it could to cut its discharges, EPA officials pointed out that other manufacturers were doing better.[21] And when the company asserted that its discharges were legal, the agency pointed to the Refuse Act of 1899.

Finally recognizing that the regulators were serious this time, Toms River Chemical tried to head off a prosecution by promising improvements, but by 1972 it was too late. Utilizing the Refuse Act, the United States Attorney's Office in Newark had just wrapped up a successful civil prosecution of more than a dozen New Jersey towns that had been dumping raw sewage sludge into the ocean.[22] Now the same federal prosecutor, Carl Woodward III, was turning his attention to the only privately owned ocean outfall line in the state. Subpoenas soon arrived at the offices of Toms River Chemical, summoning its executives to testify before a grand jury in Newark. "It was a very big case, but it wasn't difficult; the company had documented everything it did and we subpoenaed those records," Woodward remembered many years later. On July 13, 1972, the grand jury handed down an indictment charging Toms River Chemical with 206 violations of the Refuse Act for discharging chemical waste into the Atlantic Ocean and the Toms River. Each of the 206 counts (one for each day in 1971 and 1972 the EPA had detected a violation) was a misdemeanor carrying a maximum fine of $2,500; if Toms River Chemical was convicted on all counts, the maximum fine of $515,000 would be the largest environmental penalty in New Jersey history up to that point.

The indictments were not what they seemed, however. The EPA and the U.S. Attorney's Office could have gone to court to try to get an injunction forcing Toms River Chemical to clean up its discharges immediately. But even as early as 1972, the EPA was having second thoughts about its initial strategy of treating violations of laws like the Refuse Act as straightforward crimes. At the end of the year, Congress overrode President Nixon's veto and gave the EPA authority to regulate wastewater discharges via a new law that came to be known as the Clean Water Act. Like the Refuse Act, which it essentially superseded, the new law banned discharges into navigable waters without a permit, but the permitting process was much more complex and bureaucratic. So, instead of seeking an injunction, the EPA began years-long negotiations with Toms River Chemical over the terms of its Clean Water Act permit. All the while, the prosecution was held in abeyance, and the company continued to operate as it always had—only more so. In 1973, a year after the indictments, production and emissions were at record levels.

The obvious solution was for Toms River Chemical finally to replace its unlined lagoons with a real wastewater treatment system—specifically, an activated sludge plant of the design the company had been refusing to build ever since the late 1950s. Activated sludge plants, which used microbes and oxygen to reduce the organic content of sewage, had been in use for more than fifty years, but they were also expensive: Toms River Chemical's would eventually cost more than $15 million. Properly designing and building such a plant and customizing it to the factory's unique wastes would take three years, company executives insisted. They said essentially the same thing to state regulators, who were finally pushing for a new incinerator and a lined landfill.

When the EPA broke off its negotiations with the company in 1974 and issued a draft permit that would, for the first time, limit the company's discharges into the river and the ocean, Toms River Chemical immediately appealed. The proposed limits were too strict and the two-year phase-in was too short, the company argued, eventually getting its way.[23] The indictments were moved to inactive status and finally dismissed without a fine or admission of guilt. On a rainy September morning in 1977, in the presence of Governor Brendan Byrne and a bevy of local politicians who heaped praise on the company, Toms River Chemical dedicated its new wastewater treatment plant. It began operating the following year, as did the initial section, Cell One, of the company's first-ever lined landfill. It had only taken twenty-five years.

By and large, the people of Toms River reacted with a collective yawn. Most considered the ocean to be an acceptable place to deposit pollution, especially now that their own sewage was also going into the Atlantic via municipal wastewater pipes and sludge barges. Back in 1972, the news that the largest private employer in Ocean County had been indicted on criminal charges inspired just one article in the *Ocean County Observer,* the local daily newspaper. That article, which emphasized the "astounded" reaction of company officials, was quickly followed by an editorial defending Toms River Chemical as a "good neighbor" that "adds heavily to our locality's economy."[24] Many people in town were at least vaguely aware of the company's lackadaisical approach to pollution control, but there was still no public evidence that anyone in the surrounding neighborhoods had been hurt as a result. (The contamination of the Holly Street wells in the mid-1960s was still a secret.)

The situation was very different a little more than a mile northeast of the factory, however. There, the residents of a struggling, semi-rural neighborhood optimistically named Pleasant Plains were drinking industrial chemicals in their tapwater. And this time, it would not remain a secret.

Unlike many other parts of the fast-growing town, Pleasant Plains was not full of subdivisions and shopping centers catering to the hordes of affluent newcomers. Its residents were mostly laborers and farmers; they were old-timers who got their water the old-fashioned way. There were no water mains in Pleasant Plains, but no one minded because you could dig your own well and find plenty of groundwater just twenty feet down—forty at the most—and use as much as you wanted, free of charge.

The first Pleasant Plains residents to notice, in early 1974, that their water tasted and smelled bad were the owners of three properties on Route 9. There was little doubt about the cause: The three parcels were just a few hundred feet south of Reich Farm, the old egg farm Nick Fernicola had used as a dumping ground in 1971. Almost two years had passed since the work crews hired by Union Carbide had supposedly cleaned up the site, carting away five thousand leaky drums and bulldozing the trenches that Fernicola had used as crude repositories for toxic waste. The neighbors had not forgotten, however. "We all knew it was coming from the Reich farm," recalled Ernest J. Nagel, who owned a butcher shop on Route 9 with his father. The shop's water smelled, as did the water in several other nearby businesses. In the spring of 1974, the county's health coordinator, Chuck Kauffman, decided to check the entire neighborhood, sending the collected water samples to the only local facility with the technical know-how to analyze them for organic pollutants: the Toms River Chemical Corporation. The company, which for once was not the suspected cause of a local pollution problem, was happy to assist. Its chemists found unnaturally high levels of organic compounds in almost every Pleasant Plains drinking water sample they tested. The EPA confirmed the results, identifying two chemicals in the drinking water: toluene and styrene, both of which had been dumped at Reich Farm. Following up on a tip, state and county officials also made another trip to the dumpsite and found plenty of evidence that Union Carbide's two cleanups in 1972 had been shoddy. There were still fifty-one drums in plain sight, and the odor was still so strong that the state in mid-1974 ordered Union Carbide to come back a third time and remove a thousand cubic yards of chemical-soaked earth along with the additional drums.

That same summer, state environmental regulators took a drastic step: They directed the town to condemn 148 wells within a semicircle extending about a half-mile south, east, and west of Reich Farm. At the time, it was the biggest well closure in the history of the state. Residents were told to plug their wells with cement and either to buy bottled water or to get it from six portable water tanks dispatched to the neighborhood by the Army National Guard. A school, North Dover Elementary, was just outside of the condemnation zone, but the principal decided not to take chances and ordered her kitchen staff to stock up on bottled water in time for the start of classes in September. The crisis did not ease until November, when the Toms River Water Company, after weeks of wrangling with the town over costs, finally extended its pipes into Pleasant Plains, allowing affected homes and businesses to tap in.

For the people of Pleasant Plains, 1974 was a bewildering year. Accustomed to getting their water for free from their own backyards, they were told they would have to abandon their wells and start paying monthly bills to Toms River Water. Many simply refused, defying the town and state. Others, especially if they could actually smell the chemicals in their tap water, made the switch but spent the rest of their lives wondering whether they had been poisoned in 1974. "All of our wells were contaminated, and we did drink it for a period of time until we got city water," recalled William Hyres, whose auto body shop was about two thousand feet from Reich Farm. "Did anyone get sick because of it? I don't know. It did stain our clothes, I know that."

The county and town health departments made a halfhearted attempt to find out whether an unusual number of people had gotten sick in Pleasant Plains that summer, but it was quickly dropped. In June, just after news of the contamination appeared in the local newspapers, the county health department conducted a cursory health survey, questioning twenty-three randomly selected families in Pleasant Plains. Fifteen reported at least one recent illness involving the kidneys, stomach, liver, or gallbladder, but that group included several whose water was unaffected; meanwhile, some other families known to have consumed tainted water reported no illness at all. Since there was no obvious correlation, the county dropped the investigation without any follow-up to look for longer-term health problems. "We didn't have the expertise to do it, and I had almost no staff at all at that point," Chuck Kauffman explained many years later. At the time of the Pleasant Plains contamination, Kauffman was still new on the job, having been appointed the county's first health coordinator in 1973.

Twenty-five years later, when the entire town was in an uproar over cancer and pollution and a multimillion-dollar health study was under way to examine links to tainted water, no one ever tried to look back at what had happened to the residents of Pleasant Plains, despite the well-documented contamination there. Informally, the talk of illnesses would persist for years. At the butcher shop, Ernest Nagel's father, also named Ernest, died of leukemia in 1976. The senior Nagel had drunk copious amounts of water from a shallow well behind the shop—and kept doing so even after it started to smell in 1974. The following year, the Nagels joined a class-action suit against Union Carbide that was eventually settled out of court with modest payments from the company—barely enough to repay residents for the expense of connecting to public water mains. "The lawyer said there was nothing else we could do," Nagel said, "so the whole thing was just dropped."

Elsewhere in Toms River, the tribulations of Pleasant Plains provoked sympathy but very little anxiety. The articles on the topic that appeared in the local papers during the summer of 1974 invariably described the contamination as a localized problem, one that would go away as soon as the neighborhood was linked up to the Toms River Water Company's distribution system. The subtext could not have been clearer: Backyard wells were dangerous; city water was safe. The news stories avoided the awkward fact that six of Toms River Water's newest and most important supply wells—three of them just 125 feet deep—were only one mile south of Reich Farm and therefore in the path of the contamination plume as it crept southward from the illegal dump. Nor was there any talk about how the speed and direction of that underground plume might be influenced by another uncomfortable fact: Those six new public wells in the Parkway well field were running at full capacity, sucking in more than two million gallons of groundwater every day, as the Toms River Water Company labored to keep up with burgeoning demand.

In truth, there was already reason to worry. In June of 1974, Union Carbide tested a water sample from one of the Parkway wells and found petrochemical concentrations of between three and ten parts per million—at least four times higher than the state's informal guideline of no more than 700 parts per billion. The state Department of Environmental Protection then did its own tests and found solvents in four Parkway wells, though at lower levels. The highest detection the DEP found was forty-two parts per billion, which was about sixteen times less than the state's guideline of that era but still more than eight times higher than today's health standard.[25]

There was something in the Parkway wells, but exactly what it was and where it came from was anyone's guess. The typical state testing protocols of the time were only sensitive enough to indicate broad groupings like "petrochemicals," "phenols," or "organic compounds." "The tests that we had back then were very primitive," recalled Herb Roeschke, who became the town's first full-time health director in 1978. "We could not look for very specific compounds or lower concentrations." Many years later, hydrologists would guess that the chemicals found in Parkway wells in 1974 were almost certainly industrial solvents, but that was still not proof that they had come from Reich Farm. Solvents like trichloroethylene, perchloroethylene, and benzene—all now classified as known or likely carcinogens—were important components of the chemical waste Nick Fernicola had dumped at Reich Farm in 1971, but they were also used by gas stations, machine shops, and dry cleaners, among other places. Those solvents could have come from almost anywhere; Reich Farm, almost a mile away, was only one possibility.

No matter where they came from, solvents were showing up in one of the town's most heavily used wells, and that worried Chuck Kauffman at the county health department. He asked the DEP to order Toms River Water to install a carbon filtration system on the Parkway well where the contamination levels were highest. But the water company balked at the half-million-dollar cost, and the state—fatefully, it would turn out—did not insist. There is no record that anyone ever proposed shutting down the well or using less water or paying for sophisticated testing that could have identified specific pollutants even at low levels. Nor did anyone inform the thousands of Toms River Water customers who got their drinking water from the well. In 1974, the water company was still being touted as the savior of Pleasant Plains—its mains were being extended into the beleaguered neighborhood—not as a fellow victim of Fernicola's dumping.

The news that a public well was tainted finally broke in January of 1975, when an article in the *Asbury Park Press* disclosed the test results from the previous summer. The DEP and the water company moved quickly to discredit the story.[26] In a follow-up article, a top DEP official dismissed any concern: "We sure wouldn't let people drink anything that would be dangerous," he said.[27] Local officials did not challenge him; in fact, they seemed eager to forget about the whole thing. In pollution matters, they had a very long tradition of deferring to Toms River Water and Toms River Chemical. "You have to remember that except for the people at Ciba, no one in Ocean County had any knowledge that these kinds of chemicals moved through soils," recalled Kauffman, the county health coordinator. Before Kauffman's arrival in 1973, in fact, there were no full-time county health officials. There were town health inspectors, but just barely. Many Ocean County towns did not bother to do inspections at all; the Toms River inspector had a high school education and spent most of his time looking at plumbing. Kauffman's background was in agriculture: He had been an egg farmer before getting licensed as a sanitary inspector.

The Toms River Water Company was essentially left to police itself—and its focus, as always, was on meeting the water demands of the fast-growing town. As the pollution spread south from Reich Farm, the water company's officers did not waver in their public assertions that there was nothing to worry about. The company's laissez-faire attitude did not change even after phenols were discovered in the spring of 1976 in thirteen more backyard wells—this time a full mile south of Reich Farm and just a few hundred feet west of the Parkway well field, at levels as high as six parts per million. By now everyone knew the drill: The thirteen wells were ordered closed, and the National Guard dispatched a water truck to the area (filled with water from the Parkway wells), while water mains were extended to serve the two affected streets. Was Reich Farm, a mile to the north, to blame for the newly discovered contamination? There was no proof, insisted the state and the water company. Was it finally time to add carbon filters to the Parkway wells, which were just a few hundred feet from the newly discovered contamination? No, that would not be necessary; continued monitoring was sufficient.

Whatever had happened in Pleasant Plains was over, as far as the guardians of Toms River's water were concerned. The state DEP, which had sued Union Carbide and Nick Fernicola in late 1975, quietly settled the case in 1977. Union Carbide agreed to pay $60,000 to reimburse the state for its groundwater testing costs. Fernicola agreed to pay just $100 and stay out of the chemical waste business for good.

In return for the minuscule payments, the state dropped its pollution charges against both without requiring either Fernicola or Union Carbide to admit guilt.

No one had done anything illegal, and there was nothing to worry about. The water was perfectly safe.

During the first trimester of Linda Gillick's pregnancy in the summer of 1978, the Parkway wells were the source of 25 to 50 percent of the drinking water the Toms River Water Company pumped to her neighborhood, which was known as Brookside Heights. Another 10 to 25 percent of the neighborhood's water came from the Holly Street wells, which were still being used ten years after the Toms River Chemical Corporation had secretly contaminated them. Other neighborhoods in Toms River, especially in the more northerly parts of town, got at least 90 percent of their water from the Parkway wells, while some of the areas to the south got almost all their water from Holly Street. The Gillicks lived near the geographic center of Toms River, so their water was a mixture of many sources, with the Parkway and Holly well fields supplying the lion's share. Linda Gillick, of course, did not know any of this at the time. No one knew, including the water company, which operated the interconnected system without knowing which wells supplied which neighborhoods. Those facts would not be established for another quarter-century, and even then, no one would be sure what any of it really meant.

Back in the late 1970s and early 1980s, unless they were unlucky enough to live in Pleasant Plains and to rely on a backyard well, few people in Toms River knew or cared where their drinking water came from. They turned on the tap, and it was there. Certainly Linda Gillick's concerns lay elsewhere. The health of her youngest child was the focus of her life, and when Linda Gillick focused, she was like a magnifying glass on a sunny day. A fierce, formidable protector of Michael's interests, she knew more about the intricacies of his condition than most of the doctors and nurses who treated him, and she was not reluctant to say so. She also made a point of trying to bring as much pleasure as she could into the lives of both of her sons. When the local papers started publishing stories about Michael, she became savvy about how reporters did their work—a skill that would later be crucial. By encouraging or at least tolerating publicity, she discovered, she could be an example to other families while also making it possible for Michael to have some unique experiences. The Gillicks got to visit Yankee Stadium, where Reggie Jackson gave Michael one of his bats, and then Giants Stadium, where Michael pretended to lift weights with star running back Joe Morris. They even got to watch the Giants win their first Super Bowl in 1986, meeting actor Michael J. Fox there and attending the victory party, where the thrilled Michael Gillick, a devoted Giants fan, was "passed around like a human football from one player to another," his mother later wrote.[28]

For Michael, it was a twilight existence. Smart and observant, despite impaired vision, Michael knew how other people reacted to his appearance and was acutely aware that he was chubby and half the size of most children his age. Too weak for school, he was tutored at home. Most of the kids he knew were fellow patients he met during hospital stays. Many had died, and the ones who recovered often did not want to be reminded of their ordeal by staying friends with Michael. Some of his closest relationships were with adult celebrities like Michael J. Fox and local lawyer John F. Russo, who had arranged the Super Bowl trip shortly after being elected president of the New Jersey State Senate. For as long as Michael could remember, death had been omnipresent. "When I was very young, my mom gave it to me straight," he remembered years later. "She said there was a battle going on inside my body, and she didn't know who was going to win." Sometimes he wondered what it would feel like to stop fighting. Once, at age eight, in the midst of another dire health crisis, he had what he later described as an out-of-body experience. Looking down on his hospital bed from above, he saw his mother weeping inconsolably. "When I saw that, I couldn't do it. I couldn't go," he remembered.

Michael Gillick was not the only beneficiary of his mother's devotion. The long days and nights in pediatric oncology wards left plenty of time to get to know other families, and Linda Gillick, unlike her reserved husband, was a born talker. She did not wait in hospital waiting rooms, she *occupied* them—filling every square foot of fluorescent-lit space with nonstop conversation. By the time Michael was five, she had had heart-to-heart talks with dozens of parents waiting to find out whether their children would survive another medical crisis. She learned more about childhood cancer than she ever thought she would. Besides neuroblastoma, there was Wilms' tumor, which struck kidney cells, and astrocytoma, which began in the brain. There was osteosarcoma in bones and Hodgkin's disease and lymphoma in white blood cells. And seemingly everywhere was acute lymphocytic leukemia, which began in bone marrow and was responsible for about one in four cases of pediatric cancer, far more than any other type. There were so many sick children, so many frightened parents. They deserved the same support she was giving Michael, Linda Gillick decided. She took a job at a local cancer charity and threw herself into the task of raising funds to support research and assist stricken families.

Without intending for it to happen, Linda Gillick became the hub of information about childhood cancer in Ocean County. By virtue of her assertive personality, her new fund-raising job, the news articles about Michael, and the many hours she spent at hospitals, Gillick knew just about every local family with an afflicted child. Sometimes her phone would ring and a parent she did not know would be on the line, full of questions. Just one case was heartbreaking enough, and now she was hearing about dozens. Ocean County, especially Toms River, seemed to have more than its fair share of misery, she thought. But Gillick had no way to judge whether the number of cases was truly unusual. She knew nothing about cancer epidemiology; she was just a mother who kept her eyes open.

As for Michael's neuroblastoma, the Gillicks rarely speculated on its possible causes. Why dwell on the past when there was work to be done? They did not argue when doctors told them there was no medical explanation for why neuroblasts run amok in about 250 American babies each year, while staying somnolent in four million others. Devout Catholics, the Gillicks considered Michael's illness to be god's will; this acceptance helped them get through their ordeal, one day at a time.

Late one night at New York Hospital, however, Linda Gillick heard something that got her thinking. She was, as usual, talking to another parent, a man who had been a construction worker in Toms River in the 1960s. He told her a story that at first sounded outlandish, explaining that he had helped to lay a pipeline that carried toxic waste right through the middle of town. The pipe started at Toms River Chemical and passed near the Gillicks' home in Brookside Heights, where many of the factory executives lived, too. The story startled Gillick, but she was too busy caring for Michael to try to learn more. It did, however, make her wonder what else she did not know about her town.

≈

CHAPTER SEVEN

On Cardinal Drive

Seen from the front, Cardinal Drive could be almost anywhere. The houses, mostly built in the boom years of the 1960s, are split-level or ranch-style, with oversized garages, manicured hedges, and handsome picture windows set into painted brick or shiny white siding. To see what makes the street unique, you need to walk around the back. Instead of another row of houses on another gently curving street, the west side of Cardinal Drive backs up against the last large undeveloped space in Toms River. You can sit in a lawn chair or float in a backyard swimming pool and imagine you are at the very edge of suburban civilization, gazing west into the forest primeval. You are not, of course. You are looking at the forest of Ciba-Geigy. Just seven hundred feet away, completely hidden by a thick curtain of trees, is the closest of a series of waste pits and lagoons extending to the north and west. For almost as long as there has been a Cardinal Drive—at least fifty years—those leaky pits have sent toxic chemicals coursing through the ground beneath Cardinal Drive and into the Oak Ridge neighborhood.

The extra privacy on Cardinal Drive was a special attraction for the families who moved there. The McVeighs already lived in Oak Ridge but moved to Cardinal Drive in 1977 because William McVeigh loved the idea of having a forest in his family's backyard. "It was so peaceful," remembered his wife, Sheila. "You'd be back there and you'd feel like you were in the country. You couldn't see anything or hear anything, it was just trees." She first heard about the problems with the local groundwater when a neighbor came over to report that the county health department had refused to let her use well water to fill her swimming pool. That sounded disturbing, but what did it really mean to have chemicals in the earth a few feet underneath your property? Sheila McVeigh was not sure, but she was glad that she and her husband, seeking more play space for their two young daughters, had replaced a backyard vegetable garden with sod when they moved into their house. The previous owner, an avid gardener, had died of cancer.

Ray and Shelley Lynnworth had lived on Cardinal Drive even longer—so long that when they first moved into their split-level brick home in 1968, there was no back fence. The yard just trailed off into the woods, giving the quarter-acre property a certain bucolic majesty. Even after the fence went up in the early 1970s, the Lynnworth children—Jill, born in 1967, and Randy, in 1969—did not regard it as an inviolable barrier. "Did Randy climb over the fence sometimes and go into the woods of Ciba-Geigy? Of course he did. All the neighborhood kids did that," Ray Lynnworth remembered. The children swam in the nearby river, too. The Lynnworths knew that the land belonged to Toms River Chemical, but they did not consider that a bad thing. Quite the contrary, they loved the privacy. There were unpleasant odors at times—usually at night, since that was when the factory's smokestacks, out of sight but not out of range, were busiest. Their home's west-facing windows, looking toward the hidden smokestacks, were a bit grittier to the touch than the other windows, as if they had been finely etched by dust particles. But these were small inconveniences, easily overlooked.

There was talk on Cardinal Drive about unexplained illnesses, just as there was more than a mile away in Pleasant Plains, in the shadow of another toxic waste site at Reich Farm. But unlike the residents of Pleasant Plains, the families who lived on Cardinal Drive and in the rest of the Oak Ridge subdivision got their water through the pipes of the Toms River Water Company, not from their own backyard wells. The neighborhood had been hooked up to public water since the early 1960s, and it tasted fine most of the time. Some homeowners still used their old backyard wells or drilled new ones to save money on their water bills. They used this backyard water to irrigate their lawns and gardens or fill their swimming pools, though rarely for drinking or showering because the well water had a faint but unpleasant odor, a bit like paint thinner.

Toms River Chemical had said nothing to its neighbors about the risk of well contamination, even though on the other side of the fence, the aquifer beneath the factory property had been so contaminated for so long that the company had resorted to drilling a well more than two thousand feet deep in its neverending search for unpolluted water for the factory's own use. The company was acquiring so much expertise in groundwater testing, in fact, that in 1980 two of its executives, Jorge Winkler and David Ellis, bought a local water-testing firm and set up their own private testing business, staffed by their wives, both of whom also had some scientific training. The firm's clients included the Toms River Water Company and several homeowners in Oak Ridge. Winkler and Ellis did not consider this to be a conflict of interest because no one could say for certain if Toms River Chemical was responsible for the contamination in Oak Ridge and also because their wives—not they—did the analytical work at the firm, known as J. R. Henderson Labs. "It was very clear that eventually one day Henderson Labs potentially would find things that Ciba-Geigy was responsible for. Until that happened, I didn't see any reason to change course," Winkler recalled. To Winkler, the arrangement made perfect sense: Who in town had more expertise with groundwater contamination than they did?

At first, the test results from the Cardinal Drive irrigation wells were comforting: Henderson Labs conducted the county health department's standard battery of tests for bacterial contamination and found nothing. Starting in 1982, however, the lab started urging its clients to pay for a more expensive analysis capable of detecting toxic chemicals, not just bacteria. The new tests found dozens of hazardous compounds. All of a sudden, the same wells that neighbors had been using for years to water their lawns and gardens—and occasionally as sources of drinking water—were regarded as so contaminated that the county health department declared that they had to be immediately abandoned and plugged.[1] Of course, no one could say for certain how the chemicals had gotten in those backyard wells, or whether they had made anyone sick.

By the time Randy Lynnworth was twelve years old, he could outrun his equally athletic father in a five-mile race; sometimes, just to keep things close, he would spin around and run backward until his father caught up. Randy was bright and funny, an excellent student who seemed destined to become an equally accomplished adult. He was almost never ill, so when he got a splitting headache and nearly collapsed during a relay run in late 1982, his parents were worried. They got their answer three weeks later, from a brain scan: Randy Lynnworth, at thirteen, had an advanced case of a highly malignant cancer known as medulloblastoma. As with Michael Gillick's tumor, Randy's was a blastoma, which meant that it began with the malignant transformation of precursor stem cells—in his case, the cells were located at the base of the brain beside the cerebellum. The tumor formed by those rogue cells was large and virulent, and Randy might not even survive the operation to remove it. A few days after the surgery, he lapsed into a coma. After several agonizing weeks during which Randy was unresponsive, the family was advised by a physician to stop feeding him and let him die in his own bed. They did so for less than a day before changing their minds and resuming his meals. Shockingly, after ten weeks in a coma, Randy abruptly awoke, as talkative and quick-witted as ever, albeit with a speech impediment, memory lapses, and a battery of physical disabilities.

That was an electric time on Cardinal Drive and in the larger community. If Randy Lynnworth could wake up, then anything seemed possible. His sudden and horrific illness had been widely publicized, and now it seemed like the whole town was participating in his recovery. The former long-distance runner could not walk, so his parents installed a set of parallel bars in the living room and Randy began working to strengthen his upper body so that he could confidently pilot his wheelchair. In the fall, he made a triumphant return to school (by now it was high school), steering his wheelchair through the halls with the same spirit of reckless abandon he had brought to his running. Camp was not an option, so during the summer of 1984 his parents organized a day camp on Cardinal Drive for all the neighborhood kids and Randy, too. Life was far from normal, but Randy was writing poems and cracking jokes, and that was enough.

As the months passed, the Lynnworths dared to wonder if the doctors might be wrong and Randy might beat cancer after all. For the first time, they even allowed themselves the freedom to wonder why he had gotten sick in the first place. It was impossible to live on Cardinal Drive and have a child with cancer and not think about Toms River Chemical. Initially, Ray and Shelley Lynnworth had a hard time believing that there could be a connection. "The idea that there might be a cancer cluster, that was something that built very slowly," Ray Lynnworth recalled. However, he did remember how the oncology surgeon at New York Hospital reacted after first hearing about Randy: "Another one from Toms River," the doctor had said. There certainly did appear to be an unusually high number of local families touched by childhood cancer. Was there really a cluster of cases, perhaps even one that was somehow caused by exposure to manmade chemicals like the ones that Toms River Chemical had vented into the air and dumped into the ground? Would it ever be possible to find out? The Lynnworths wondered.

The triumphs of John Snow and William Farr during the cholera epidemics of the mid-nineteenth century finally proved what had been suspected by observers as far back as Hippocrates: Infectious diseases waxed and waned across populations in nonrandom patterns that could be mapped in space and time. In other words, they clustered. But what about diseases like cancer that took years to develop and generally were not infectious? Percivall Pott, with his wretched chimney sweeps, and John Ayrton Paris, with his Cornish copper smelters, had described what they thought were associations between those occupations and scrotal cancer. Other physicians noticed apparent clusters of scrotal cancer among workers in factories that manufactured paraffin wax derived from coal tar.[2] But their observations, involving just a handful of cases, lacked the scope and precision of Farr's statistical ledgers and Snow's maps. Skeptical peers were mostly unconvinced, their doubts bolstered by the work of Siméon Poisson and other pioneering statisticians who showed how easy it was for apparent clusters to be caused by nothing more sinister than random chance, especially when only a few cases were involved.

But why *shouldn't* cancers cluster? If Rudolf Virchow was correct that all cancers begin when an external event disrupts a healthy cell and triggers a frenzy of cell division, it stood to reason that the external trigger—whatever it was—would have similarly lethal effects for many other individuals exposed to it, just as the cholera bacterium did. The patterns of cancer incidence deduced by a well-designed epidemiological study might even provide crucial clues about the identity of those triggering events, just as Snow's maps implicated a waterborne contagion for cholera.

Cancer was immensely more difficult to study than a cholera epidemic, however. Through his microscopic analysis of cancerous cells, Virchow had helped to prove that cancer was not one common disease but many rarer ones, each with its own pathology and place of origin inside the body. Besides uncontrolled cell division, just about the only characteristic that most types of cancer had in common was that it usually took years for a tumor to grow large enough to be noticed. So a John Snow–style study of potential environmental causes of cancer—incorporating columns of figures, maps, questionnaires, physical exams, environmental measurements, and all the rest—stood a chance of being fruitful only if a researcher was lucky enough to stumble upon an island of misery whose population had just the right characteristics. The population had to be definable and stable, without too many people coming or going over the years. It had to be affected by an unusually intense and easily measured form of pollution, one that might be carcinogenic. And it had to be afflicted with so many cases of a particular type of cancer that the cluster could not reasonably be dismissed as an unlucky fluke.

Where were those unhappy but epidemiologically fecund places? In the late nineteenth century, one of the best candidates was a region Paracelsus had passed through 350 years earlier: the Erzgebirge

Mountains, which straddle the border between the German state of Saxony and Bohemia in the Czech Republic.[3] The Erzgebirge (German for "metal ore mountains") had been continuously and heavily mined since 1410. The silver deposits discovered in the sixteenth century near the Bohemian town of Joachimsthal (Saint Joachim's Valley, in English) were so bountiful that *thaler* came to be a synonym for coinage; in English-speaking countries, it was translated as *dollar*, since *dale* was a synonym for valley.

Even in Paracelsus's day, the Erzgebirge miners were known to die young, often from the same set of debilitating symptoms that came to be known as "mountain sickness" or "miners' exhaustion," among many other descriptive terms. In fact, mountain sickness was lung cancer, although Paracelsus did not fully recognize it as such.[4]

In the late nineteenth century, the largest complex of mines was in Schneeberg, on the German side of the mountain range. It was a desultory region of impoverished villages, and there was almost no work to be had outside of the mines, which by the late 1800s were mostly producing bismuth, nickel, and cobalt. The latter two metals were extracted from a gray crystalline mineral called smaltite that consisted of cobalt, arsenic, iron, and nickel. For centuries at Schneeberg, smaltite had been mined by pickax, but by the 1870s miners were blasting it out with dynamite, which Alfred Nobel had commercialized in 1867. It mattered little to the health of the miners, because both techniques, by ax or by blast, generated thick clouds of toxic dust in the poorly ventilated mineshafts, some of which were more than two thousand feet deep. The conditions were so brutal that miners who managed to survive to middle age were often so incapacitated that they could no longer work in the mines and instead scratched out a living carving wooden toys and nutcrackers like the one in Tchaikovsky's ballet.

Into this pathetic tableau entered a young doctor named Walther Hesse in 1877. As the newly appointed district physician in the Schneeberg region, Hesse was responsible for eighty-three villages, which he set about visiting, traversing the mountain roads via horse and buggy and occasionally on foot. At thirty years old, he was already a man of the world, having trained in Dresden and Leipzig and fought the

French as a battlefield surgeon in the Saxon Army.[5] He was appalled by what he saw in the villages and mines of the Erzgebirge. "As a rule," he later wrote, "the miners marry early and leave behind a large number of children in pitiful circumstances, who naturally must look for [a] line of business as soon as possible. . . . Like their fathers, they age quickly, and with the greatest probability face the same previous mentioned fate of their fathers."[6]

There was one other physician in the region, Friedrich Härting. The medical officer of the largest mine in Schneeberg, Härting kept track of the number of deaths at the mine and had been quietly urging his employer to improve working conditions. To build a more persuasive case, the two physicians decided to conduct a study together to measure the prevalence of "mountain sickness" and perhaps even identify its likely cause. Like John Snow and other pioneers of the nascent field of epidemiology, Hesse and Härting attacked the problem on multiple fronts. They counted cases, autopsied dead miners (confirming for the first time that mountain sickness was, in fact, lung cancer), searched for similarities among the ill, and measured environmental conditions in the mineshafts.

In two manuscripts, a short article Hesse wrote in 1878 and a longer one he coauthored with Härting the following year, the two physicians disclosed their findings. Between 1869 and 1877, they reported, 145 of the approximately 650 miners in Schneeberg had died of lung cancer, most of them under age fifty. Excluding accidents, lung cancer accounted for 75 percent of all deaths. It was a shockingly high percentage, even compared to miners elsewhere. Clearly, the Schneeberg mines were a particularly fertile environment for lung disease and early death, but why? The doctors' autopsies of twenty dead miners suggested that they had inhaled something highly toxic: Their lungs had shrunk to the size of two fists, were riddled with bronchial tumors, and contained almost no air. What could have done all that damage?

Hesse and Härting tried mightily to find out, but could not. They were looking for something that was present in the Schneeberg mines but not elsewhere, since they had written to the managers of other mines and were assured that there were no lung cancer epidemics there. Smaltite, the arsenic-infused mineral, was their leading suspect. It was much more common in Schneeberg than elsewhere and was known to be a lung irritant. Another clue: Schneeberg miners told the two physicians that mountain sickness seemed most commonly to afflict laborers who spent their time extracting minerals, instead of blasting tunnels or doing other jobs. Since nickel and cobalt miners worked most directly with smaltite, while bismuth miners did not, the physicians tried to compare cancer rates in the two groups but were thwarted because the miners constantly changed jobs and because the smaltite dust was everywhere, even in the bismuth-mining areas. Hesse and Härting looked for other ways to make comparisons, too, including the use of wax paper dust-collectors to see if some tunnels were dustier than others and of filter-equipped facemasks to try to measure how much ore dust miners inhaled.

In their 1879 report, however, they conceded that each of those attempts failed to pinpoint a specific cause for the cancers. After two years of investigation, they were stymied. They wrote up their inconclusive results, called for improved conditions in the mines (again, to little effect), and never again published on the subject. Härting returned to his quiet work as a mine physician, and the peripatetic Hesse moved to Berlin and embarked on what he doubtless regarded as a much more fruitful area of medicine than cancer epidemiology: infectious disease.[7]

For all his progressive impulses, Walther Hesse, who died in 1911, never fully grasped the importance of what he and Friedrich Härting accomplished in Schneeberg in 1878. For the first time, someone had taken the tools of infectious disease epidemiology—case counts, geographic and temporal patterns, interviews, environmental measurements, physical examinations—and applied them rigorously to cancer. The result was a confirmed cancer cluster, the first one that could withstand doubters' scrutiny. Later, other German scientists—most notably Ludwig Rehn, who first linked bladder cancer to aniline dye manufacture in Frankfurt in 1895—would be inspired by the Schneeberg work. True, Hesse and Härting had failed to identify the specific cause of the lung cancer epidemic in Schneeberg, but as subsequent generations of epidemiologists would affirm in Toms River and elsewhere, it was extremely difficult to complete even the first step of confirming a nonrandom, true cancer cluster. The second step, determining a true cluster's likely cause, was just about impossible in most cases—but not in Schneeberg, as it turned out.

What Hesse and Härting did not know was that the mineshafts of Schneeberg were veritable shooting galleries of high-energy gamma rays and fast-moving particles emitted by the unstable nuclei of radioactive elements like radon, bismuth, cobalt, nickel, and uranium in smaltite and other local minerals. Experimenting in the 1890s with minerals from the Erzgebirge, Henri Becquerel and Marie Curie, among others, identified the basic processes of radioactivity. Soon after, researchers returned to the mines with photographic plates, Geiger counters, and other tools to measure radiation. Most of the sources of radioactivity in the mines, it turned out, were metallic elements, but one—radon, formed by the radioactive decay of uranium and radium—was a gas that could be easily inhaled. Soon after the Nazis occupied Bohemia in 1938, they discovered that lung cancer was an epidemic among miners in Joachimsthal as well as Schneeberg. The Germans launched an ambitious study of the two mining centers, conducting autopsies on dead miners, experimenting on animals, and taking extensive radon measurements in the tunnels.[8] By 1939 they were certain that radon was the cause of the lung cancer epidemics in Schneeberg and Joachimsthal, and they were right. That knowledge, however, did not end the abusive conditions at both locations, especially after the prospect of atomic weaponry made uranium the most precious natural resource in the world. After the war, Joseph Stalin used slave labor at Joachimsthal to build the Soviet Union's atomic stockpile. The U.S. government, meanwhile, downplayed similar evidence of widespread lung cancer among miners on the Colorado Plateau as it rushed to secure fissionable uranium for nuclear weapons. Even now, radon remains one of the few environmental carcinogens over which there is little debate—it is second only to cigarette smoking as a cause of lung cancer in the United States, responsible for an estimated twenty-one thousand deaths each year thanks to its ubiquity in many types of soil.[9]

Using every technique and tool he could think of, Walther Hesse had managed to find a cancer cluster in one of the most obvious places in the world to look for one. His successors even succeeded in identifying its cause. For the nascent science of cancer epidemiology, it was a misleadingly promising start.

The slow awakening on Cardinal Drive in the early 1980s was not happening in a vacuum. The world was changing, too, and some parts of it were changing much faster than Toms River. In 1977, a *Niagara Gazette* reporter named Michael Brown began investigating reports of illnesses and pollution in a working-class neighborhood of his hometown of Niagara Falls, New York. The neighborhood came to be known as Love Canal for its namesake feature, a never-completed canal that from 1942 to 1953 was used as a clay-lined dumpsite for drummed and uncontained waste chemicals, including aniline derivatives and benzene. In 1953, Love Canal was covered with dirt and sold by the Hooker Chemical Company to the local school board for the token sum of one dollar, plus the inevitable liability waiver. The members of the Niagara Falls Board of Education knew what was in the canal and had even toured the site with Hooker officials, who drilled holes into the clay so that school officials could see the chemicals beneath. But the school board members, like their counterparts in Toms River ten years later, were intent on finding cheap land for another school to serve their fast-growing community, so they built an elementary school right beside the old dump. Hundreds of new homes soon followed, in an area that had been mostly open land during the years when Hooker was dumping. Homebuyers were not told that they would be living beside (and in some cases on top of) an old hazardous waste dump.

By the 1970s, the Love Canal neighborhood was rife with reports of strangely colored water in basement sump pumps, dead backyard vegetation, foul odors, and illnesses including miscarriages, birth defects, and cancer. In some parts of Love Canal where the earthen cap atop the old landfill had collapsed, the tops of fifty-five-gallon drums poked through the surface, especially after heavy rains. Michael Brown's stories in the *Niagara Gazette* galvanized the community, prompting a charismatic young housewife named Lois Gibbs to organize street protests and attempt some do-it-yourself epidemiology, including mapping the locations of homes where there were health problems. Her son, Michael, had developed epilepsy and a low white blood cell count shortly after beginning kindergarten at the elementary school beside the dump. In the summer of 1978, after studies confirmed high levels of airborne chemicals in dozens of nearby homes, New York State took the unprecedented step of declaring a health emergency at Love Canal, closing the school and evacuating the residents of 239 nearby homes. Five days later, President Jimmy Carter followed with a federal disaster declaration—the first one ever prompted by a non-natural disaster.

Seemingly overnight, the long-slumbering issue of hazardous waste became a national crisis, with Love Canal its epitome. In Niagara Falls, the ensuing maelstrom of heavily publicized protests and worrisome medical studies—at one point, panicked residents briefly held two Environmental Protection Agency officials hostage at the headquarters of the local homeowners' association—eventually forced the state and federal governments to expand the evacuation zone to include up to nine hundred additional homes in 1980.

Love Canal had laid bare the limitations of the EPA's clout on hazardous waste cases, and Congress resolved to do something about it. Four years earlier, it had passed the Resource Conservation and Recovery Act, which gave the agency authority to compel emergency investigations and cleanups at waste sites. But the law's reach was limited, and the short-staffed EPA used its authority only at the most egregious dumpsites. In most cases, the old dumps were left to the states to address or ignore as they saw fit. After the Love Canal crisis, however, Congress rushed to enact a powerful new law designed specifically to deal with hazardous dumps. Signed into law by President Carter in December of 1980, it was called the Comprehensive Environmental Response, Compensation, and Liability Act, but everyone knew it as Superfund. It directed the EPA to compile and rank a list of the most dangerous toxic waste sites in the country and then oversee their cleanups, which would be funded by the dumpers or, if they could not be found, by a special "superfund" raised from fees on the chemical and petroleum industries and from general tax dollars.

Under the doctrine of retroactive liability, companies could be held responsible for cleanup costs even if their dumping was legal at the time. If a dumper refused to comply with a cleanup order, the EPA could tap the Superfund to pay for the remediation and then take the company to court and recoup up to three times as much as the agency had spent on the cleanup.

The EPA set to work compiling its list of the worst waste dumps. Finalized in 1983, the first "National Priorities List" included 406 dumpsites (later additions would eventually quadruple that number). Sixty-five sites on the original Superfund list were in the undisputed capital of hazardous waste dumping in the United States: New Jersey, which had twenty-four more sites than its closest rival, Michigan. With nine dumps on the list, Ocean County alone had more Super-fund sites than thirty-six states. Two of them were in Toms River: Reich Farm and Toms River Chemical, which ranked 105th and 134th, respectively, on the nationwide list. (Love Canal ranked 116th.)[10]

As far as the people of Toms River knew at the time, Reich Farm and the chemical plant were merely potential threats, but many local residents were already becoming aware of the actual perils of living near a Superfund site for other reasons. Just a few months after Love Canal became national news, in November of 1978, state officials notified 150 families living near the Jackson Township Landfill that their well water was unsafe to drink; most had to use bottled water for almost two years. The landfill, which was just ten miles northwest of Toms River Chemical, was an old strip mine that had been used as a municipal landfill since 1972, after pollution problems forced the closure of Jackson's previous dump. When a state health survey found that residents were suffering from rashes due to the contaminated water, the families sued the town, winning a $16 million jury award in 1983 that was reduced on appeal. The case was one of the first to establish that a polluter could be forced to set aside funds to be awarded later if a victim developed cancer or some other latent disease—a precedent that would later be important in Toms River.

The advent of Superfund, and the lawsuits it helped to spawn, was bad news for Toms River Chemical—and for Union Carbide, too, since it was on the hook for the Reich Farm contamination. (Nicholas

Fernicola certainly could not afford to clean up the mess he had made dumping Union Carbide's drummed waste.) Until then, both companies had been managing quite nicely in their dealings with the New Jersey departments of health and environmental protection. Except for the partial cleanups at the Reich Farm site, the state agencies had done very little to require Union Carbide or Toms River Chemical to remove waste from the soil or pump it out of groundwater. Instead, the state's efforts had been focused (without much success) on the much less ambitious goals of preventing new spills and keeping existing pollution out of drinking water wells. But the federal EPA was less susceptible to political pressure than its state counterparts, and the Superfund law gave the agency broad discretion to insist on comprehensive cleanups even if a dumpsite did not pose an immediate health threat. Before Superfund, a corporation responsible for a leaking dump might face a liability of a few hundred thousand dollars, or perhaps several million dollars for the largest spills. After Superfund, those costs would multiply tenfold or even a hundredfold for the biggest and most complicated cases, including the two in Toms River. Love Canal was exhibit A. Thanks to retroactive liability, Occidental Petroleum, which bought Hooker Chemical in 1968, ended up paying the state and federal governments $227 million to cover the costs of cleaning up the dump and relocating more than one thousand families. Occidental also paid slightly less than $20 million to settle a lawsuit filed by more than thirteen hundred residents of the neighborhood, including Lois Gibbs.

Thanks to Superfund, toxic liability was now a major debit on the balance sheets of even the largest corporations. For more than a century, hazardous waste had generally been left wherever it was dumped, whether in Basel or Toms River or thousands of other places. The passage of Superfund and its counterparts in other countries at last raised the prospect of meaningful cleanups. But the potential costs were so vast that Superfund sites became legal battlefields, attracting legions of pugnacious lawyers, engineers, toxicologists, biostatisticians, epidemiologists, and other advocates-for-hire whose conflicting assertions slowed cleanups to a crawl and made health studies vastly more contentious and complex. Environmental epidemiology had always been an extremely difficult pursuit, rife with uncertainty. Now, with tens or even hundreds of millions of dollars hinging on each outcome, it would be even tougher.

There was no storybook ending on Cardinal Drive for the Lynn-worths. In 1986, they sold their house of eighteen years and moved to a different part of town. Their new home was on a large lot, and its backyard was so lushly landscaped that they no longer had to rely on someone else's forest for the privacy and space they craved. The family needed a change of scenery because Randy Lynnworth was dying. A few months earlier, he had woken up unable to feel his leg, which had gone numb. After more than two years of remission, the medulloblastoma was back. This time, the tumors had spread from the base of his brain to his spine. The year that followed was an extended nightmare of agonizing chemotherapy and inexorable decline, but Randy continued to write poems when he had the strength. One was read at his funeral in the spring of 1987, the year he should have celebrated his high school graduation (he was awarded a diploma posthumously). The boy who loved to run never regained the ability to walk more than a few steps, despite all of his hard work on the parallel bars and in physical therapy. Instead, more than six hundred residents of Toms River walked for him, behind his casket.

By the time Randy Lynnworth died, Toms River Chemical and its successor, Ciba-Geigy, were no longer revered names in Toms River. A company that had been venerated for so long had fallen fast and far, thanks to a series of shocking events starting in 1984 that no one could have predicted. Through it all, the Lynnworths were interested but uninvolved observers. They were not comfortable in the public eye the way that Linda Gillick was. They had met the Gillicks at Memorial Sloan-Kettering Cancer Center shortly after Randy's cancer returned and had gratefully joined the network of families Linda Gillick was building. A few weeks before Randy died, the Lynnworths filed a lawsuit against Ciba-Geigy and eventually got a modest settlement from the company, which did not admit any liability.

Unlike many Toms River parents who lost their children to cancer, Ray and Shelley Lynnworth never claimed to be certain that pollution was to blame. In fact, the more Ray Lynnworth learned about the complexities and controversies associated with health studies, the more convinced he was that no one would ever be able to find out why Randy had gotten sick. "I can't base something like that on intuition or what's in my heart, you have to be more scientific about it," he explained. "I'm not a political activist, that's not where my heart and soul are. For me, it's not about waging a battle for justice or revenge. There are others who feel differently, and that's fine. It's a very personal choice." What the Lynnworths wanted most of all was to honor their son's memory in their own quiet way. They did so by turning their do-it-yourself summer camp into a formal organization, called Team Randy. More than twenty-five years later Team Randy was still organizing summertime travel activities for Toms River teenagers, with a special emphasis on serving those who are physically disabled or in financial need.

The Lynnworths never went back to Cardinal Drive. If there was going to be a battle for justice or revenge, others were going to have to wage it.

PART II

BREACH

CHAPTER EIGHT

≈

Water and Salt

The first cracks in the long peace between Toms River and its chemical plant appeared on the morning of April 12, 1984, at the intersection of Bay and Vaughn avenues, right in the middle of town. Overnight, the road surface had buckled, and a county road crew was sent to investigate. The crew used a backhoe to dig out the cracked asphalt and discovered that the soil underneath was deep black instead of sandy brown and saturated with a liquid that had a strong chemical smell.

Roden Lightbody drove over to take a look, too. He went because he was a traffic engineer for Ocean County, but that was not the only reason. Red-faced and gruff, Lightbody was an important political figure in the county, and also in Toms River. In the grand tradition of the Ocean County Republican machine, he was both a public employee *and* a public official. In fact, he was the mayor of Toms River (officially, of Dover Township). By the time Lightbody arrived at the scene, a television news crew was there, and the reporter was asking questions the mayor could not answer.

"When they dug out the road surface, they found all kind of mucky black material. I thought maybe it came from Ciba-Geigy, but at first we didn't know," Lightbody remembered many years later. "We called them and said something's happening here, and Ciba-Geigy at first said, 'It's not our problem.'" Lightbody knew that the company operated a pipeline that ran through Toms River, but he was not sure where, and it did not appear on the town's utility maps. The pipeline was not quite a secret, but it was close. It had been in the local newspapers in 1965 and 1966, when Ciba completed its construction and started pumping five million gallons of partially treated wastewater every day into the Atlantic Ocean instead of the river. But that was a generation earlier, and the leaders of Toms River had long since perfected the art of forgetting unpleasant nuggets of local history, especially those that concerned Ciba-Geigy. Like everyone who mattered in Toms River, Lightbody had strong connections to the factory. His brother had worked there since the 1960s. The mayor had heard plenty of stories over the years about burial pits, strange smells, and unexplained illnesses—who in town had not? But it was not the job of local government to check up on Ciba's environmental practices, in Lightbody's view. After all, the company seemed to know what it was doing. "We were not in a position to demand anything from Ciba-Geigy," he remembered. "We didn't have the knowledge."

In many ways, the company had never been more popular in Toms River than it was at the beginning of 1984. The factory's environmental problems seemed to be in the rear-view mirror, growing ever distant with the passage of time. Its solid waste was finally going into a lined landfill and its wastewater to a new treatment plant and then the ocean. Moreover, there was less waste overall than there had been during the peak 1970s because of a shift in the factory's product mix. Starting in 1983, the company no longer made anthraquinone vat dyes, its original product, and concentrated instead on more profitable azo dyes, resins, and specialty chemicals. The 1972 indictments were ancient history, the river pollution of the early 1960s was forgotten entirely, and the secret contamination of the town's drinking water in 1965 was still a secret. There was a bitter labor strike at the factory in late 1980—several strikers were arrested for throwing nails under managers' cars—but tempers had since cooled. With one thousand workers and managers and a $35 million payroll, the chemical plant was still the largest private employer in Ocean County, even if the total workforce was a few hundred below the peak employment before the strike. As if to formalize a clean break with the past, in late 1981 the Swiss changed their factory's name. Henceforth, it would be known as the Ciba-Geigy Toms River Plant, instead of the name that had meant so much to the town for so long: Toms River Chemical.

The name change also symbolized the company's aspirations to move beyond chemical manufacturing in Toms River. Its plans became public in January of 1984, when Ciba-Geigy announced with great fanfare that after a two-year transition period it would move its United States pharmaceutical manufacturing operations from Cranston, Rhode Island, to Toms River. What Ciba-Geigy did not announce was the reason for the relocation: a decade-long battle with Rhode Island environmental officials over the company's wastewater discharges—one and a half million gallons per day—into the Pawtuxet River and Narragansett Bay. The discharges ended in 1983 when a new municipal treatment plant opened, but by then the company had become deeply unpopular in Cranston and was the target of protests and lawsuits.

In Toms River, however, Ciba-Geigy in early 1984 was still at the center of the community, with strong bonds to the political elite and, via its union, to the citizenry. So on the afternoon of April 12, when Ciba-Geigy belatedly acknowledged that its waste pipe was the source of the leak and a white-suited crew from the factory arrived at the corner of Bay and Vaughn to take over the effort to repair the breach and clean up the mess, Roden Lightbody and the other public officials at the scene were happy to give way and let the trusted experts take over. The leak was quickly traced to a corroded hole the size of a golf ball in the waste pipe. The factory's highly acidic wastewater had chewed through steel and two layers of supposedly leak-proof enamel, made from coal tar.

By the end of the second day, the crew from Ciba-Geigy had finished the repairs and was carting away the soaked soil. In a few more days, the road could be repaved, and life would be back to normal on the corner of Bay and Vaughn. Another bit of unpleasantness would be buried and forgotten in Toms River. Any other result was too far-fetched even to contemplate. Certainly no one could have imagined that a two-inch hole would eventually be the undoing of mighty Ciba-Geigy in Toms River.

Don Bennett's desk was always the most cluttered in the offices of the *Ocean County Observer,* which made sense because he was, by decades, the longest-serving employee in the newsroom. The heaping piles of paper on his desk were the strata of small-town news: charity auction notices, county legislative agendas, press releases touting high-achieving middle-schoolers, and the like. The longer he stayed at the *Observer,* the more space there was in the newsroom for his piles and files, because his colleagues often were not replaced when they departed. When new reporters were hired, they were often young enough to be his grandchildren.[1]

In 1984, Don Bennett was already an anachronism in the newsroom, but that did not bother him. He and his wife raised six sons in Toms River, and he never dreamed of moving on to a bigger newspaper in a bigger town. He was a proud member of the community, a former officer in the Junior Chamber of Commerce and the son of the former superintendent of schools. Although his ambitions never extended beyond Ocean County, he took his job very seriously. He wanted to uncover news, not just cover it. "I wanted to keep this community informed. If I left to work somewhere else, then the community wouldn't be getting important information it needed," he recalled. "I felt I was needed here." On his desk he kept a quotation from investigative reporter Bob Woodward: "All good work is done in defiance of management." That quotation, he explained, "has just sort of been my spirit."

Not far from the clutter of his desk Bennett kept two large file cabinets crammed with material on the Ciba-Geigy chemical plant, a place that had fascinated him ever since he arrived in Toms River in 1958, when he was seventeen, just six years after the factory began operating. As a teenager, Bennett swam in the cloudy, smelly river water just downstream from the factory's outfall pipe. As a young father, he coached hockey at a nearby ice rink and watched his skaters struggle to breathe through the evening stench. And as a reporter, he had occasionally heard stories from union employees about dangerous environmental practices at the plant.

Turning that information into articles in the *Observer,* however, was very difficult. The factory "was like a closed empire out there in the woods. It was thirteen hundred acres, and the production area was in the middle of it, so there were these buffers on all sides," Bennett remembered. Company executives generally refused to answer questions about health and safety at the plant. The county legislators and town committee members never talked about Ciba-Geigy except to extol its contributions to the economy and to local charities. The plant's blue-collar workers were potentially the best sources for stories about conditions there, but the only time Bennett ever heard from them was when it was contract negotiation time and the union was trying to put pressure on the company. As soon as there was an agreement on a new contract, his phone stopped ringing. For all of Ciba-Geigy's power and influence in Toms River, its factory remained somewhat of a mystery, rarely appearing in the pages of the *Observer,* the *Asbury Park Press,* or any of the local weekly papers unless it was for sponsoring a Boy Scout troop or a golf benefit or some other charitable act.

Don Bennett wrote those good-news stories, too; it was part of the job. But he always kept an eye out for other kinds of articles, stories about the safety of the Ciba-Geigy plant and the impact of its smokestacks, discharge pipes, and landfills on the surrounding community. Over the years, Bennett managed to get a few of those stories into the paper. In the mid-1960s, he covered the controversy over the construction of the ten-mile ocean pipeline, for example. But his reporting never seemed to have much impact; memories were short, and Ciba-Geigy was very important to the local economy.

In early 1984, however, Bennett managed to find a story that he figured would make a splash. His tipster this time was his own mother. She was in the beauty salon one morning and heard some wives of Ciba-Geigy employees talking about an interesting piece of information their husbands had brought home: The factory's wastewater treatment plant was not only treating its own effluent, it was also taking in waste from other factories around the region and had done so for years. That was news, Bennett thought, since the company was pumping all of that wastewater through the pipe and into the ocean just three thousand feet off Ortley Beach, a popular area for swimming.

The story got even better when Bennett spoke to Edward Post, who worked for the state Department of Environmental Protection and was in charge of enforcing industrial wastewater regulations around the state. Ciba-Geigy, it turned out, had been violating its state-issued ocean discharge permit by taking in all kinds of outside wastes: caustic soda from the Nestlé instant coffee plant in Freehold, cleaning wastes from Sandoz in East Hanover, aluminum salts from the Wickhen antiperspirant factory across the border in Huguenot, New York. In all, at least a dozen companies were trucking liquid waste to Ciba-Geigy's wastewater treatment plant. Post and his state colleagues believed that the company was in clear violation of its discharge permit, and in 1983 put Ciba-Geigy on notice that if it continued to import waste it would be subject to prosecution. The company's lawyers deflected the state's demand by arguing, without any apparent irony, that none of the outside waste was as toxic as what Ciba-Geigy was sending to its own treatment plant and that mixing in the outside waste was actually helpful because it neutralized the company's own highly acidic dye and resin wastes.

Bennett knew it was a juicy story. It was one thing for Toms River to tolerate discharges from a company that employed several thousand locals, sponsored scout troops, and owned the country club. It was quite another for the local shoreline to be a dumping ground for factories from all over New Jersey. So on the morning of April 12, when he heard that the company's waste pipe had ruptured at the corner of Bay and Vaughn, Don Bennett was already prepared to make life uncomfortable for Ciba-Geigy.

The presence of a television news crew at the site of the leak on that first morning was the first clue that this cleanup might not go as smoothly as usual for Ciba-Geigy, even before Don Bennett had written a word about its waste-handling practices. Toms River was no hotbed of environmentalism, but by 1984 local attitudes were beginning to evolve. Publicity over Love Canal had raised awareness about the risks of toxic chemicals, and so had the forced closure of many local backyard water wells, in Pleasant Plains and elsewhere, that were legacies of the town's rural past. Linda Gillick was not yet an activist—she had her hands full coping with five-year-old Michael's unending medical crises—but she and her neighbors were starting to pay attention to the town's water problems, and they did not like what they were hearing. In 1983, New Jersey newspapers had also been filled with stories about an unfolding scandal at the Environmental Protection Agency over mismanagement of the new Superfund program. Agency chief Anne Gorsuch Burford was forced to resign; her subordinate Rita Lavelle was convicted of lying to Congress. The scandal was a big issue in New Jersey, the state with the most Superfund sites. One of Burford's chief antagonists was James Florio, a tough-talking congressman from South Jersey who had coauthored the Superfund law in 1980 and now pursued Burford with a prosecutor's zeal.

There was an even more important reason why the rupture of the Ciba-Geigy pipeline was a big story in Toms River. For the first time, a pollution problem in town was literally impossible to overlook. In the past, the community's environmental fiascoes had always occurred out of sight: in the back corner of an egg farm, or on the grounds of a hidden factory screened by a thick forest of oak and pine, or forty-five feet below the surface of the Atlantic at the end of a pipe. This time, it was at a busy intersection smack in the middle of town, just down the road from the new Ocean County Mall, which had displaced the riverfront downtown as the true hub of Toms River and the entire region.

"Until the leak, many of these residents never even realized there was a pipe behind their houses," Jorge Winkler, the Ciba-Geigy executive, would ruefully recall years later. "It somehow brought the pipeline and the company into the public eye, that's what the leak did. And that's when the whole climate changed." A local lawyer who became a prominent opponent of the chemical plant, Dan Carluccio, agreed. "So many people were new in town that there was no inherited knowledge about the area. People just came here and assumed that everything was fresh and clean and nice. They were so shocked to find out there was a pipeline leaking chemical waste," he remembered. That the waste was being dumped into the Atlantic—the wellspring of Ocean County's identity and the keystone of its economy—was especially galling. "People would come down to the Jersey shore and they'd say, 'You have a chemical plant here? You find those in Linden [near Newark], not here in paradise,'" Carluccio said. "It was a shock to many people."

A steady stream of negative news stories began to erode Ciba-Geigy's virtuous public image. What made the stories so devastating was not the news of the leak itself but the opportunity it created for reporters like Don Bennett to bring up broader issues. Even after tests showed that the effluent had not contaminated any water wells near the site of the rupture, there were plenty of other avenues for reporters to explore. Bennett was writing unflattering articles about Ciba almost every day, including stories about its leaking landfills and disputes with state regulators as well as its acceptance of wastes from other factories. Many in Toms River were shocked to learn that Ciba-Geigy and Reich Farm were on the Superfund list of the country's worst hazardous waste dumps, alongside Love Canal and the industrial sites that local residents associated with decaying cities like Newark and Camden, not their own shiny hometown.

An even bigger shock, especially to residents of the beach communities, was that Ciba-Geigy had been pumping its treated wastewater into the ocean for almost twenty years. The economies and identities of shore towns like Lavallette and Seaside Park were wholly dependent on clean beaches and a clean ocean. Few people there worked at Ciba-Geigy; summer tourism, not the chemical industry, was the engine of the shore economy. After losing their fight in the mid-1960s to prevent the construction of the ocean pipeline, the leaders of the beach communities stopped talking about it because they were worried about driving tourists away. But the flurry of articles about the leak in the spring of 1984 made it impossible to keep pretending that the pipe did not exist. Ortley Beach was part of Dover Township—its mayor was Roden Lightbody, who was close to Ciba-Geigy—but the community just to the north of it, Lavallette, had its own government and an outspoken mayor named Ralph Gorga. Now that the pipeline was back in the news, Gorga started attacking the company as an out-of-control polluter.

Ciba-Geigy was suddenly facing its first all-out public relations crisis in a county where it had always gotten its way. The company responded with all the finesse and humility of Marie Antoinette on the eve of the French Revolution. Ciba-Geigy's designated spokesman, at first, was Jorge Winkler, who by now was director of production and environmental affairs at the Toms River plant. An engineer, Winkler had a strong German accent that to American ears made him sound even more arrogant than he actually was. As he would explain years later: "I had an excellent education in science; I never had a course in public relations." He told reporters, politicians, and anyone else who would listen that the leak was a trivial matter and that anyone who claimed otherwise should not be taken seriously because only Ciba-Geigy's chemists were qualified to assess the risk. And then Winkler made an assertion that would become notorious in Toms River. The five million gallons per day of treated waste that the company was pumping into the ocean, he declared, were harmless because the effluent was composed of "ninety-nine percent water and a little salt." In other words, all the company was doing was putting salt and water into a saltwater ocean.

The statement was not only misleading, it was self-defeating. A frank explanation—that the treated wastewater contained relatively low levels of toxic byproducts of the factory's dye and resin manufacturing operations—might have tamped down the controversy before it could explode into a full-blown conflagration. Instead, Winkler's preposterous description and his refusal to spell out exactly what was in the "salt" (Ciba-Geigy claimed that the identities of the chemicals in its wastewater were trade secrets) only served to motivate people like Don Bennett at the *Observer* and Mayor Gorga of Lavallette. They zeroed in on the obvious questions: Which chemicals were in the company's waste? What harm were they doing? A company so secretive and disingenuous, they figured, must be hiding something important.

Until the twentieth century, the evidence that industrial chemicals could cause cancer rested almost entirely on the wobbly pillar of observational epidemiology. These were the anecdotal reports of people like Bernardino Ramazzini, Percivall Pott, and John Ayrton Paris, later bolstered by the more methodical analyses of Walther Hesse and others. They noticed unusually high rates of particular cancers in certain groups of people—scrotal cancer in chimney sweeps, lung cancer in cobalt miners, bladder cancer in aniline workers—who also had been exposed to unusually high levels of hazardous compounds, usually in complex mixtures. But their reports merely demonstrated an *association* between exposure and disease—a correlation, not a causal relationship. And their observations provided almost no help in determining which particular chemicals might be responsible for triggering cancer, from the hundreds of possible suspects that a chimney sweep or a miner or a dye worker might encounter.

To strengthen the case for environmental carcinogenesis, research would have to move from the outside world to the laboratory, from observation to experiment. For centuries, in dangerous places like the mines of Schneeberg, humans had been conducting uncontrolled experiments in the induction of cancer. The question now was whether those conditions could be replicated in the tightly controlled setting of a laboratory. Infectious disease research had made a similar transition in the late nineteenth century, and the results were spectacular. Louis Pasteur and Robert Koch were national heroes in France and Germany, respectively, for their microbiology work. Building on the real-world observations of John Snow and other epidemic-trackers, they grew pathogenic bacteria in Petri dishes and infected a menagerie of animals, from mice to chickens. Through his lab experiments, Koch isolated the microbes responsible for cholera, anthrax, and tuberculosis, laying the groundwork with Pasteur for vaccines and other strategies that would finally bring those ancient plagues under control in Europe and America. In the wake of Pasteur's and Koch's successes, many of their disciples turned their attention to cancer, hoping for similar results.

Cancer had other ideas. As usual, what worked for infectious diseases did not work for cancer. Microbes like *Vibrio cholerae* could be cultured on a plate; cancer cells could not. (That would change in 1951, with the successful propagation of HeLa cancer cells into a perpetual cell line; until then, attempts to culture cancer cells had failed.) Koch and Pasteur had shown that some pathogenic bacteria could be transplanted from one mouse to another or even from human to mouse without losing their potency, but tumors proved to be extremely difficult to transplant, despite many attempts.[2] By 1900, it was clear there would be no "eureka" moment when the mysteries of cancer would be dramatically revealed in the laboratory of an indomitable hero scientist, as cholera's mysteries had been revealed to Koch.

The failure of the transplant studies cast doubt on the widely held belief that cancer was caused by living parasites in the human body, perhaps by the same bacteria implicated in infectious diseases. If not a mystery parasite, then what? Victorian-era scientists developed at least three alternative hypotheses for tumor formation, and all of them were linked to Rudolf Virchow, the pivotal figure in a great age of medicine—the "professor of professors," as he was known.[3] Virchow's own "irritation" theory suggested that almost any kind of external trauma could induce a tumor as long as it occurred repeatedly over a long period. He asserted, but could not prove, that chronic irritation could lead to inflammation, cell damage, and then tumors in the epithelial cells that lined the cavities, organs, and other surfaces of the body—the cells where most cancers originate. A rival theory was advanced in 1875 by Virchow's former assistant, Julius Cohnheim, who proposed that tumors arose from clumps of "embryonal rest"—fragments of embryonic tissue that persisted in adults and were composed of stem cells that never differentiated into mature cells capable of performing specialized functions. If those long-dormant cells were activated by an outside stimulus, tumors would follow, asserted Cohnheim, who based his theory in part on the fact that fetal tissue looked like cancerous tissue under the microscope. His idea was later modified into the theory of dedifferentiation, which suggested that cancer began when specialized cells abruptly regressed into their more primitive state and began dividing rapidly.

All of those competing theories would eventually be shown (after major modification) to have at least some validity for certain types of cancer, but unless some yet-undiscovered microbe was to blame, they all failed to address the big question: Why does cancer begin? What type of catalyst could trigger Virchow's cell irritation, activate Cohnheim's embryonic rest, or cause specialized cells to begin regressing? As the twentieth century began, no one could say. Like an apparition, cancer seemed to arise out of nowhere, with no discernible initiator. Moving cancer research from field observation to laboratory experiment had yielded nothing but frustration and intense disagreement among the champions of various theories of tumor formation.

Katsusaburo Yamagiwa of Japan knew about those conflicts firsthand.[4] Born in 1863 into a noble family of the samurai class that had lost most of its wealth, he proved to be such a brilliant medical student that the Japanese government in 1892 sent Yamagiwa and two other young scientists to Robert Koch's lab in Berlin to study tuberculosis. They were not made to feel welcome. The precise reasons are obscure: The delegation may have shown up on Koch's doorstep uninvited, Koch may have blamed them for the Japanese government's mistreatment of a friend, or they may have found the work uninteresting—or perhaps all three. It surely did not help matters that the young scientists came from a nation viewed as uncivilized by many Europeans of that era, though there is no proof that attitude was shared by Koch, who later visited Japan and posed for photographs in a kimono. Whatever the reason, the Japanese quickly dispersed to other laboratories, with far-reaching consequences for cancer research because of where Yamagiwa ended up: at the Berlin laboratory of the great Virchow, the champion of cellular irritation theory and no friend of the upstart Koch and his focus on microbes.

Yamagiwa thrived under Virchow's tutelage and fully embraced his mentor's irritation theory of carcinogenesis. Virchow knew about the case studies from the Schneeberg mines and the aniline factories suggesting that some pollutants might be carcinogenic. He also recognized the possibility that chemical exposures might provoke the irritation he believed led to malignancies. But Virchow had never tested that idea experimentally. It is likely that his new protégé Yamagiwa, during his sojourn in Berlin, resolved to try if he ever got the chance. Returning to Tokyo in 1894, Yamagiwa was initially assigned to study diseases that were hindering the expansion of the Japanese Empire. But his service ended in 1899 when he contracted tuberculosis, a disease Yamagiwa surely thought he had left behind when he departed Koch's lab seven years earlier. For the rest of his life, he suffered from a hacking cough, shortness of breath, and chronic exhaustion. Confined to his Tokyo laboratory after a long and intermittent recovery, Yamagiwa returned to the consuming passion he had developed during his time with Virchow: the search for catalysts capable of initiating cancer via cellular irritation.

The suspected carcinogen Yamagiwa decided to study was, in many ways, an obvious choice. It was coal tar, the original industrial pollutant—bane of chimney sweeps and chief elixir of the chemical revolution ever since William Perkin used it to create the first synthetic dye in 1856 in his parents' attic. Tar was already a leading suspect based on a long string of observational studies, from Pott's chimney sweeps to reports of cancer in workers in the dye and kerosene industries, both of which relied on tar derivatives. In 1907, before there was any laboratory evidence that coal tar was carcinogenic, the British government formally recognized scrotal cancer as an industrial disease, declaring that "men engaged in handling pitch or other tarry products" would qualify for workers' compensation if they developed scrotal cancer.[5]

Yet by the time Yamagiwa began his coal tar experiments in 1913, European scientists had given up trying to use it to induce cancer in animals. Several had reported failures after painting tar onto the ears of dogs or injecting it into rats; others had smeared azo dyes and kerosene onto the skin of rats. Most of those experiments had triggered lesions in the affected areas but no cancerous tumors. Coal tar research seemed to be at a dead end, especially after a dramatic announcement in 1913 from Copenhagen: A former student of Koch's named Johannes Fibiger declared that after six years of research, he had discovered a parasite that caused cancer. While dissecting wild rats infected with tuberculosis, Fibiger had noticed that many had stomach tumors. He eventually concluded that the cause was a microscopic nematode worm he called *Spiroptera carcinoma*, which lived in the stomachs of cockroaches that were then consumed by rats. (The nematode originated in South America and the Caribbean but was carried to Europe in the cockroach-infested holds of sugar ships.) Fibiger fed those infected cockroaches to mice in his laboratory and reported that he could reliably induce tumors in the rodents' stomachs and esophagi. Fibiger even claimed to have transferred the stomach tumors from one mouse to another. The Danish scientist had seemingly won the race to confirm the first carcinogen—and it was a cancer microbe, not a chemical pollutant.

Word of Fibiger's apparent breakthrough reached Tokyo after Yamagiwa and his assistant, Koichi Ichikawa, had already embarked on their coal tar experiments. Mindful of his mentor Virchow's ideas about chronic irritation of epithelial cells, Yamagiwa made two choices that distinguished his work from that of his failed predecessors: He chose to experiment on rabbits because the insides of their long ears provided plenty of accessible epithelium, and he decided to paint those ears with coal tar over many months, not just a few weeks. Beginning on September 1, 1913, Ichikawa painted tar on the animals' ears every two or three days. (Yamagiwa was too weak for the laborious work.) After 112 days, tumors appeared. On April 2, 1914, Yamagiwa excitedly reported the results to the Tokyo Pathology Society. But his audience, aware of the failed European experiments, was skeptical. The growths Yamagiwa saw on the rabbits' ears were merely inflammation, not malignancies, several of his colleagues asserted.

Undaunted, Yamagiwa and Ichikawa secured a grant to purchase sixty more rabbits and began the process of repeating their lengthy experiment. Things were going well until, during the summer rainy season, an infectious disease swept through the cages and killed almost all of the rabbits. Among the few survivors, however, two developed the same tumors the researchers had seen before, so Yamagiwa decided not to abandon the project. Instead, he bought more rabbits. By the time he formally presented the results to the Tokyo Medical Society on September 25, 1915, Ichikawa had painted coal tar on the ears of 137 rabbits over 250 days and had documented the presence of cancerous tumors on the ears of seven of those rabbits. This time, Yamagiwa got a much more favorable reception. To celebrate his success, Yamagiwa penned a haiku poem, famously translated into English as: "Cancer was produced! Proudly I walk a few steps." He wrote up his results in a paper he dedicated to the memory of his mentor Virchow, who had died in 1902.

It was a historic moment, but few scientists were paying attention.[6] While Johannes Fibiger's fame spread quickly, Yamagiwa's achievement got much less attention because of Japan's remoteness. By the early 1920s, however, scientists in Europe and the United States (including Fibiger) were replicating Yamagiwa's coal tar experiments in mice, rats, and dogs, confirming the carcinogenicity of tar and developing a template for testing other suspect compounds. Fibiger's nematode experiments, meanwhile, could not be replicated. Even so, they were acclaimed as the synthesis of the two major schools of thought on carcinogenesis: Fibiger's nematodes were microbial but allegedly caused tumors via irritation of the stomach and esophagus of their animal host, which is why his results excited disciples of both Koch and Virchow. Fibiger was repeatedly nominated for a Nobel Prize, finally receiving it in 1926, after the prize committee rejected splitting his award with Yamagiwa.

It was, in hindsight, one of the biggest errors in the history of the Nobel Prize. As was already becoming clear in 1926, Fibiger's nematode was *not* the direct cause of the growths he saw in rats. By the 1930s, research had demonstrated that the stomach lesions Fibiger saw in rats were benign and appeared only in animals fed a diet deficient in vitamin A, which was the crucial cause.[7] The nematodes were, if anything, merely a contributing factor because they caused tissue irritation (a realization that belatedly lent more support to Virchow's irritation theory). Microbial carcinogens, both parasitic and viral, would not be confirmed until the second half of the twentieth century.[8] By the 1950s, Fibiger's name had disappeared from many histories of cancer research. When his work was included, it was often as an illustration of experimental error.

Yamagiwa's discovery, on the other hand, launched modern experimental cancer research, setting the stage for the identification of hundreds of chemicals that, at sufficient doses, cause cancer in lab animals. After Yamagiwa, cancer could no longer be dismissed as a vague threat confined to dangerous places like mines and factories, or as an uncontrollable illness that struck randomly and without apparent cause. The era of the carcinogen had arrived.

When Stephanie Wauters heard Ciba-Geigy's Jorge Winkler describe the chemical composition of his factory's wastewater as "ninety-nine percent water and a little salt," she was furious. A former high school science teacher who had just started law school in 1984 (she would eventually become a prosecutor and then a judge), Wauters was no shrinking violet. She and her husband, John, an accountant, lived on Tunesbrook Drive, about a quarter-mile west of the site of the pipe leak. Like many of their neighbors, the Wauters had their own water well in the backyard, and Stephanie was upset when she learned that her children's drinking water had come from the same shallow aquifer the leak had contaminated. She wondered how long the pipeline had been leaking. Then she heard Jorge Winkler's comments, which made no sense to her at all. "Our motive initially was just to find out what was going on and to protect our children," she recalled years later. "Then we heard that the waste was from Ciba, and that Ciba was claiming it was just diluted effluent. But how could it be just water and salt, especially if they were accepting waste from all those other companies in violation of their permit? They weren't being honest. I felt we had to get more involved."

She and her husband, along with some of their neighbors, invited four science-minded friends over for coffee. Three were high school science teachers: William Skowronski and Peter and Susan Hibbard. The fourth, Stephen Molello, worked at the Oyster Creek nuclear power plant south of town. They decided to form a group, which they later named Ocean County Citizens for Clean Water, with Stephanie Wauters as the leader. The group's mission would be to push for stricter water testing and for full disclosure by Ciba-Geigy of the types and quantities of chemical waste it was discharging into the ocean, burying in landfills, and sending up its smokestacks.

These were daunting goals, considering Ciba-Geigy's local clout and penchant for secrecy. Within a few weeks, the group got even more ambitious. From Don Bennett's articles in the *Observer,* Wauters learned that Ciba-Geigy's ocean-dumping permit had expired and that the state Department of Environmental Protection would be conducting hearings to decide whether to renew it, and if so, under what conditions. That permit was absolutely crucial to Ciba-Geigy because the company had no other way to get rid of the almost two billion gallons of wastewater it discharged every year—a total that would surely rise after the pharmaceutical manufacturing operation moved from Rhode Island to Toms River. Without the permit, there would be no pipeline to the ocean; without the pipeline, the Toms River factory would have to close.

Stephanie Wauters and her friends realized that the expiring permit created a rare opportunity. Ciba-Geigy had little incentive to pay attention to a few science teachers, but the company would have to listen to the state Department of Environmental Protection. If the citizen's group could force the agency to take a tough line with Ciba-Geigy, the company would have to compromise or risk its entire Toms River enterprise. No one in Ocean County Citizens for Clean Water had a clear idea about how to accomplish that, but they were ready to try. The group's core members were all Democrats accustomed to outsider status in Toms River, so they took an outsider's approach that bypassed Roden Lightbody and rest of the local power brokers. Instead, they published letters in the local papers and sought allies among reporters. At the *Observer,* Bennett obliged with detailed coverage of their activities. In the beach communities, meanwhile, Mayor Gorga spread the word about the new group. Soon, a few dozen people were crowding into the Wauters' living room for meetings.

Cancer was not yet a key point of attack for Ciba-Geigy's critics, but it was always lurking in the background. Many people in town had heard murmurings about illnesses on Cardinal Drive and in other neighborhoods near the chemical plant, as well as among its workers, but the general feeling was that nothing could be proved. The company would not even divulge the names of the chemicals in its waste, so how could anyone hope to connect the plant to a specific pattern of illnesses? "From Day One we were concerned about cancer in the community, but we didn't have the resources to look into it," Wauters remembered. "We were taking on the whole social structure of the company and the politicians and agencies; we felt we couldn't fight the cancer fight, too."

Even so, new information about cancer was trickling out, and none of it was good for Ciba-Geigy. Back in 1981, the same year the company applied for a new ocean discharge permit, a team of inspectors from the U.S. Environmental Protection Agency had visited the plant, collected samples of treated wastewater, and brought them back to an agency lab, where saltwater tanks filled with tiny bug-eyed sea creatures were waiting. These were mysid shrimp, which were often used in tests to see whether polluted water was harming marine life.[9] For years, Ciba-Geigy had refused to conduct those mysid tests, preferring instead to test its effluent on a much hardier animal, the sheepshead minnow. (The company's critics would later call the minnow the "cockroach of the sea" for its ability to survive even in highly toxic environments.) The reasons for the company's reluctance became clear as soon as EPA technicians poured various amounts of Ciba effluent into the mysid tanks and waited the requisite four days to see what would happen to the animals. Ciba-Geigy's effluent, the agency concluded, was "highly toxic" to the mysid, more than half of which died in every mixture the EPA tested—even the most diluted, which consisted of just 6 percent effluent and 94 percent salt water.[10]

The agency also conducted a second test that Ciba-Geigy had long refused to undertake, one that was directly relevant to the cancer issue.[11] It was aimed at seeing whether the company's treated wastewater triggered genetic mutations in bacteria. That was important because mutations, alterations in DNA, frequently lead to the uncontrolled cell growth of cancer; mutagenic chemicals often, though not always, are carcinogens, too. The results from the second test were even worse for Ciba-Geigy. Its wastewater was more damaging to DNA than any effluent that had ever been tested at the EPA's New Jersey lab, according to a summary of the tests prepared by the state Department of Environmental Protection. Ciba-Geigy, the report concluded, should not be permitted to continue to discharge "such a clearly mutagenic wastewater."[12]

Ever since there had been a chemical plant in Toms River, its executives had refused to disclose what was in its waste. With the new EPA tests, agency officials finally knew enough to put the lie to the claims that only water and salt were going into the ocean and that the company's new treatment plant was removing any hazardous compounds before discharge. Now regulators knew that even after treatment, and even in diluted form, the more than five million gallons of wastewater Ciba-Geigy discharged every day into the Atlantic was still hazardous enough to kill living things and scramble their DNA. That was a shock because, under the traditional approach to regulating pollution—one chemical at a time—the company's effluent had always been acceptable, or nearly so. Now it was clear that the mixture, even when diluted, was toxic.[13] By the time officials realized how misguided their assumptions had been, the company had already sent more than thirty billion gallons into the ocean—and billions more into the river, before 1966.

The 1981 test results were a disaster for Ciba-Geigy, but its managers could at least console themselves that no one outside the regulatory agencies knew. There was still a chance that the company could get a new permit for its pipeline without public scrutiny and thus keep making chemicals in Toms River. But on the morning of April 12, 1984, with the discovery of the leak at the corner of Bay and Vaughn avenues, all hope of a quiet accommodation vanished. A state source tipped off Don Bennett about the results of the 1981 mysid and mutagenicity tests, which he promptly splashed all over the *Observer*. By the summer of 1984, Bennett was writing articles about *all* of the chemicals used at the factory. Thanks to the information the company was forced to provide on its application for a new permit, the composition of its waste was no longer a trade secret. Instead, the people of

Ocean County were reading about it every day. Bennett even wrote a two-day series featuring detailed descriptions of twenty-two hazardous chemicals in the company's wastewater and 109 others used at the plant.[14]

The list was devastating, but by the time it was published, Ciba-Geigy had bigger worries than another damaging story in the local paper about cancer-causing chemicals.

≈

CHAPTER NINE

Hippies in the Kitchen

There were hippies in Rose Donato's kitchen. She was seventy-one years old and stood less than five feet tall, and she liked to keep a clean kitchen. It was the brightest room in her beachfront bungalow, the place where she sipped her morning coffee and watched the sun rise over the surf as she paged through the morning's *Ocean County Observer*. And now hippies were there—and on her front lawn, too. Their sleeping bags were crowding her perennials.

It was her own fault. She had invited them, sort of. It had started with an article Donato noticed in the *Observer* three months earlier, in April of 1984. She read newspaper stories all the way through if they were about environmental topics, and this particular one hit close to home. The article, by Don Bennett, described a rupture in a pipeline that carried wastewater from the big chemical plant in Toms River, which was ten miles inland, to a discharge zone a half-mile off-shore from Ortley Beach—a location well within view from Rose Donato's kitchen, since Lavallette was just north of Ortley Beach. Donato had summered in Lavallette for many years, long enough to remember the controversy over Ciba-Geigy's plans to build the pipeline back in the mid-1960s. She had considered joining the opposition back then but was mollified by assurances from public officials that the pipeline would be carrying "a clear, harmless water," her daughter Michele remembered.[1] So eighteen years later, when Rose Donato read in the *Observer* that hazardous chemicals had just leaked from the pipe in a Toms River neighborhood, she was "absolutely infuriated," as Michele Donato put it. "I think she felt very, very betrayed by her government officials, who had lied." Rose Donato could be tough when the situation called for it. A divorcée, she had run a demolition business in New Brunswick before retiring, and she was used to pushing back if someone tried to take advantage of her. And so she resolved, as soon as she finished the article in the *Observer,* to push back against what she considered a grave offense to the Atlantic Ocean.

Rose Donato was not an activist, but she knew what she loved: her garden, her view, her beach, and her ocean. She had been involved in an effort to stop Rutgers University from expanding into a forest near New Brunswick, where she lived when she was not in Lavallette. And every year since the 1970s, Donato had sent a check for twenty dollars to an environmental group she especially admired: Greenpeace. Newsletters from the organization would arrive in her mailbox, and they were always filled with accounts of the derring-do of plucky young activists who somehow managed to disrupt all kinds of nefarious activities on the high seas, including nuclear weapons testing in Polynesia, seal clubbing in Newfoundland, whaling off the California coast, and industrial waste dumping in the North Sea. The last one is what caught Rose Donato's attention. If Greenpeace was fighting dumping off the coast of Scotland, why not off the coast of New Jersey, too?

She wrote a letter to Greenpeace asking for help. And then the adventure began.

The news that a chemical company was discharging its waste near a New Jersey beach was just what Dave Rapaport wanted to hear—not because he supported ocean dumping but because he wanted to end it. He worked for Greenpeace U.S.A. and was somewhat of a prodigy there. Bearded and intense—he looked like a young Al Pacino in *Ser-*

pico, but with wire-rimmed glasses—Rapaport, at age twenty-five, had just been put in charge of Greenpeace's new toxics campaign in the Northeast, which was supposed to target companies that discharged waste into lakes, rivers, and oceans. It already had a donated boat: a forty-foot, steel-hulled ketch named *Aleyka.* Rapaport and other staffers had fixed the boat up the year before. Now, in the spring of 1984, he was looking for destinations on the East Coast. The trick was to find creative ways to use the *Aleyka* that would generate news coverage for Greenpeace and build support for its campaign. To Rapaport, the discharge at Ortley Beach sounded ideal.[2] New York City media were nearby, and the dumper, Ciba-Geigy, was one of the largest chemical manufacturers in the world. The fight in Toms River, he thought, could turn into a big deal.

The first thing Rapaport did was to call Stephanie Wauters and arrange to meet with her and the other founders of Ocean County Citizens for Clean Water. They were suspicious of Greenpeace and worried it was far too radical for conservative Toms River. But they also recognized that they needed help to rouse the community, and the young man from Greenpeace looked reassuringly respectable, with his neatly trimmed beard and glasses. "I went down there to Stephanie's house and talked to them and listened to their stories," Rapaport recalled. "From everything I heard, it seemed like as good a place as any to bring the boat. It all kind of added up, so we put it on the itinerary."

For the next few weeks, in May, Rapaport scouted out the area, especially Ortley Beach and next-door Lavallette. The beach towns on the Barnegat Peninsula were odd but endearing; they were blue-collar resorts full of quirky characters. Rapaport needed to make some friends. If the *Aleyka* was going to come to New Jersey, its crew would need a base of operations on shore. In Lavallette, he heard about Rose Donato's beach house, which was perfectly situated just off the beach, within easy range of ship-to-shore radio. He asked her if she would be willing to allow her house to be used for a good cause, and Donato agreed. She had written Greenpeace a letter, after all. If she had any worries about the fact that Rapaport was vague about

how many people would come and how long they would stay, the sep-
tuagenarian did not show it. "She just said, 'OK, you can stay at my
house,'" Michele Donato remembered, "and that was how it all
started."

On one of his first mornings in Lavallette, Rapaport rented some
snorkel gear and swam out from the beach to try to find the pipeline
discharge area. He quickly recognized that he was out of his depth, in
every sense. The first vents did not appear in the half-buried pipe until
it was twenty-five hundred feet offshore; by then, it was under forty
feet of murky water. (The vents then continued for another thousand
feet to the end of the line.) Clearly, if he wanted to anchor the *Aleyka*
directly over the pipeline, Rapaport would need help from the locals,
preferably ones with boats, scuba equipment, and knowledge of the
underwater topography. Cautiously, he began to look for allies. He
met with an ex–Navy Seal at a topless bar (he was a diver who had
done work for Ciba) and hung out with several charter boat captains.
"It was almost like espionage," he recalled, "because I was trying to
find people who were sympathetic, while also trying to avoid giving
too much information to people who might turn out to be sympathetic to Ciba."

Rapaport already had a pretty clear idea of the type of aggressive
mischief he wanted to undertake, though he kept the exact nature of
his plans a secret. What he had in mind was what was known in the
environmental movement as monkeywrenching, illegal but nonlethal
actions aimed at disrupting a company's harmful environmental
practices.[3] (Critics of monkeywrenching had another term for it:
criminal sabotage.) He wanted to plug up the discharge vents in
Ciba-Geigy's pipeline, at least temporarily. It was a bold idea, technically difficult to accomplish in deep, cloudy water and also politically
risky, since public opinion in Ocean County might turn against the
Greenpeace activists if they were perceived as reckless vandals. That
would be the exact opposite of the citizen uprising against Ciba-Geigy
that Rapaport hoped to inspire. "The whole purpose was to find a
way to dramatize what was going on with Ciba, to expose it, and to
inspire people to take action," Rapaport remembered many years

later. "We wanted people to think, 'If those crazy guys from Green-peace can do it, so can we.'"

Worried that the pipeline-plugging operation might fail or back-fire, Rapaport also identified a second target, one that was visible above the treetops as he drove by the Ciba-Geigy factory complex on Route 37. It was the company's water tower, one of the tallest objects in Toms River at 160 feet. It looked like a giant aspirin tablet on stilts. The tank on top was white except for the words "CIBA-GEIGY" in six-foot black letters. A narrow catwalk with a railing ringed the lower edge of the tank; below the railing, the tank rested on six stilts that were each about as tall as a ten-story building. One stilt was equipped with a ladder surrounded by a protective cage. To a Green-peace veteran like Rapaport, an industrial tower on private property was like a fire hydrant to a dog: It was territory that had to be marked. In 1982, five daring activists had garnered nationwide publicity for Greenpeace by simultaneously climbing power plant smokestacks in three states and unfurling banners protesting acid rain. Ciba-Geigy's smokestacks were too short and narrow for that type of stunt, but its water tower was a climber's delight—assuming he or she could scale the factory's perimeter fence, evade the security guards, and sprint all the way to the tower ladder without being caught. Rapaport had just organized a similar climb in Baltimore, so why not in Toms River?

As June became July, Rose Donato acquired a clearer sense of what it meant to host a Greenpeace action. A stream of shaggy young men—and a few women, too—appeared at her doorstep and pitched their sleeping bags on her floor. "We just kind of took over her house," Rapaport remembered. "More and more people started showing up." Initially, Donato was still at her winter home in New Brunswick, hav-ing left extensive instructions for Rapaport and his fellow campaign-ers about how to take care of the bungalow. When she arrived in Lavallette to check up on them, she realized how futile those instruc-tions were. The beach house got even more crowded when the *Aleyka* showed up and anchored offshore. Using rubber rafts, crew members went back and forth from the boat to the house, all day and night. One Friday evening in mid-July, Michele Donato showed up for what she thought would be two weeks of lazy vacationing at her mother's place. "I was shocked to see them all there," she remembered. "When I showed up, the house was loaded with twenty-five people, and my mother was sitting there in her kitchen, looking kind of bewildered. I mean, they were sleeping *all over the place*." After briefly kicking them all out, Rose relented and instead left for New Brunswick, mollified by the gift of a bottle of scotch from her new friends at Greenpeace. Her daughter, a lawyer and self-described "old hippie" at thirty-five, stayed and joined the campaign. "It was all very, very exciting," Michele Donato remembered. "I wanted to be part of it."

She was not the only one caught up in the impending drama. As the Greenpeace contingent grew and its activities became more visible, the people of Lavallette started to take notice. There was a cultural chasm between the middle-class families who summered there and the young activists, but they found common cause in their opposition to the pipeline. "We might have been dirtier than they wanted their own kids to be, but we were good kids and the people in Lavallette recognized that," Rapaport said. A few trusted locals began visiting the *Aleyka* and sitting in on Greenpeace strategy sessions, which were always more like parties anyway. Lavallette mayor Ralph Gorga, still the only public official in the county willing to attack Ciba-Geigy in the newspapers, helped by allowing Greenpeace to leave its inflatable boats on the beach, even though a borough ordinance forbade it. Lavellette, he declared, would be a safe haven for Greenpeace.

The aura of illegality only added to the romance of the enterprise. Twice, Rapaport scaled the perimeter fence and led clandestine nighttime scouting missions onto the factory property to assess Ciba-Geigy's security, map the fastest route to the water tower, and collect soil samples for later testing. Meanwhile, off Ortley Beach, scuba divers (some from Greenpeace and some local) jumped off the *Aleyka* and searched the cloudy ocean bottom for the pipe. After many failed attempts, they managed to collect a few bottlefuls of the dark liquid that was shooting out one of its discharge vents. The bottles of brown water found a home on the front porch of Rose Donato's increasingly crowded beach bungalow. When they wanted inspiration, Dave Rapaport and his band of monkeywrenchers-in-waiting would hold the

bottles up to the light and wonder what kinds of chemicals were swirling inside.

The 1915 triumph of Katsusaburo Yamagiwa in inducing cancer in
rabbits exposed to coal tar had a galvanizing effect on cancer researchers around the world, especially after his former student Hidejiro Tsu-
tsui in 1918 used the same "tar painting" technique to produce tumors
in mice, which were much easier to handle in a laboratory than rabbits. One of the most determined of those scientists was a tall, be-
spectacled, and rather severe Oxford-trained pathologist named
Ernest Laurence Kennaway.[4] Joining the small research staff of London's Cancer Hospital in 1922, Kennaway immediately embarked on
a search for the specific ingredients of coal tar responsible for causing
cancerous tumors in animals and, presumably, humans. Yamagiwa
had confirmed the first carcinogenic substance, but Kennaway wanted
to find the first carcinogenic *molecule*. What made the quest so difficult was coal tar's composition: It was not one chemical but several
thousand. The high-temperature combustion of coal generates a
complex and highly variable stew of all sorts of compounds, largely
but not exclusively hydrocarbons. Most of these combustion products
are airborne gases, others are solids in ash, and the rest comprise the
viscous liquid known as coal tar. This thick, dark-brown goo was
carcinogenic—Yamagiwa had proved that—but Kennaway wanted to
know why. Identifying the particular ingredients responsible would be
a huge step toward understanding and preventing cancer.

He had only a few clues. The biggest was that the tars produced
when coal was burned at higher temperatures were more carcinogenic
to mice than those generated by less heat. So Kennaway started testing
pure chemicals such as anthracene that were produced only when coal
was burned above 700 degrees Celsius, or about 1,300 degrees Fahrenheit. None of those pure compounds induced tumors in his mice,
but he did find that when he burned organic substances—including
human skin—at the same high temperatures, he could produce mixtures that also gave mice cancer. Kennaway realized that the combustion products he created from his experiments were mostly members
of a large class of chemicals called polycyclic aromatic hydrocarbons,

or PAHs. The mystery carcinogen in coal tar, he reasoned, was probably a PAH.

By the late 1920s, Kennaway was certain he was getting close to his goal, but he may also have begun to doubt that he would be the one to reach it. He had Parkinson's disease; he had noticed its symptoms—shaking hands, a shuffling gait, and slightly slurred speech—a few years earlier, and they were getting progressively worse. Like the tubercular Yamagiwa, Kennaway increasingly relied on assistants and colleagues to undertake the laborious work of analyzing potential candidate chemicals and testing them on mice. He had always been fastidious about recording each day's work in the lab, but now it took him an hour or more to write just a few pages.

When the marathon search finally ended in 1932, Kennaway was too incapacitated to run its final leg.[5] Instead, members of his lab team distilled and tested PAHs extracted from two tons of coal tar pitch donated by a nearby London gasworks, finding one molecule that had all of the characteristics they were looking for. It was a previously unknown compound called 3:4-benzpyrene, now known as benzo(a)pyrene, and its identification marked the first time a pure chemical found in the real world, and not just synthesized in a laboratory, had been shown to cause cancer in a lab animal. The ambiguity was over: The combustion of organic compounds—in boilers, engines, chimneys, and all the other fires of industry—produced at least one molecule capable of causing malignant tumors in a living organism. Soon Kennaway's lab would find more than a dozen other carcinogenic PAHs, all of them present wherever organic compounds are incinerated—from the coal-fired power plants of inland China and the American Midwest to the kettles and boilers of Ciba-Geigy's Toms River plant.

The significance of what Kennaway achieved went far beyond a particular molecule or class of chemicals, however. If Katsusaburo Yamagiwa launched the era of the carcinogen, then Ernest Kennaway turned the search for cancer-causers into a methodical, fruitful science—one that would often be undertaken by teams of collaborating chemists, physicists, biologists, and physicians emulating the work of Kennaway's own team. Thanks in large part to Kennaway and his colleagues, dozens of carcinogenic pollutants would be identified by the 1940s, and environmental causation—the effects of toxic chemicals on living cells—would gain wide acceptance as an important (if largely unproven) explanation for many human cancers. But those developments would also transform the social conditions in which environmental cancer research is conducted. Now that specific products of commerce—including the detritus of dye manufacture—had been directly implicated as causes of deadly disease, the discoveries of Kennaway's successors would no longer be greeted with acclamation. As governments took their first steps toward meaningful regulation of the chemical industry, science would become both a weapon and a target.

Anyone in Toms River who did not already know Greenpeace was in town found out during the last week of July of 1984, when the group staged a press conference in Lavallette and called for the immediate closure of the pipeline.[6] Exactly how that would happen went unsaid, but Ciba-Geigy executives did agree to meet the following day with Dave Rapaport and other Greenpeace representatives to talk about it. By now, the company was beginning to recognize what it was facing, so Jorge Winkler and other executives adopted a conciliatory tone. Yes, they told Rapaport, Ciba-Geigy agreed that ending ocean discharges was a worthy goal. No, they added, it was not feasible with current technology. "Of course they weren't going to meet our demands," said Rapaport, "and we told them if they didn't, we would do what we had to do"—whatever that meant.

Dave Rapaport knew what it meant. He would have to talk his friend Sam Sprunt into climbing the water tower. He already had one confirmed climber, a serious-minded twenty-three-year-old woman from Boston named Beverly Baker who worked full-time for Greenpeace. But it would take two volunteers to carry all the banners and supplies. Rapaport wanted Samuel Sprunt to be the second. Sprunt, who was twenty-six, did not work for Greenpeace. He was a graduate student at the Massachusetts Institute of Technology, studying physics. Sprunt cared about environmental issues—he had participated in some anti–nuclear power demonstrations as an undergraduate at

Stanford University—but he was in Toms River mostly because he was Rapaport's friend and because he was pursuing a young woman with the group.

As it turned out, the young woman had already left Toms River by the time Sprunt arrived, but Rapaport urged Sprunt to stay because he possessed a characteristic Rapaport thought would come in handy on a Greenpeace mission: He was not afraid to try something crazy in the pursuit of adventure. Sprunt had proved that the previous summer in Newport, Rhode Island, at the America's Cup yacht race, where the upstart *Australia II* was about to win a historic victory, breaking the United States' perennial hold on the most famous sailing trophy in the world. Stealthily swimming up to the yacht in the middle of the night, Sprunt and a friend managed to photograph its super-secret winged keel, avoiding detection by a squad of Australian military guards on the pier just a few yards away. "The whole world wanted a picture of that keel, and we had it," Sprunt remembered. "We walked all around Newport that night, soaking wet, trying to find a reporter who would believe our story." Alas for Sprunt, by the time his film was developed, *Australia II* had won the final race, and its gloating owner had revealed the keel to the world. Sprunt's hard-earned photo was worthless, but he had a great story to tell his friends, including Rapaport, who thought it sounded a lot like a Greenpeace stunt.

Ciba-Geigy's security guards would surely be child's play for a man who could outsmart the Australian Air Force, Rapaport thought. Sprunt was not so sure. Even if he made it up the water tower, Sprunt knew that he would eventually be arrested for trespassing. He worried about what his parents would think, and whether the university would cut off his doctoral stipend. But Sprunt liked his Greenpeace friends and believed in their cause. So, a few hours before dawn on Monday morning, July 30, he and Bev Baker put on hard hats (so that they would look like workers from a distance) and slipped over the fence and onto the vast factory grounds. Thanks to the earlier reconnaissance missions, they had a hand-drawn map and knew how to maneuver through the road grid to reach the water tower. They ran in a crouch, stopping only to duck behind some drums when they saw the headlights of a patrol car. After what felt like a long time to Sprunt, they made it to the center of the complex and saw the shadow of the tower looming in front of them. First, though, they had to cross an open area beside two buildings. The lights were on, and they could hear the voices of some night shift workers. The two paused to consider their options. "We had the element of surprise," Sprunt said, "and I figured that even if they saw us they wouldn't know what to do, so we decided to continue the crouch run. We just had to hope there wasn't something locked, like a gate, at the bottom of the ladder."

They were in luck. There was no lock. The pair started climbing the ladder, which was more than 120 feet high. Their backpacks would not fit underneath the protective cage that encircled the ladder, so they decided to try to carry their sleeping bags, food, and a large banner up by hand. That proved disastrous when, halfway up the ladder, Baker dropped the banner. "I watched it sickeningly spiral down the tube and hit the bottom, and I thought, 'Oh God, that's the one thing we can't lose,'" Sprunt said. He waited to make sure that no one was coming to investigate, then climbed down to grab it while Baker continued to the top. A few minutes after beginning his second ascent, Sprunt dropped his sleeping bag and had to go back down a second time. By the time he finally made it all the way up he was exhausted, and collapsed on the narrow catwalk that would be his home for the next two and a half days.

At the eight o'clock shift change, several hundred Ciba-Geigy workers looked up in the sky and saw something shocking: two figures silhouetted against the water tower, and just below them a giant white banner with a spray-painted message: "Reduce It, Don't Produce It."[7] Some factory employees were amused; most were angry. John Talty, who would later be the president of the union, was only half-joking when he later recalled, "I had to talk some of the chemical operators in Building 102 or 103 out of shooting them." Up on the catwalk, Sprunt and Baker viewed the hubbub below with a mixture of curiosity and anxiety. "We could see it was a crisis down there, and that no one was sure what to do," Sprunt remembered. "I figured we would last through the morning and then they would send the cops up to get us."

That did not happen, because the executives at Ciba-Geigy feared a confrontation. Forcing the two activists back down the long ladder would be dangerous, and the company did not want to turn them into martyrs. Instead, "we offered them lunch," remembered Ciba-Geigy executive Jorge Winkler. A foreman met Sprunt halfway up the ladder and gave him blankets, a flashlight, and safety ropes as well as some peanut butter sandwiches. Sprunt was especially grateful for the sandwiches because the only edible items he was carrying were nuts and berries packed by a granola-loving volunteer. The workers below, meanwhile, were furious about the coddling. "We couldn't believe it when they gave them blankets and food," remembered Talty, the union leader. "We told the managers that was a mistake, that these Greenpeace guys were bullies and that the story would get bigger the longer they stayed up there. And that's exactly what happened. It got bigger."

When the first news helicopters arrived in the afternoon, their crews were rewarded with eye-catching video of the water tower, the banner, and the two figures at the railing. The protest was featured that evening on newscasts in Philadelphia and New York City, marking the first time—but not the last—that the Ciba-Geigy factory was the subject of a big-city television report. The residents of Toms River were not yet accustomed to living in a notorious town, and they didn't like it. The factory in the woods was no longer their secret.

Sam Sprunt and Bev Baker stayed up on the water tower for two days and two endless nights. Life on a narrow, iron catwalk ten stories up was not comfortable. They were cold, and they were wet. Worried about slipping through the railing while lying down in their sleeping bags, Sprunt and Baker used the Ciba-provided ropes to tie themselves down, but it didn't help much. "The whole thing was pretty scary," Sprunt remembered. They took heart, though, in the chaos they were provoking down below.

At the front gate of the factory, workers and poster-waving activists traded insults and even a few shoves. Things got completely out of hand on the second day of the protest, when a Ciba-Geigy worker broke the window of a car in which two activists were sitting, reached in, and struck a woman from Greenpeace. He was charged with assault. Another employee was charged with criminal mischief for destroying a Greenpeace sign. Inside the factory, a Ciba-Geigy worker made his own sign, which he waved at the water tower. It read: "Jump, Bev, Jump." The news cameras caught it all. The story kept building on television and in the New Jersey papers, *The Philadelphia Inquirer,* and eventually *The New York Times.* Worst of all for Ciba-Geigy, the Associated Press filed numerous dispatches that were published all over the United States and in Europe, the company's home base. "It was really an enormous amount of coverage," Rapaport remembered. "It was, in many ways, the best-staged sequence of a protest that I was ever involved with. We basically laid siege to the plant."

On the third day, August 1, after long negotiations with Ciba-Geigy and the town police, Rapaport finally radioed Sprunt and Baker and told them to come down—this time wearing safety harnesses provided by the company. They were exhausted and more than ready to call it quits. They had not slept for more than two hours at a time and had to use a plastic bag in lieu of a toilet. (The bag burst when they tossed it down, which pleased no one.) That afternoon, Sprunt and Baker were taken to the police station and issued summonses for trespassing. Sprunt then went out for the hamburger he had been craving for days and ended the day with an ocean swim. In a few days he would leave Ocean County and never return.[8] Sprunt would eventually become a physics professor at Kent State University in Ohio, where he studied liquid crystals and occasionally pondered his unique role in the Toms River saga.

Greenpeace, however, was just getting started in Toms River. As soon as Sprunt and Baker were safely sprung from the police station, a spokesman for the group announced to reporters that three scuba divers that morning had plugged thirteen of the fifty discharge vents in Ciba-Geigy's pipeline, using foam gaskets with wooden plugs. It was not quite true. The jury-rigged plugs were not watertight and did not actually block the discharge; in any case, divers hired by Ciba-Geigy quickly removed them. But Rapaport managed to turn the ineffectiveness of the attempted monkeywrenching to Greenpeace's advantage by pointing out that the organization was careful to leave most of the vents open because it did not want to cause another rupture farther up the pipeline, like the one near Stephanie Wauters's home that had started everything back in April. To Rapaport, nothing was more important than winning the goodwill of the town's residents. He wanted Ciba-Geigy to be the bad guy, not Greenpeace.

By all accounts, his strategy was working. The leaders of Ocean County Citizens for Clean Water were still suspicious of Greenpeace's radicalism but were happy to capitalize on the group's highly publicized actions to boost their own membership. By early August, the citizens' group was drawing more than 150 people to its meetings, which were now held in town hall instead of Stephanie Wauters's living room. The union sometimes sent Ciba-Geigy workers over to try to disrupt the proceedings, but the environmental group's supporters were usually able to outshout them. In the beach communities, meanwhile, the Greenpeace activists were treated like conquering heroes. The group set up a booth on the boardwalk where volunteers collected hundreds of signatures on a petition calling for the closure of the pipeline. By that time, Greenpeace had gotten hold of the damaging reports that regulators had produced about Ciba-Geigy's wastewater. As a result, they were able to hand out "information sheets" detailing what the company was dumping. There was even a comic book–style pamphlet featuring drawings of fish skeletons in Barnegat Bay and spills of black liquid next to suburban homes. Much of the information was exaggerated; all of it was effective.

If all that was not enough to catch the attention of beachgoers, there was also the coup de grâce: a large jar filled with what was, by now, a familiarly opaque liquid the color of brown shoe polish. It was, according to Greenpeace, another sample of Ciba-Geigy effluent collected from one of the discharge vents. "It was really, really dark-brown and ugly," recalled Michele Donato. "People would see it and say, 'Oh my God, *that's* what's coming out?'" To passersby, the implication was clear: You are swimming in this stuff. Never mind that no one had any direct evidence of the discharge harming marine life (although the tests on mysid shrimp suggested it might). Never mind that for humans, the risk was very close to zero because the effluent was being discharged more than a half-mile offshore and was drastically diluted by the surrounding ocean. None of that seemed to matter, especially when people read in the *Observer* that many of the discharged chemicals were carcinogens. "Even if there was a chance of just one child swimming out there and developing a cancer ten years later or, God forbid, a pregnant woman, I just knew we had to stop this," remembered Nancy Menke Scott of Lavallette, who was so inspired by the tower-climbers and pipe-pluggers of Greenpeace that she spent the next seven years fighting Ciba-Geigy on an almost full-time basis.

The crew from Greenpeace stayed in Lavallette for a few more weeks, enjoying what was essentially a carnival of environmentalism on the boardwalk. Rapaport knew the group was scoring big when, one night during dinner at Rose Donato's bungalow, a man who was later identified as a prominent member of the Ocean County Chamber of Commerce ran up to the front porch and grabbed one of the bottles of brown wastewater. He dashed back to his car, but Rapaport and several others from Greenpeace were right behind. They chased him down the borough's main road until he was pulled over by the police, who forced him to return the jar of precious effluent.[9] There were more press conferences, too, including one in which Greenpeace announced that a soil sample the group collected on one of its nighttime raids onto Ciba-Geigy property had tested positive for cyanide. The residents of Ocean County may not have recognized many of the tongue-twisting chemicals that Don Bennett was writing about in the *Observer,* but they knew what cyanide was. A company spokesman responded that cyanide had not been used at the factory for years but that same week had to acknowledge that Ciba's own tests had just revealed that the inside of its pipeline was corroding and needed extensive repairs.[10] It was more evidence that that effluent the company had described as "ninety-nine percent water and a little salt" was, in fact, strong enough to eat through steel.

By mid-August, things were winding down. Many members of the Greenpeace team had already left town with the *Aleyka,* though they promised to return the following summer. Rapaport was in New York

City, where on August 6 he and three cohorts generated another bonanza of publicity for Greenpeace by climbing the scaffolding surrounding the Statue of Liberty (it was being restored) and hanging a banner proclaiming, "Give Me Liberty from Nuclear Weapons—Stop Testing."

The Toms River that Greenpeace left behind at summer's end was very different than the one Rapaport had first encountered three months earlier. For one thing, the shoreline communities were radicalized. To Rose Donato, Nancy Menke Scott, and hundreds of thousands of other New Jerseyans, the ocean was sacred; it was what made their state something other than the butt of jokes by New Yorkers and Philadelphians. The outrage they felt was deeply personal, and they were ready to keep on fighting. On the other side of Barnegat Bay, the residents of Toms River were more conflicted. They, too, loved the ocean, but their ire over Ciba-Geigy's discharges was tempered by their connections to the company and its importance to the local economy. Most were not yet ready to turn on Ciba-Geigy, but they were unhappy that their town had been labeled as polluted. Reputation meant everything in Toms River, and now their community was being portrayed on television and in the newspapers as no different from Newark, Trenton, and the other run-down industrial cities they had scorned for so long.

A threshold was crossed in Toms River during the summer of 1984, thanks to people like Don Bennett, Stephanie Wauters, Dave Rapaport, and even Rose Donato. The chemical plant was no longer something to be proud of, and it never would be again. Its executives were no longer controlling events, and they never would again. And for the people of Toms River, there were suddenly many reasons to question their town's two-generation romance with the chemical industry. Soon there would be more.

≈

CHAPTER TEN

The Coloring Contest

On the morning of July 30, 1984, a few hours after Samuel Sprunt and Beverly Baker of Greenpeace snuck onto the grounds of the Ciba-Geigy plant and climbed the water tower to begin their protest, three state officials showed up at the factory's front gate for a rare surprise inspection and another hellish trip into the landfill pit.

For Jim Manuel, a newly hired hazardous waste specialist at the state Department of Environmental Protection who would make many such descents into the pit starting later in 1984, a visit to Ciba-Geigy was a thoroughly unpleasant experience—and one that was awash in déjà vu. Like almost every other science-minded boy from Ocean County (Manuel had grown up in next-door Lakewood), he had applied for a job at Ciba-Geigy after graduating from college. "It was one of the stellar employers of the area," he recalled. "Nobody ever questioned anything they did." Now, returning to the plant as an inspector, Manuel was getting an inside look at the company's operations.

He was not impressed. Some buildings at the complex were clean and modern, but others looked like they had been contaminated for decades. Treeless areas near the production buildings had been used as dumps as far back as the 1950s and still bore the scars of their misuse. Manuel knew that the soil was sandy and that an aquifer lay beneath it. He wondered where all those long-buried chemicals had gone once they reached groundwater. But those older dumps, which totaled more than fourteen acres, were outside the state's purview. The U.S. Environmental Protection Agency would have to deal with them through the Superfund process, which the EPA was still organizing in 1984. Manuel and his DEP colleagues were focused on the new, two-acre dumpsite their own agency had sanctioned: the double-lined landfill where the company was burying ten thousand drums and nine thousand cubic yards of sewage sludge every year.[1]

By the time Manuel got there, the DEP had known for more than five years that the landfill's inner lining was leaking. The dump's first section, Cell One, had opened in 1978, and within a year state inspectors had documented a leak of about one drop per second. By mid-1982, Cell One was filled to capacity, and the leak had swelled to approximately one gallon per day. The company then started dumping in a new section, Cell Two, which promptly began leaking about *forty* gallons per day.[2] The landfill's inner lining, made of polyvinyl chloride plastic, could not withstand the solvents and other corrosive chemicals in Ciba-Geigy's waste. An identical plastic outer liner was all that was preventing the waste from breaking through and again polluting the groundwater, as it had during the decades of open-pit dumping at the factory. In between those two liners was a two-foot layer of sand, portions of which were now soaked with leaking waste.

That was not supposed to happen, because hazardous and liquid wastes were not supposed to be buried in the landfill. To reduce the chance of leaks, the operating permit issued by the state DEP specified that only dry, nonhazardous waste could be dumped there. All hazardous waste was supposed to be trucked offsite to specialized facilities. By shirking that requirement and dumping almost all of its waste in its own landfill (everything but the wastewater it was pumping to the ocean), the company was probably saving about a million dollars per year.[3] By mid-1984, the evidence was overwhelming that Ciba-Geigy was violating the permit requirements, potentially a criminal offense. DEP inspectors had made more than 130 visits to the landfill since 1979, documenting leaks each time. The state even tested the leaking liquid, confirming that it included indisputably hazardous chemicals such as toluene. Still, the DEP did not try to fine or prose-cute Ciba-Geigy or revoke its landfill permit, which would have forced the company either to shut down or to ship all of its waste and sludge offsite at huge expense. Instead, the state let Ciba-Geigy keep dump-ing, day after day and year after year.[4]

It was a sorry example of environmental enforcement and an ab-solutely typical one—then and now. Compliance in New Jersey and everywhere else still depends almost entirely on voluntary reporting and negotiation. At their meager staffing levels, the oversight agencies have no alternative. The EPA, for example, is supposed to monitor hundreds of thousands of polluting facilities but has never had more than eighteen thousand employees, with not even one in ten directly involved in enforcement. So the system still relies on self-reporting by companies, with only sporadic direct oversight by overworked agency inspectors.[5] When there is a violation, agencies usually lodge civil charges that almost always end in negotiated agreements, called con-sent orders, in which the offending company promises to change its behavior and sometimes pays a fine.

Criminal convictions in environmental cases are very rare, since they generally require prosecutors to prove beyond a reasonable doubt that a polluter *intended* to violate the law—a formidable hurdle be-cause environmental rules are complex and open to varying interpre-tation. Even in the rare instance when an enforcement action begins as a criminal matter, it often ends in a civil settlement, which is what had happened with the indictments of Toms River Chemical in the 1970s. Union Carbide had followed a similar path with a 1977 consent order in which the state dropped its charges of groundwater pollution in Pleasant Plains in return for a settlement of $60,000. In the 1980s, criminal prosecutions of polluters were so rare that in 1984, the EPA referred just thirty-one cases to the Justice Department for prosecu-tion, while more than three thousand cases were handled administra-tively or in civil court.[6] Since enforcement depends on self-reporting and the distant threat of fines, a polluter's decision on whether to comply becomes a business calculation: What are the chances of being caught? Would paying the fine be cheaper than the cost of complying in the first place? Many waste handlers simply conclude that compliance doesn't pay.[7]

No state in the 1970s and 1980s had more trouble controlling polluters than New Jersey, which was still a hub of the chemical industry. In 1976, when Congress passed the Resource Conservation and Recovery Act, which for the first time tried to define what hazardous waste was and how it would have to be handled, the price of hauling toxic waste quickly quadrupled in New Jersey. So did the incentive for dumpers to evade the new rules. Some dumpers, it turned out, were already comfortable operating illegally because they had connections to organized crime.[8] In 1979, New Jersey lawmakers finally made it a felony, punishable by five to ten years in prison, to dispose of toxic waste illegally. But the difficulty of proving criminal intent remained a huge hurdle to getting convictions.[9] The situation was particularly frustrating for the state Department of Environmental Protection because the agency depended on the threat of tough enforcement, including criminal penalties, in its uphill struggle to try to get polluters to comply with its permits. That threat was especially hollow for a huge entity like Ciba-Geigy, which could afford top legal and engineering talent and had impeccable connections to politicians. The company had already shown that it could beat a criminal prosecution, back in the 1970s. No agency was eager to tangle with Ciba-Geigy again, and the company knew it.

By the early summer of 1984, however, the calculus was changing in Toms River, thanks to the pipeline leak and ensuing public outrage. Greenpeace was in town, and the beach communities were riled up. Ocean County Citizens for Clean Water was in the newspapers almost every day, and so was Don Bennett, who was writing one critical story after another in the *Observer*. It was obvious to Jim Manuel and his fellow inspectors that the status quo was no longer acceptable. Their bosses in Trenton were going to have to do something about Ciba-Geigy—or face the political consequences.

A day at Ciba-Geigy was always an ordeal for Manuel and his colleagues. First they would spend hours in a cramped office, poring over manifests that purported to show the disposition of every drum of waste the factory produced. That was a walk in the park compared to what came next: The inspectors would climb down into the landfill pit and examine the drums up close. The scene was straight out of Dante's *Inferno*. The stench was awful, and the drums, lined up in rows and lashed together by ropes, sat on a thick bed of jet-black sludge trucked over from the wastewater treatment plant. "It was a nasty place," Manuel remembered. "The sludge they put in there was almost like axle grease. When it rained, everything would get extremely goopy. Sometimes you couldn't get anywhere close to the drums so you had to kind of circle them from a distance."

Under those nightmarish conditions, it was extremely difficult for the inspectors to figure out what was actually inside the drums—and impossible once they were buried in sludge, which happened every few weeks. Even so, state inspectors kept catching Ciba-Geigy in the act of dumping hazardous waste and liquids into the landfill, thus violating its permit. "It probably happened three, four, five times," Manuel remembered. Sometimes, the inspectors would follow their noses, seeking the distinctive odors of liquid solvents. Other times, the company was tripped up by its own paperwork. The manifest sheets, which identified every drum by its serial number, showed that some drums that were supposed to be shipped offsite were dumped into the pit instead.

When caught, Ciba-Geigy managers did not always fess up. Often, they claimed that the waste in question did not meet the legal definition of "hazardous." There was, for instance, the "filter cake" residue scraped off the filters used in resin production. The company sent hundreds of drums of filter cake to its "dry, nonhazardous" landfill every year, even though the cake was drenched in liquid toluene. Managers claimed that the cake chemically bonded with the solvent, rendering it nonhazardous. Manuel considered that nonsense. Years later, when the state was pulling drums out of the landfill, he would prove his point by reaching inside a long-buried drum, grabbing a chunk of filter cake and squeezing it, and then watching the toluene pour out onto the ground.

To Manuel and the other state inspectors, the evidence was clear: Ciba-Geigy had violated the terms of its permit and was continuing to do so. The question now was whether anyone was finally going to do anything about it. By mid-1984, there had been so many violations at the plant that the inspectors had dropped their customary practice of giving the company advance notice that they were coming. Now they just showed up unannounced, as the three inspectors did on July 30. Once again, they discovered evidence that drums containing liquid solvents were going to the landfill.

Two days later, on the morning that Sprunt and Baker climbed down from the water tower and Greenpeace's divers made their symbolic attempt to plug the outfall pipe, Ciba-Geigy executive Jorge Winkler was out of town. He was making an emergency trip to Trenton to meet with top officials of the state Department of Environmental Protection. As the director of production and environmental affairs, Winkler was the point man in the company's clumsy attempts to defend its waste-handling practices. Now he was in Trenton admitting that Greenpeace was at least partially correct and so were the state inspectors. Ciba-Geigy had made a "mistake" by continuing to bury liquid waste, Winkler told the officials. This was not a gray area, like the semisolid filter cake. Instead, the drums at issue were filled with pure liquid, which was "a clear violation," Winkler acknowledged. The state officials sent him out of the room to discuss the matter in private and then called him back in and dropped a bomb: They told Winkler that the company would have to shut down the landfill immediately and remove all fourteen thousand drums that were buried in Cell Two.

Shocked, Winkler pleaded for a partial reprieve that would allow Ciba-Geigy to keep using the landfill for its wastewater sludge only. Otherwise, he said, the factory would have to close down and start furloughing workers. The state officials agreed, and for a few more weeks the factory continued to operate normally. The only difference was that instead of burying its waste drums, Ciba-Geigy stored them in a temporary holding area as it continued to negotiate with the state. Now Winkler was making more concessions, having belatedly realized the precariousness of the company's position. On August 10, Ciba-Geigy finally told the state that its wastewater treatment plant would no longer accept waste from other factories—more than a year after the state had demanded it and four months after reporter Don

Bennett's articles in the *Observer* made it a high-profile issue. Winkler also organized some quiet meetings with local politicians, urging them to stand by Ciba-Geigy as a pillar of the region's economy and assuring them that the company would, as usual, reach a quiet accommodation with the state DEP.

Things seemed to be calming down, especially after Greenpeace sailed away. The company's well-connected legal team—led by Matthew Boylan, who happened to be a former director of the state Division of Criminal Justice—was negotiating with the DEP. At the end of August, Winkler felt confident enough to take his usual vacation on Upper Saranac Lake in New York. Even after all the craziness of the summer of 1984, he was sure that the company would be able to cut a deal, just as it always had. The Adirondacks were always beautiful over Labor Day, and Winkler did not want to miss seeing them.

In 1932, the same year that Ernest Kennaway's decade of research at the Cancer Hospital of London culminated in the identification of benzo(a)pyrene as the first confirmed chemical carcinogen, another iconoclastic pathologist obsessed with unearthing the causes of cancer reached a turning point in his career. His name was Wilhelm Hueper, and like so many of his predecessors in cancer research, he had a Paracelsian knack for irritating almost everyone he knew.

Raised in a liberal, comfortable home in northern Germany, Hueper interrupted his medical studies in 1914, at age twenty, to volunteer for military service at the outbreak of World War I, despite being a pacifist who railed against the "stupid adventures" of war.[10] He fought in the trenches in Belgium—one of his tasks was to haul huge tanks of poisonous chlorine gas to the front lines to use against the French whenever the winds were favorable—and then switched to field hospital work. He triaged the wounded at what he called "an orgy of mass murder" during the five-month Battle of the Somme, in which there were more than five hundred thousand German casualties. During the final year of the war, Hueper was a prisoner in France, sleeping in a stable and surviving on a daily ration of two cups of coffee and a small piece of bread.

These were the types of experiences that would leave any man short-tempered and willful, and Hueper was both. Immigrating to America in 1924 after getting his medical degree, he worked as a pathologist at a Chicago hospital before quitting in 1930 in a dispute over money. He then accepted an offer to join the University of Pennsylvania Cancer Research Laboratory in Philadelphia. Like cancer researchers all over the world at that time, he tried to emulate Katsusaburo Yamagiwa's success in inducing tumors in test animals, but Hueper's attempts to match the feat by injecting mice with arsenic compounds succeeded only in killing the mice. He did make progress on another front, however. Irénée du Pont, one of the richest men in America and a former president and longtime board member of the industrial behemoth founded by his great grandfather, bankrolled the research lab where Hueper worked. As a result, Hueper made several trips to the du Pont mansion in Wilmington, Delaware. Hueper was even invited to see the company's massive Chambers Works complex across the river in Deepwater, New Jersey. Hueper's lab experiments on chemical carcinogenesis were not going particularly well, but he had a natural affinity for workplace investigations. Perhaps a tour of one of the world's largest chemical manufacturing plants might provide fresh inspiration.[11]

DuPont, the company, and du Pont, the man, would come to regret the 1932 invitation, but their association with Hueper began innocently enough. During the tour of the factory complex, several chemists told Hueper that as part of its aniline dye production process DuPont was making large quantities of two chemicals Hueper knew well: benzidine and beta-naphthylamine. (The same two compounds were also being produced by the Cincinnati Chemical Works, a rival facility run by DuPont's Swiss competitors.) Hueper knew about them because they had been identified as the likely causes of the bladder cancer clusters that had been detected in dye factories in Germany, Switzerland, and England as far back as 1895, the year Ludwig Rehn first took note of "aniline tumors" among dye workers in Frankfurt. Those European studies suggested that bladder cancer appeared in factory workers ten to fifteen years after they began handling the two chemicals; the Chambers Works had been producing aniline dyes since 1917, and it was now 1932. To Hueper, it was obvious what was about to happen: DuPont was going to face an epidemic of bladder cancers. In fact, it might have started already. Returning to Philadelphia, he dashed off a strongly worded note and sent it directly to Irénée du Pont. Hueper waited for months but got no answer, and when he raised the issue with his supervisor at the laboratory, du Pont's close friend and personal physician Ellice McDonald, he was told there were no bladder cancer cases at the Chambers Works.

A few months later, however, the medical director of DuPont turned up at the doorstep of Hueper and McDonald's laboratory with alarming news: There *were* bladder cancer cases at Chambers. In fact, the company had identified twenty-three cases among its aniline workers. Hueper immediately wrote another letter to Irénée du Pont, this time proposing that the company create its own toxicological institute to study chemical risks to employees and consumers. Again, there was no response, until a few months later when Hueper accompanied McDonald on another house call to the Wilmington mansion. Pulling Hueper aside for a confidential chat, Irénée du Pont told him the timing was not right to fund an institute because the country was in the depths of the Great Depression and DuPont could not spare any more funds for health research.

The following year, in 1933, Hueper felt the effects of the economic collapse more directly: McDonald fired him. The two men had clashed—Hueper regarded McDonald as an unethical self-promoter, and we can only guess what McDonald thought of Hueper—and now Hueper was told that the laboratory could no longer afford his salary. Desperate for work, Hueper returned to Germany with his wife and son to search for a medical position. Spaces had opened because so many Jews had fled after Adolf Hitler's rise to power, but Hueper got no offers. The family returned to the United States, and Hueper finally landed a humble post at a small hospital in a Pennsylvania mining town. Then, in 1935, something surprising happened: Hueper received an offer from the director of the newly created Haskell Laboratory of Industrial Toxicology. The Haskell lab, under construction in Wilmington, was about to become the in-house health research arm of the DuPont Corporation. Irénée du Pont had taken Hueper's advice after all, and now Hueper was being invited to be part of the team. He quickly accepted.

Arriving in Wilmington, Hueper found that the bladder cancer epidemic he had predicted was in full swing, with more than fifty confirmed cases among dye workers. He quickly launched a long-term study of dogs exposed to beta-naphthylamine, the aniline dye ingredient known as BNA. DuPont had previously tested its other key dye ingredient, benzidine, on lab animals and had never found tumors, but the company always halted its experiments after less than a year. It was the first example of what Hueper would later adopt as a truism: Manufacturers could not be trusted to run carcinogenicity tests.[12] He knew it took more than a decade of exposure to dye chemicals for humans to develop bladder cancer, so he thought that animal tests should last for several years at least—especially since both Yamagiwa and Kennaway had exposed animals to carcinogens for more than a year before seeing tumors. In his new lab, Hueper set up a much more thorough experiment in which sixteen female dogs, large enough to have their bladders checked with a cystoscope, would be given BNA with their daily chow.[13] The dogs would be monitored at least two years, twice as long as previous experiments.

While he was waiting to see what would happen to the dogs, Hueper decided to take another trip across the Delaware River to tour the BNA operations at the Chambers Works—this time as a DuPont employee. When he got there, Hueper was surprised to see that the BNA manufacturing area was extremely clean, with no telltale powder strewn on any surfaces. Spotting a foreman, Hueper told him, "Your place is surprisingly clean." The reply came back: "Doctor, you should have seen it last night; we worked all night to clean it up for you."[14] He decided to make an impromptu visit to a separate building where benzidine was made. This one had not been cleaned up, and the loose powder was everywhere. "With one look at the place, it became immediately obvious how the workers became exposed," he later wrote. Fifteen years earlier, in 1921, the Geneva-based International Labour Office had examined the accumulating evidence about dye manufacture and bladder cancer and had identified benzidine and

BNA as the most likely suspects, urging manufacturers to adopt "the most rigorous application of hygienic precautions."[15] Judging from Hueper's visit to the dye production buildings at the Chambers Works, that message had apparently never reached the DuPont Corporation. Angry about what he regarded as an attempt to mislead him about conditions in the factory, Hueper reacted in typical fashion: He dashed off yet another note to Irénée du Pont, this one complaining about the "deception." As usual, there was no reply, but Hueper was never again allowed to tour the dye operations at the Chambers Works.

Hueper's effective banishment from the factory floor was a sign that something fundamental was changing at DuPont—and in the chemical industry as a whole. Manufacturers had always downplayed the health consequences of their business practices, but their tactics tended to be subtle because nothing more aggressive was required. Governments lacked the popular mandate and medical evidence needed to challenge them. But by the mid-1930s, the balance was shifting. Kennaway's 1932 confirmation that benzo(a)pyrene was carcinogenic energized the worldwide search for additional industrial chemicals capable of inducing malignancies. By 1936, more than a dozen had been shown to be carcinogenic, and DuPont's initial enthusiasm for Hueper's aniline research program was flagging by the day. The company published its last study on bladder cancer at the Chambers Works in 1937, identifying eighty-three cases—fifty-six of them diagnosed in the previous three years.[16] Henceforth, DuPont would keep the growing body count to itself, although in a private letter in 1947 the factory's medical director observed that *every one* of the workers who had handled BNA in the early years of the chemical's production at Chambers had developed bladder cancer. Public disclosure of that fact might have helped workers at the Cincinnati Chemical Works, where a similar bladder cancer cluster was emerging in the late 1940s, but DuPont did not publish what it knew.[17] The era of scientific openness was ending. A new era of regulatory warfare was beginning.

Combat, of course, was nothing new to Wilhelm Hueper. He was not about to abandon his work, especially when, twenty months into his dog experiment, cystoscope examinations showed that many of them were developing tumors in their bladders. By the time the experiment ended after thirty-six months, thirteen of the sixteen dogs had bladder lesions—the same fast-spreading growths Hueper saw in his examinations of dye workers at DuPont.[18] More than forty years after Ludwig Rehn first took note of "aniline tumors" in German dye workers, a specific dye chemical—beta-naphthylamine, BNA—had finally been shown to be a carcinogen. Hueper had added another important industrial chemical to the growing list Kennaway had started five years earlier. A few months later, in November of 1937, the inevitable ax fell: Hueper was fired again. This time, he was told that he could not discuss any of the work he had done at DuPont without the company's consent, which he would not receive.

There would be no more publicity about cancer clusters in the dye industry for a very long time. The new rules of contested science would not permit it.

On the night of the most notorious murder in the history of Toms River, September 7, 1984, Dane Wells was left alone at the Ciba-Geigy landfill to stand watch over hundreds of unburied drums. They stood in long rows, like an army of shabby toy soldiers, glistening faintly in the light of a nearly full moon. "It reminded me of some sort of science fiction lunarscape," Wells remembered many years later. "As someone who had concerns about toxic waste, it was frightening to me. I wanted to get out of there."

A very unlikely set of circumstances had brought Wells to the landfill on that eerie night. At the time, he was a young investigator in the Ocean County Prosecutor's Office. Wells and his boss, Dick Chinery, who ran the office's special investigations unit, had spent the previous summer trying to build a criminal case against Ciba-Geigy. It was exciting stuff for a couple of local cops who had grown up in Ocean County—"pineys," they called themselves—and were accustomed to investigating burglaries. In reality, Chinery was less of a boss to Wells than a co-conspirator, in an entirely noncriminal sense. The two saw themselves as Jersey versions of frontier lawmen: incorruptible, adventurous, and not overly concerned with the niceties of legal procedure and office hierarchy. Like reporter Don Bennett at the *Observer,* Chinery and Wells were close observers of the local power structure but not a part of it, and they shared Bennett's fascination with the contradictions of Ciba-Geigy. How could a place so familiar, where thousands of people had enjoyed the best wages in the county for two generations, also be so mysterious? "Ciba-Geigy was kind of like a closed book," remembered Chinery. "No one really knew what was going on there. We wanted to find out."

They pursued the case with gusto, running down every lead that they could find. Operating under the loose supervision of County Prosecutor Edward Turnbach, Chinery and Wells arranged secret roadside meetings with informants who worked at the factory and made a surreptitious alliance with Greenpeace to exchange damaging information about Ciba-Geigy. (As a memento, Wells even saved one of the bowl-shaped wooden plugs that Greenpeace had used to try to block the ocean discharge pipe.) In July came the electrifying news that the state DEP had secretly referred the Ciba-Geigy case to the state Division of Criminal Justice for possible criminal prosecution. Wayne Smith, the chief investigator of the division's Environmental Crimes Bureau, would be coming to Toms River, and Chinery and Wells would be working with him on a joint investigation.

Excited, the two county investigators told Smith everything they had learned about Ciba-Geigy. But they soon found out that the state-county investigation would not be much of a partnership. Smith and the lawyers from the Environmental Crimes Bureau were the experts, and they were not interested in collaborating with a couple of amateurs who probably thought toluene was a brand of toothpaste and filter cake was for dessert. "They didn't understand the complexities of environmental crimes," Smith remembered. "We did." Smith told the pineys to be patient while he built a criminal case that Ciba-Geigy executives had intentionally, repeatedly violated the terms of the company's state-issued landfill permit. To Wells and Chinery, it sounded like an airtight case. They were sure that a raid on the factory, followed by mass arrests, was imminent.

But weeks passed, and there was no raid. In Toms River, meanwhile, the rumors were flying. There was talk that Ciba-Geigy had asked local politicians to lobby state officeholders on the company's behalf and that the company's well-connected attorneys had even appealed to the state's Republican governor, Tom Kean, who would be up for election the following year. Wayne Smith would later maintain that there was never any serious doubt that the state would eventually charge Ciba-Geigy, though the timing was uncertain because it was a complicated case. But Chinery and Wells did not believe that. They were sure that the big shots from Trenton were looking for any excuse either to drop the criminal case or to cut the county investigators out of it. It was a question of money, not just pride, because if the state brought the prosecution by itself, any penalties Ciba-Geigy paid would go exclusively to Trenton and would not be shared with Ocean County.

Convinced that their case was slipping away, Chinery and Wells decided to try something outlandish. At the end of August, Chinery picked up the phone and called the state Environmental Crimes Bureau. He informed the unhappy prosecutor at the other end of the line that the Ocean County Prosecutor's Office had decided to raid Ciba-Geigy alone and pursue its own prosecution. The county cops would be going in, Chinery said, just as soon as a judge issued a search warrant. That was a lie. There was no warrant request, no plan to raid the factory. The county investigators—whose first try at environmental enforcement had been a farcical attempt to arrest a belligerent cesspool dumper in 1980—were utterly unequipped to tackle a site as huge as Ciba-Geigy. In fact, Wells and Chinery were unequipped to tackle a site of any size. "We had no idea what to do. We didn't even own a white chemical suit," Wells recalled. "We were just so disgusted by the foot-dragging that we wanted to make something happen." It was a gamble, but Chinery thought the state would not dare to call his bluff. If the local cops went in alone, the state would miss out on all the news coverage that would come from busting Ciba-Geigy and would miss out on any fines, too. Even worse would be the perception that the state had caved in to political pressure to go easy on the company, in contrast to the plucky county prosecutor. It was, Chinery decided, a foolproof plan. The state would have to act.

He was right. A few hours after Chinery's phone call, a carful of prosecutors from Trenton showed up at his office in the old county courthouse, ready to do almost anything to head off a unilateral county raid. Before the meeting was over, the state officials had agreed to share any fines with the county and to swear in Chinery and Wells as special deputies in the attorney general's office so that they could monitor the investigation from the inside. The state prosecutors promised to raid the plant quickly so that Ciba-Geigy would not have time to clean up the site or destroy any evidence. In return, County Prosecutor Turnbach conceded the obvious: The state would take the lead on the prosecution.

The following Saturday, September 1, was the start of a long holiday weekend. Jorge Winkler was relaxing at his lake house in the Adirondacks, blissfully unaware of what was about to happen at the factory he had helped to run for seventeen years. In Trenton, Wayne Smith was hard at work. He spent the weekend combing through all the evidence the state had collected at the landfill and writing a request for a search warrant, which a judge issued September 4. Two days later, at eight-thirty in the morning, more than a dozen investigators from the state Division of Criminal Justice showed up at the factory complex with an escort from the town police. Two county investigators were there, too: Dick Chinery and Dane Wells were finally getting a chance to storm the castle. By then, they were expected. Matthew Boylan, the Ciba-Geigy lawyer and former director of the state Division of Criminal Justice, was waiting at the front gate. But there was nothing he could do to stop his former colleagues from entering. Many times over the past few years, DEP inspectors had been kept waiting for an hour or more at the gate. This time, thanks to the warrant, the prosecutors and police drove right in.

And that is how Dane Wells ended up watching moonlit barrels of toxic waste. The drums were critical evidence in the nascent criminal investigation. Even after spending two full days confiscating records and taking chemical samples at the factory, state investigators were not finished collecting evidence. Someone had to stay all night to make sure the drums were not tampered with.

Wells was alone at the landfill that night because every other cop on the graveyard shift in Ocean County was working on an extraordinary murder case. Robert Marshall was telling the police a terrifying story: He and his wife, Maria, were heading home from a night of dinner and gambling in Atlantic City when they pulled off the Parkway south of Toms River and parked their Cadillac Eldorado at a darkened picnic area to fix a rattling tire. While crouching by the tire, he told police, he was struck from behind and knocked out. When he regained consciousness, his wife was sprawled in the passenger seat of the Cadillac, dead, the apparent victim of an assailant who stole her money and shot her twice in the back. The murder was a stunner in Toms River. A handsome couple, Rob and Maria Marshall had three teenage sons, were prominent members of the Toms River Country Club, and lived in Brookside Heights, the same neighborhood where most of the Ciba-Geigy executives lived. What made the case even more of a sensation was that within days of the murder, there was talk in town that the Marshalls had had a troubled marriage and that the police doubted Rob Marshall's account of what had happened.

With the raid on Ciba-Geigy and the growing suspicion that Rob Marshall had arranged his wife's death, the people of Toms River felt the ground shifting beneath their feet. Just a few months earlier, Marshall had chaired the annual gala of the Ocean County United Way; Ciba-Geigy had, as usual, been the biggest donor. Now both were suspected of conspiracies that directly challenged the community's smug self-image as an island of safety and morality in a dangerous, immoral world. In a town where no one ever asked many questions, suddenly everything was up for vigorous discussion. The pages of the *Ocean County Observer* became a chronicle of civic strife: "Doctor Seeks Ciba Boycott," "Ciba Fails to Search for Reported Cyanide," "Ciba-Geigy Workers' Bags Searched," "Chemicals Found in Private Wells near Ciba Plant," "Greenpeace Cites Clam Kill," "Ciba Hints Dead Clams Were 'Planted,'" "Health Chief Is Told: 'Don't Drink the Water.'"[19]

No one had a rougher September than Jorge Winkler, the factory's director of production and environmental services, when he returned from the Adirondacks. On September 10, he was called into the plant manager's office and told he was losing his title because of the violations at the landfill. A few days later, the local papers reported for the first time that a water-testing laboratory that back in 1981 had failed to detect toxic chemicals in a backyard well on Cardinal Drive was co-owned by Winkler and another Ciba-Geigy executive, David Ellis. Winkler's supervisors had known about this potential conflict of interest, but the public had not. Soon afterward, Winkler, Ellis, and four others were suspended with pay. They continued to insist that any problems at the landfill were innocent errors, but to the Republican power brokers in Ocean County, the message was unmistakable. The suspended managers had been the public face of Ciba-Geigy on environmental matters; local politicians had dealt with them for years. Now they were gone, and the politicians were not sure whom or what to believe.

Everything in Toms River seemed to be falling apart now. On September 26, a grand jury indicted a Louisiana man in what a judge described as the "killing for hire" of Maria Marshall. Four days later, the people of Ocean County got two more jolts with their Sunday morning paper. Rob Marshall, now openly suspected of arranging his wife's murder, had checked himself into a psychiatric hospital after taking an overdose of sleeping pills. And on three full pages of the *Observer,* reporter Don Bennett laid out in excruciating detail exactly what Ciba-Geigy was sending into the ocean off Ortley Beach. The word *cancer* appeared eight times in the story, which described twenty-two toxic chemicals in the company's waste stream.[20] Two weeks later, county legislators voted unanimously to urge the state Department of Environmental Protection to investigate the environmental impact of Ciba-Geigy's discharge pipe. It was still a long way from the shutdown that an increasing number of county residents were demanding, but it was an epochal step for a board that had never before challenged Ciba-Geigy.

The civil war in town was starkly apparent in October of 1984, when more than eight hundred people attended a DEP public hearing on the company's request for a new permit for its ocean pipeline. The crowd was split down the middle. Sign-waving critics of Ciba-Geigy sat in front; the back rows were filled with hundreds of employees in blue windbreakers bearing the company logo. One of the quieter audience members that night was Linda Gillick. She may not have said much, but she seethed when a union official, Thomas Dooley, took the microphone and declared: "There is no greater devastation than people losing their jobs."

Linda Gillick was just beginning her transition from kinetic fundraiser for children's cancer causes to tough-minded political activist, but in a letter the *Observer* published a week later, she gave an early flash of her talent for passionate persuasion in her response to Dooley. "Do you have so little regard for humanity that monetary values are foremost?" Gillick wrote. "Try the devastation of pain, disfigurement and blindness of a child from cancer. Try paying the medical bills that wipe out your income even if you have a job. Try taking a child out in public so tortured by the disease physically that ridicule and silent accusations kill a part of his spirit every day."[21] Michael Gillick, almost six years old at the time, had been undergoing chemotherapy and radiation treatments for all but the first few months of his life.

Even events on the other side of the world were resonating in Toms River. On December 2, 1984, forty-two metric tons of methyl isocyanate gas leaked from a tank at a pesticide plant in Bhopal, India. Within two weeks, an estimated eight thousand people were dead, and tens of thousands were seriously injured. It was the worst industrial accident in history, and the majority owner of the plant was a subsidiary of Union Carbide, source of the waste that Nick Fernicola had dumped at Reich Farm in 1971. But the connection drawn by most Toms River residents was to Ciba-Geigy, especially after Don Bennett reported that the company was storing up to six tons of phosgene gas on the factory grounds. The sweet-smelling poison gas had been used in dye production at the Toms River plant starting in 1959, and there had been at least three leaks since then, including the one that killed a factory worker in 1974.[22] In Bhopal, Union Carbide used phosgene to make methyl isocyanate. Now, Bennett was reporting that there was no plan to evacuate the town in the event of a Bhopal-style leak in Toms River. It would take another seven months and a dozen more newspaper stories before Ciba-Geigy succumbed to public pressure and stopped using phosgene in Toms River.

A week before Christmas of 1984, Robert Marshall was finally arrested and charged with solicitation to commit murder. Prosecutors told the press that Marshall's motive in arranging the murder of his wife of twenty-one years was to collect a $1.5 million life insurance policy, pay off his debts, and move in with his girlfriend, a former vice principal at an area high school.

Ciba-Geigy had an apparent motive, too: By the early 1980s, it was saving about a million dollars a year by claiming that its hazardous waste was not really hazardous and thus could be buried on the factory grounds. Whether anyone in Toms River had been harmed as a result was a question that had not yet been asked, much less answered.

From the vantage point of her backyard on Cardinal Drive, the property her husband had insisted on buying for its sylvan tranquillity, Sheila McVeigh had an up-close view of the tumultuous events of 1984 and 1985 in Toms River. The more she learned about what was happening behind the trees at Ciba-Geigy, the more upset she got. When a friend who was active in Ocean County Citizens for Clean Water called to ask if she had developed any rashes from living so close to the plant (she hadn't) McVeigh started going to meetings and soon became the group's secretary. McVeigh taught sixth grade at a school on the other side of town, but her two daughters had attended West Dover Elementary. That was the same school where, in November of 1984, the Board of Education had responded to parent demands by hiring an air-testing firm to make sure that no fumes were drifting over from Ciba-Geigy, whose property bordered the playground. The plume of contaminated groundwater beneath the entire neighborhood was another matter; there was nothing to do about it except to make sure that the schoolchildren were drinking only the water supplied by the Toms River Water Company—water everyone in town assumed was clean.

One morning in the spring of 1985, an envelope from Ciba-Geigy arrived in McVeigh's mailbox at school and the mailboxes of the other teachers. The company was sponsoring an endangered-species coloring contest and wanted teachers to encourage students to enter. "So I told the kids, 'OK, you know what, we're going to draw ourselves. We are the endangered species because of all this pollution,'" McVeigh remembered. "After that, the superintendent invited me down to have a chat. He told me that what Ciba was doing was no big deal, and that it was like spilling a can of soda in a swimming pool. I told him I really don't believe this company should be sponsoring anything, and that I was going to organize a boycott of the contest." She got about twenty other teachers to join in and even met with Ciba executives, visiting the factory behind the trees for the first time. "I was scared to death, but I went there, and wouldn't you know it, the plant manager said the same thing to me, that it was like spilling a can of soda in a swimming pool. They all had the same line."

Similar acts of rebellion were breaking out all over town. In a sense, the entire county was engaged in a coloring contest, with each side attempting to paint the other in the most unflattering shades possible. By now, Ciba-Geigy had smartened up and hired a former Federal Bureau of Investigation agent as head of factory security and a public-relations man who was much better than Jorge Winkler at communicating with reporters and the public. The company still faced a hostile press, determined prosecutors, and a skeptical public, but at least its executives as far away as Basel finally understood the stakes. They were fighting for the factory's survival and could no longer count on the public's unquestioning support. Whatever the summer of 1985 brought, they would be ready—and after the craziness of the previous summer, they were sure it would bring something.

They were right, of course. Greenpeace had made its reputation by saving whales, not fighting chemical companies, but the Toms River operation had been so successful that there was never any doubt that the activists would return the following summer. In fact, Dave Rapaport had promised they would. In 1985, he was scheduled to sail up the California coast for Greenpeace, so the new leader would be Jon Hinck, a seven-year Greenpeace veteran and New Jersey native who as a boy had gone to Boy Scout camp in Toms River.

In mid-April, the double-masted wooden schooner *Fri* (Danish for "free"), chartered by Greenpeace from a supporter, sailed into Ortley Beach and anchored at a now-familiar spot: about three thousand feet offshore, directly above the Ciba-Geigy pipeline, which lay half-buried in the sediment forty feet below. Greenpeace was going to make a second attempt to plug the pipe's discharge vents, and this time the activists were ready to make a more credible effort. The *Fri* was almost three times the size of the *Aleyka,* with enough room for a team of filmmakers to record the action, a full crew, and four experienced divers plus two advisers who plugged pipes professionally for industry. The wood-and-foam plugs Greenpeace used in 1984 had flopped, so this year the group had devised a superior alternative consisting of semi-spherical metal caps that could be attached to the discharge vents with bolts and waterproof cement.

In the early morning hours of April 20, 1985, Jon Hinck placed a call to the Ciba-Geigy security office. Worried about creating a dangerous situation, he wanted to make sure that the company knew what Greenpeace was up to. Hinck told the incredulous shift supervisor that unless Ciba-Geigy shut down its pipe voluntarily, Greenpeace would again take "direct action" to block it. "It was pretty funny," Hinck remembered. "The guy said to me, 'Let me get this straight. You would like us to shut down our discharge to facilitate your plugging our pipe?'" To no one's surprise, the supervisor declined the request, so Hinck sent his divers into the chilly, choppy Atlantic. Attaching each plug was a slow and hazardous process because the water was so cloudy. There was another reason for the divers' struggles, but Greenpeace did not know it until months later. After getting the telephone call from Hinck, Ciba-Geigy's pipeline operators began pumping huge volumes of river water into the line, doubling the water pressure. At that pressure, attaching a metal cap to a spurting vent was like trying to fly a kite in a hurricane.

Greenpeace's attempt to monkeywrench the pipeline was so difficult that when a state police patrol boat showed up before lunch, Hinck half-hoped that they would all be arrested so that he could call off the operation. Instead, the police merely told them to stop. When Hinck refused, the cops took down the names of everyone on board and left. By the end of the afternoon, the divers had managed to plug fifteen of the fifty discharge vents. They were back at it the following day after getting their scuba tanks refilled. But in the afternoon, after Ciba-Geigy's own divers confirmed the blockages and the company got a judge to issue arrest warrants, the same patrol boat again pulled alongside the *Fri*. This time the police arrested twelve people, charging them with disorderly conduct and criminal mischief. Ciba-Geigy then sent a crew out to remove the plugs.

Sprung from jail the following day on bail of $250 each, the young activists immediately headed back out to the outfall site, where they collected more water samples and then headed down to Trenton. On the front steps of state Department of Environmental Protection commissioner Robert Hughey's office, the activists set up a wading pool, filled it with brownish water collected from the outfall, tossed in a couple of dead fish and a beach ball, and invited the commissioner to take a dip as reporters watched. He declined.

There would never be any direct evidence to support Greenpeace's central contention that Ciba-Geigy's offshore discharge was a threat to ocean swimmers. By any standard, the company's wastewater was not safe to drink, but no one was drinking it or swimming anywhere nearby—except for the divers from Greenpeace or Ciba-Geigy who kept checking the vents and collecting more water samples. In fact, the dilution effect of the ocean was so extreme that water samples taken ten yards away from the discharge vents were unable to detect any hazardous chemicals at all.

To Greenpeace, however, the literal truth of its arguments was less important than their moral resonance: Ciba-Geigy was dumping in the ocean, and the ocean belonged to the public. Besides, Greenpeace was providing compelling street theater and great fun for everyone except the company. On a cloudy day in May, two hundred people waving flags and singing "God Bless America" attended a rally on the Lavallette boardwalk that served as a valedictory for Greenpeace. In October, charges against the twelve activists were dismissed after a local judge ruled that he lacked jurisdiction for offenses committed offshore. Greenpeace would not return to Ocean County; it did not need to.[23] "In Toms River, we were a catalyst. The community got excited and motivated and worked together," Hinck recalled. "We knew it wasn't going to fall apart after we left."

As bad as things looked for Ciba-Geigy, its leaders could at least take comfort in the fact that they had not been criminally charged. Six months after the raid, prosecutors issued grand jury subpoenas to more than a dozen managers, but there were no indictments. Soon the conspiracy theories again were flying in Toms River, just as they had the previous summer in the weeks before the raid. Had the company found a way out?

Ciba-Geigy certainly seemed to be assuming that if it played ball with state regulators, the criminal case might go away. On April 25, 1985, after more than two years of on-and-off negotiations and less than a week after Greenpeace embarrassed the company by partially plugging its outfall pipe a second time, company executives and Commissioner Hughey reached a sweeping deal. The company agreed to pay the largest penalty for an environmental violation in New Jersey history—$1,450,000—in return for being allowed to operate its ocean pipeline five more years. Ciba-Geigy's actual costs would be much higher because the company also agreed to remove all fourteen thousand drums it had buried since 1982 in the active section of the landfill, Cell Two. (The removal ultimately cost more than $4 million.) Those drums would have to be trucked to hazardous waste landfills in other states; Ciba-Geigy's own landfill from now on could be used only for sludge from the wastewater treatment plant. The company's wastewater also would have to meet stricter standards, including the use of sensitive mysid shrimp as a test species. Since Ciba-Geigy already knew that its effluent killed shrimp and mutated DNA, the company would have to improve its treatment process or risk the state shutting down the pipeline and closing the plant.

With the settlement agreement in place, the company tried to turn the page. In a full-page advertisement in the *Observer,* Ciba-Geigy promised to "focus on the future" while conceding that it "should have done more to respond to the legitimate concerns voiced by our neighbors."[24] It was the closest the company would ever come to apologizing to the people of Toms River, though there would be many more opportunities to do so later. In July, a contractor began the massive job of removing all the drums that Ciba-Geigy had dumped in Cell Two. Pressured by Ciba-Geigy's executives and union, the county legislature returned to its traditional stance of servile obsequiousness, rejecting a proposal to hold a countywide referendum on whether the pipeline should be shut down and approving a weakly worded substitute instead. For the first time in more than a year, things were looking up for the company. Greenpeace was gone, the war with state regulators seemed over, and the local politicians were back in line.

And then, on October 24, 1985, the other shoe dropped with a resounding thud. A grand jury in Trenton handed up indictments charging Ciba-Geigy and three executives with engaging in a decade-long criminal conspiracy to circumvent state and federal environmental laws. A fourth Ciba manager faced a lesser charge, illegal dumping. The indictments painted a devastating picture of life inside the factory gates, charging that executives had flouted the law by ordering the dumping of thousands of hazardous waste drums in the landfill. To cover their tracks, the grand jury charged, the executives doctored records, deceived state inspectors, and pursued their scheme for years—even after they knew that the landfill was leaking and that an older underground pollution plume had seeped off the factory grounds, tainting backyard water wells in Oak Ridge.

What made the charges especially shocking in Toms River is that the grand jury's indictments attributed the criminality not just to the company in general but to individual executives. Like the Marshall murder case, this was a *personal* betrayal. It was the first time in state history that the executives of a chemical company had been personally indicted for environmental violations, and the four managers who were targeted all had strong ties to the town's social and political establishment.[25] They were not naïve small businessmen or mob-connected midnight dumpers; they were highly trained professionals who wore a coat and tie to work, played golf at the country club, and sailed at the yacht club.

Two months after the Ciba-Geigy indictments, Robert Marshall's murder-for-hire trial began in a small town near Atlantic City. (It had been moved there out of concern that an unbiased judge and jury could not be found in Toms River.) After an eight-week trial, Marshall was convicted and sentenced to death by lethal injection—a sentence eventually reduced to life in prison.[26] To the humiliation of image-obsessed Toms River, the case would be chronicled in a best-selling book, *Blind Faith*. The book, and the television miniseries that followed, painted a scathing picture of a debauched society of vapid social climbers in a town where "you were what you drove, you were what you wore, you were where you lived—no matter how heavily mortgaged it was."[27]

Together, the very different crimes of Ciba-Geigy and Rob Marshall smashed to smithereens the guiding mythology of Toms River. Its residents could no longer pretend that their town was an apple pie refuge from the chaos of the outside world. Outsiders had not poisoned the water supply, polluted the ocean, and then hid their actions for years. It was not thugs from Newark or Camden who arranged to put two bullets in Maria Marshall's back. Those betrayals came from within. Their own neighbors had done it, the people they knew from the Little League and the United Way's annual fundraising gala.

As in any culture that loses its guiding myth, struggle and dislocation ensued. Instead of denial, there was open conflict, a coloring contest writ large. Jorge Winkler watched from the sidelines. Although he was never indicted, he was not allowed to return to work by the factory's new management team. He stayed in town another ten years, insisting throughout that he was guilty of nothing but inept public relations. If he had returned to Switzerland, he later explained, "it would have been a signal to the people of Toms River that I had done something wrong and had to disappear." In 1995, he finally left, after twenty-eight years. He and his wife moved to Montana, where the craggy peaks of the Bitterroot Range reminded him of the Alps. He skied, hiked, and tried not to think about his bitter experiences in Toms River or the calamities that occurred there after he left.

"It's a nice place to hide," Winkler said of Montana. "When I came out here, the first thing I did was to make sure I was not near a Superfund site."

CHAPTER ELEVEN

Cases

More than twenty years had passed since George Woolley took a final plunge into the Toms River, brushing purplish foam off his arms as he surfaced in its reeking waters. In the years since that foolhardy 1962 swim, the river had gotten cleaner and Woolley had developed an abiding interest in environmental health. He had worked at the factory since 1964, when he was twenty-two and an offer from the Toms River Chemical Corporation rescued him from a humdrum job behind a hardware store counter. As a teenager, Woolley had dreamed of being a science teacher, but there was no money for college. He had no complaints, though. By the 1980s, experienced laboratory technicians like him were getting more than $500 a week at Ciba-Geigy plus overtime. The job also allowed Woolley to indulge his interest in science by learning as much as he could about the chemicals he was handling. This self-taught knowledge had come in handy many times: After reading up on the hazards of epichlorohydrin in the late 1960s, for example, he told his supervisor that he would not work with it anymore unless the company provided properly ventilated workspaces, which it promptly did.

Like many longtime workers at the plant, Woolley's feelings toward

his employer were nuanced. He had seen working conditions gradually improve during his long tenure and was proud to have helped
hasten the changes through his activism in the union, Local 8-562 of
the Oil, Chemical, and Atomic Workers International Union. In his
early years at the factory, he had handled noxious, unfamiliar chemicals with almost no protective gear and had made dozens of trips to
the old open-pit dump to pour cans full of solvents into the "smudge
pots" that led straight to the sandy ground. By the mid-1980s, however, both practices had been banned. Although relations between the
workers and management were always tense whenever a new labor
contract was being negotiated, the two sides worked reasonably well
together otherwise. Woolley regarded Ciba-Geigy as a decent employer that was generally willing to do what the rules required—but
nothing more than what the rules required. As he explained years
later, "I was working for a chemical company that, in my opinion,
was one step ahead of the law—and I mean that in a positive way."

The uproar over the leaking pipeline, and the torrent of bad publicity that followed, had brought labor and management closer together, united by a shared sense of persecution. Now that the company
was notorious, Ciba-Geigy *needed* the support of its workers in a way
it never had before, and the workers—fearful of losing the best
blue-collar jobs in the county—were eager to be drafted. The union
represented 650 workers, and even though they were a much smaller
percentage of the town's total population than in earlier years, they
were still a formidable political force when mobilized. Now they
showed up by the hundreds at public hearings, wearing company
jackets and union baseball caps; some even drove out to the beach
communities and organized counterdemonstrations near Greenpeace
rallies. When Ocean County Citizens for Clean Water started getting
publicity, Ciba-Geigy workers tried to take it over, demanding to see
its bylaws, calling for elections, and showing up uninvited at meetings. The union president at the time, Jim McManus, was particularly
bellicose. "Swim somewhere else!" he would shout, if someone complained about the ocean outfall.

Even before the public criticism began, the chemical plant was in decline. At just under one thousand people, its workforce had fallen by one-third since its 1968 peak. The Swiss had lost patent protection for their most important dyes, allowing competing factories in Asia and Eastern Europe (some of which they owned) to use the same relatively simple manufacturing processes to make dyes much more cheaply. The new plants in places like India and Poland saved money by dumping barely treated waste into open pits or rivers, just as Toms River Chemical had done a generation earlier. An even more important economic advantage for overseas manufacturers was the cost of labor. Hourly wages at the Toms River plant were more than eight times higher than those at most foreign factories. A similarly huge disparity in wages had already devastated the American textile industry, creating yet another incentive for dye manufacturers to move overseas. Fabric was usually dyed before it was cut and made into clothing, so it made sense for dye manufacturers to be located near the booming textile factories of East Asia.

By the mid-1980s, it was obvious that high-wage, low-tech chemical plants like the one in Toms River did not have a long-term future in the United States, and the workers at the plant knew it. But that realization only strengthened their determination to hold on to their jobs for as long as they could. Besides, Ciba-Geigy had given them reason for hope by announcing plans to make Toms River the new home of its American pharmaceutical manufacturing operations, which were not as labor-intensive and therefore less vulnerable to foreign competition than dyes. If Ciba-Geigy's executives told them that the ocean pipeline was necessary to keep the factory running, then the unionized workers would fight for it.

And yet, for all their aggressiveness in challenging Ocean County Citizens for Clean Water and Greenpeace, the unionized workers of Ciba-Geigy had more in common with the plant's critics than either side acknowledged publicly. The more the environmentalists talked about the dangerous chemicals the company was dumping, the more worried the workers became about their own safety, especially after Don Bennett began listing many of the chemicals in his stories in the *Ocean County Observer*. "There was all this talk about cancer on the outside, so we said, 'What about the people who worked in the plant?' We were much more exposed than they were," remembered John Talty, an officer in the union starting in the mid-1980s.

Practically every long-term employee knew someone who had died of cancer. "It would just be people you knew from around the plant. There were a lot of them, and not all of them were listed as dying from cancer," Talty explained. Everyone had his or her own theory about where the clusters were: George Woolley was convinced that there were an unusual number of cancer cases in Building 108, where epichlorohydrin and another carcinogen, ethylene oxide, were used. Talty worried about North Dyes, where benzidine derivatives were used to make azo dyes, which were now the most important products made in Toms River. And everyone had heard about young Randy Lynnworth, the teenager who lived just outside the factory fence on Cardinal Drive and was still struggling with brain cancer in 1985. Company officials continued to deny that chemical exposures at the plant were causing cancer, but relatives of several dead workers had won large workers' compensation awards on the grounds that the cancers were work-related.[1]

The factory was a maelstrom of conflicting anxieties, and as the head of the union safety committee, George Woolley was caught in the middle. Many Ciba-Geigy workers were very worried that they were risking their health yet were terrified that the negative publicity over the company's pollution problems could cost them their jobs by forcing the closure of the waste pipeline and the factory. "Everyone was under so much stress because of everything that was going on," Woolley recalled. "We were being put through the wringer, and it was a terrible, terrible experience. We were doing everything we could to fight to keep the plant open despite all of the bad press, but we also knew there were too many cancers at the plant and that no one had studied that." Now the union had to choose. Should it push the company to conduct a cancer study, even at the risk of further spooking the workforce and providing more grist for damaging news stories? Or should it downplay the health risks and focus on trying to keep the plant open? For self-taught environmental health enthusiast George

Woolley, a hometown guy who loved his job but did not fully trust his employer, it was a difficult choice.

Indecision was not one of Wilhelm Hueper's problems. His 1937 dismissal from the DuPont Corporation—the second time in five years that DuPont had fired him for insubordination—might have been cause for introspection. Perhaps the world of industrial health research, increasingly dominated as it was by scientists in the pay of manufacturers, was not a natural fit for someone so bullheaded. But Hueper was not one for self-analysis. He had been trained in a very different tradition, one epitomized by legendarily cantankerous Teutonic physicians such as Rudolf Virchow and even the ancient Paracelsus. These very confident men were clinicians first: Their emphatically expressed conclusions about the causes of illness were based on the firm bedrock of their own hands-on examinations of sick patients. In the parlance of epidemiology, they were "case series" investigators. Hueper's experimental triumph in inducing bladder tumors in BNA-exposed dogs merely confirmed what he already believed from examining workers, touring factories, and reading old case reports from factory physicians in Europe. Hueper had already been sure that dye chemicals caused bladder cancer; now he had proof.

But science was changing, and Hueper's forceful views about what constituted proof were becoming increasingly unfashionable among cancer researchers.[2] With the coming of World War II, manufacturers like DuPont were seen as crucial national security assets. Inevitably, health concerns took a backseat in the drive to ramp up wartime production of plastics, steel, and insecticides. There were scientific reasons, too, why Hueper's ideas were increasingly out of the mainstream. The "numerical medicine" pioneered by Pierre Louis and Siméon Poisson a century earlier had come into full flower with the development of biostatistics as a legitimate academic discipline. The new biostatisticians were not only interested in tracking infectious disease, they also took the first steps to measure the extent of chronic diseases—especially various forms of cancer—across broad populations. What they learned was that cancer was increasing and seemed to be everywhere, including in many people who had never set foot in a factory or a mine. The official statistics for England and Wales showed a quadrupling of the cancer rate among men over age forty-five between 1860 and 1939 and a near doubling among women.[3]

Especially disturbing was the huge increase in lung cancer. Predictably, it was noticed first by the Germans, with their long history of effective public health surveillance. In the late 1930s, research at the University of Jena implicated tobacco use as the overwhelming cause of lung tumors. The strongest evidence came via one of the first true case-control studies, in which a German researcher named Franz Müller compared the smoking habits of lung cancer victims to those of healthy people or those with other kinds of cancer.[4] It sounds bizarre in retrospect, in light of the participation of many German physicians in Nazi atrocities during the war, but their counterparts in Britain and the United States made pilgrimages to Jena and other German cities in the years before World War II to learn about the Third Reich's vigorous efforts to curb smoking and promote public health among ethnic Germans—campaigns that had no analogue in the tobacco-loving democracies.[5]

One of those visitors from abroad was a young English medical student who would go on to have a towering influence on scientific attitudes toward pollution and cancer in Toms River and throughout the world. Richard Doll ended his week's stay in Frankfurt in 1936 impressed with the quality of German science but disturbed by the anti-Semitism he encountered in professors and students alike.[6] Doll had a good head for numbers; he was in medical school only because he had drunk three pints of ale on the night before an entrance exam and had thus failed to qualify for a scholarship to study mathematics at Cambridge. Despite his fascination with statistics, Doll wanted to be a brain surgeon, though his plans changed with the outbreak of World War II. He spent most of the war working in army hospitals and aboard ships. By the time the fighting ended, Doll had no interest in going back to school for seven years to become a surgeon.

What interested Doll instead was the growing movement, spurred in England by the creation of the National Health Service, to take chronic-disease epidemiology out of the factories and mines—Wilhelm Hueper's domains—and apply it to the whole of society. In Britain,

the first target of this new effort was the mysterious leap in lung cancer cases. In 1947, the country's leading statistical epidemiologist,
Austin Bradford Hill, was asked by a government committee to devise
an epidemiological study on an unprecedented scale, involving many
hundreds of hospital patients. Hill chose Richard Doll as his collaborator. Doll knew of the work Müller had done in Nazi Germany before the war identifying smoking as the key cause of lung cancer but
thought that the studies were too small to be convincing. He hoped to
reach a much more definitive conclusion.

The series of studies that Doll and Hill undertook would become
the most celebrated pieces of research in the history of epidemiology,
rivaled only by John Snow's work on cholera almost a century earlier.
They would also marginalize the workplace-centered research of rivals like Hueper and cast serious doubt on the usefulness of investigating cancer clusters like the one in Toms River. Doll's studies, first
with Hill and later with Richard Peto, not only established that smoking and other "lifestyle" habits were major risks for lung cancer
and heart disease, they also enshrined two relatively new forms of
research—the case-control study and the prospective cohort—as the
new gold standard for epidemiology. Everything else, including the
"case series" studies and animal tests that had been important for so
long, were now seen as suspect.

By virtue of their design, the smoking studies could withstand
skeptical analysis in ways that Hueper's case-by-case reports on his
examinations of DuPont dye workers, as well as his experiments on
dogs, could not. Doll and Hill's first major lung cancer study, published in 1950, adopted the case-control format that had been employed only sparingly in the past, and never on such a large scale.
Seven hundred and nine confirmed lung cancer patients from twenty
London hospitals were included, and their habits were compared to
an equal number of patients hospitalized for other reasons. The non-
cancer control group was selected to match the cases in age, wealth,
male-to-female ratio, and area of residence. Most of the participants
smoked, but Doll and Hill discovered that those with lung cancer
were twice as likely to smoke more than twenty-five cigarettes a day.
The cancer/smoking connection was especially strong among men:

Just *two* of the 647 male lung cancer patients did not smoke, compared to twenty-seven of an equal number of cancer-free men.[7]

It was a convincing result, bolstered by a similar study published four months earlier in America.[8] (Doll was chagrined—he wanted to be first.) But the research was not entirely bulletproof because it depended on the patients' uncertain recollections of their smoking history. Moreover, there was a possibility that Doll and Hill had overlooked some other possible cause. So they planned a second study, a prospective cohort, in which they would follow a very large population over many years, periodically quizzing them about their habits and waiting to see how many got lung cancer. Because they wanted to find a group they could count on to stick with such a long-running study, Doll and Hill chose physicians—forty thousand of them. Doll would ultimately follow the group for decades, obtaining results that were similar to the earlier case-control study but less vulnerable to challenge by the tobacco industry. Unlike the case-control study, the cohort results did not depend on participants remembering their past smoking habits, and with such a large group it was easier to be confident that smoking really was the most important factor determining a person's lifetime risk of developing lung cancer.

Wilhelm Hueper watched with alarm as Doll's "risk factor" studies soaked up acclaim that his own work had never received. In 1942, soon after his second firing from DuPont, he had committed his vast knowledge to paper, producing a massive tome—its 896 pages included citations to almost four thousand studies. Called *Occupational Tumors and Allied Diseases,* it is still considered a classic in the field. The book was so highly regarded that in 1948 the National Cancer Institute hired Hueper as the director of its new environmental cancer section. Even from this exalted perch, however, he had little influence, especially after the worldwide sensation of Doll's smoking studies. Hueper was an outspoken skeptic of what he dismissively called "the cigarette theory," questioning how an "ill-documented, simple, unitarian theory" could be a major cause of conditions as pervasive and diverse as lung cancer and heart disease.[9] His objections ran much deeper than a mere scientific dispute over the strength of the evidence. Ever since Rudolf Virchow had manned the Berlin barricades in the failed revolution of 1848, German public health investigators had identified strongly with the laboring classes. They saw themselves as fighting for safer conditions and fair recompense for the human casualties of industrial production. Many of those workers smoked, drank, and ate poorly, so the new emphasis on "lifestyle factors" as causes of chronic disease would not only draw scarce research dollars away from workplace studies, it would also, Hueper feared, allow companies to avoid taking responsibility for conditions in their factories by shifting blame to the bad habits of their employees.

Hueper was completely wrong to minimize tobacco's importance as a health risk. Cigarettes turned out to be at least as damaging as Doll and Hill had asserted in the early 1950s. Their work helped save millions of lives, including many of Hueper's beloved blue-collar laborers who gave up the destructive habit of smoking. Doll eventually became so convinced of the importance of cigarettes as carcinogens that he would argue that smoking caused 30 percent of all cancer deaths. Workplace chemical exposures, he boldly asserted, were responsible for only 4 percent and pollution just 2 percent.[10] Doll's critics attacked this claim, questioning his data and his motives: In the years just before his 2005 death, Doll, a former socialist, served as a paid consultant and expert witness for manufacturers of pesticides, industrial chemicals, and asbestos.

Despite his errors about tobacco, Hueper was right about the chilling effect that "lifestyle factor" research would have on the identification and mitigation of chemical hazards in the workplace and in neighborhoods like Toms River as well. Beginning in the mid-1950s, medical journals began publishing fewer case reports and more case-control studies. Instead of physicians reporting on their examinations of a few dozen patients and describing perceived patterns of illness, the authors of these new studies were often biostatisticians who had never actually examined a patient but instead relied on paper records from thousands of cases.[11] Hands-on physicians, who had dominated the search for cancer's causes since the days of Paracelsus and Bernardino Ramazzini, were rapidly losing influence. Young doctors increasingly looked to large epidemiological studies for information about what was making their patients sick, instead of emulating

Ramazzini and drawing their own conclusions based on what they saw and heard from patients.

In Toms River, this attitudinal shift ensured that people like George Woolley and Linda Gillick would be met with skepticism, and even condescension, when they started asking questions about all of the cancer cases they noticed. Richard Doll's smoking studies were more reliable than the old case series studies because their use of matched control groups made it easier to exclude alternative potential causes. But there was an important caveat: To reduce the confounding role of chance, the study populations needed to be large—much larger than a single factory or town, ideally. That was especially true if the disease being studied was rare. Otherwise, no one could be sure whether a detected association between a risk factor and a disease was real or merely coincidental.

Researchers responded predictably to Doll's triumphs. They gravitated toward very large studies involving thousands of people and relatively common diseases. That was very bad news indeed for proponents of factory-based health studies. Even at a huge plant such as DuPont's Chambers Works, only a few hundred workers were likely to be directly exposed to a particular suspect chemical, and that chemical might be used in only a few other places anywhere in the world. With such a small population to study, a difference of just one or two cases of a rare disease could drastically alter the results—and could easily be dismissed as nothing more than luck. For industrial workers, who had long benefited from case series reports, the rise of the new epidemiology was unwelcome in two ways: Those old case reports were now dismissed as unsubstantiated, and the new, more credible case-control and cohort studies were rarely attempted in factories—or in small residential communities like Toms River.

Wilhelm Hueper recognized the trend and railed against it, but as the years passed, fewer people listened. It was at least partly his own fault. He was as combative as ever at the National Cancer Institute—his boss once said Hueper was "usually right about what he said and what he did, but the way he was right was wrong. He had an uncanny facility for abrasiveness."[12] His past clashes with DuPont continued to plague him; he would later assert that DuPont officials had accused him, at various times, of being a secret Nazi and of having "commu-nistic tendencies"—an interesting juxtaposition of ideological ex-tremes.[13] Hueper eventually made so many enemies that his supervisors at the cancer institute told him that he could no longer do factory studies. He responded by speaking out more than ever. At the request of Rachel Carson, Hueper vetted portions of *Silent Spring* before her profoundly influential book was published in 1962; Carson cited his research approvingly in the text. He retired five years later and died in 1979, at age eighty-five, after venting his spleen in an intricately de-tailed autobiography few people have ever read. He called it *Adven-tures of a Physician in Occupational Cancer: A Medical Cassandra's Tale*. It was never published, and it exists in manuscript form only at the National Library of Medicine in Bethesda, Maryland, not far from the cancer institute.

Like Cassandra, the mythological prophetess who foretold the fall of Troy but was ignored by her father the king, Wilhelm Hueper lived to see his own unhappy prophecies fulfilled, including his prediction that research into cancer's causes would come to be dominated by studies of lifestyle choices like smoking and nutrition, with relatively little attention paid to involuntary chemical exposures like the ones in Toms River. Just like dismal Cassandra, Hueper was powerless to stop it.

Case series reports were already falling out of fashion when a forty-five-year-old man walked into Dr. Arthur Wendel's Cincinnati office in January of 1958 complaining of blood in his urine. The man worked on the factory floor of the Swiss-owned Cincinnati Chemical Works, and no one—doctor, patient, or employer—was surprised that he had ended up in the care of one of the city's busiest urologists. The men who worked at the Cincinnati Chemical Works often com-plained of trouble urinating, and if their symptoms persisted the company sent them to Wendel for a more thorough examination.[14]

Wendel met his new patient and took his history, learning that the man had worked at the factory for eight years and that one of his principal jobs was shoveling benzidine into kettles. Wendel then did what he always did in such cases: He performed a cystoscopy, an uncomfortable procedure in which a thin tube equipped with lenses is inserted into the urethra. Looking through the lens, Wendel discovered a cancerous tumor in the man's bladder. It was the fourth case of bladder cancer Wendel had diagnosed in a Cincinnati Chemical Works employee over the previous twelve months, and he was sure it was not a coincidence. "He was seeing a disproportionate number of patients from the chemical industry, and he picked up on that," recalled his son Richard, who was also a urologist and later joined his father's medical practice. The four affected workers not only worked for the same company, they also worked in the same building and performed similar jobs, handling benzidine.[15]

This was, of course, a very old story. Bladder tumors had been linked to dye manufacturing ever since Ludwig Rehn reported on three cases in Frankfurt dye workers in 1895. By 1925, so many cases had accumulated that the Swiss government officially recognized bladder cancer as an occupational disease in the dye industry, making the stricken workers eligible for special compensation. British dye workers (though not their American counterparts) received similar benefits starting in 1938 after a cluster of bladder cancer cases appeared at a Ciba-owned dye factory in Manchester.[16]

The cause was almost certainly exposure to benzidine and beta-naphthylamine, or BNA. Thanks to Wilhelm Hueper's dog studies, the case against BNA was so strong that by the 1950s DuPont and the Swiss had stopped using it at all of their plants, including in the United States. The discovery of cheaper substitutes made the decision to abandon BNA a relatively easy one. Benzidine, however, was a very different case; there were no similarly inexpensive alternatives. It was used in much greater volumes than BNA ever was, yet when dye workers at the Chambers Works kept getting bladder cancer in the 1940s and 1950s, DuPont scientists blamed past exposure to BNA, not the ongoing inhalation of benzidine dust.[17] DuPont stuck to its position even after researchers at a Ciba plant in England reported in 1951 that twenty-one of the sixty-six workers who had developed bladder cancer there worked only with benzidine, not BNA.[18] Dye makers also benefited from growing skepticism about the relevance of animal experiments. In a study published in 1950, for example, researchers at

Allied Chemical and Dye Corporation acknowledged benzidine was carcinogenic in rats but noted that the tumors occurred in other organs, not the bladder. They also pointed out that rabbits and monkeys exposed to benzidine did not develop bladder tumors and that only one of seven dogs did.[19]

The Cincinnati Chemical Works had been a national center of benzidine production since 1929, producing about three thousand pounds per day. There were large, uncovered piles of benzidine powder at the company's St. Bernard factory, and the dust was everywhere. Workers shoveled benzidine by hand, often with no protection other than gloves and sometimes masks. In the late 1940s, there was a flurry of urinary tract tumors among workers there; the company blamed BNA and instituted stricter rules for handling it. But the Swiss owners, despite the evidence from their own factories in Europe, did not mandate similar controls on the much more important benzidine. In fact, benzidine production increased in Cincinnati after 1952, after vat dye production shifted to Toms River and freed up more space in the overcrowded St. Bernard plant to make benzidine-based azo dyes. As a result, workers at the aging St. Bernard factory throughout the 1950s continued to shovel piles of benzidine powder by hand and also used their hands to clean the powder off filter presses in the open air.[20]

So when Arthur Wendel diagnosed a fourth case of bladder cancer among workers at the St. Bernard plant in January of 1958, alarms went off in the executive offices of the Cincinnati Chemical Works and its parent companies in Basel. They looked for help from the department of environmental health at the University of Cincinnati, which was housed in the independent Kettering Laboratory, a unique arrangement. Named for a longtime director of research at General Motors, the Kettering Lab was heavily funded by manufacturers and had a reputation for research that tended not to challenge industry interests. Mitchell Zavon, a Kettering physician who headed the team summoned to assist, arranged for Wendel to examine all twenty-five men who handled benzidine at St. Bernard. Despite all of the evidence from Europe about bladder cancer and dye chemicals, the Cincinnati Chemical Works had never encouraged its workers to get yearly bladder exams (a standard practice in Europe), and its medical record

keeping was poor. Now Wendel and his cystoscope would begin filling in the record.

By the summer of 1958, Wendel had completed his examinations
of the twenty-five men. They were as young as thirty and as old
as fifty-nine and had spent anywhere from three to twenty-eight
years working on the factory floor in St. Bernard. Nine of them—
36 percent—had bladder cancer, including five previously undiscov-
ered cases. It was a staggeringly high percentage. Among the nineteen
men who had handled benzidine for at least seven years, the rate was
even higher: 47 percent.

"It was just unbelievable how much cancer there was in that plant,"
remembered Eula Bingham, a young biologist at the University of
Cincinnati who was part of the research team summoned by the company to assess the situation. Bingham would go on to a distinguished
fifty-year career as an environmental cancer researcher, including a
stint as the director of the Occupational Safety and Health Administration in the late 1970s, but the extremely high case count at the Cincinnati Chemical Works made an indelible impression on her. Years
later, she even invited a retired worker from the plant to visit an industrial hygiene class she taught at the University of Cincinnati to demonstrate how he used to shovel benzidine by hand. "In my entire
career, I never saw anything else like that again," Bingham recalled. "I
never saw so many people get cancer in one place. Even back in 1958,
there were enough cases to know that something terrible was going
on."

Later in 1958, a protégé of Wilhelm Hueper's named Thomas
Mancuso, who worked for the Ohio State Health Department, confirmed Wendel's findings in a more sophisticated study that included
a control group. He found that plant workers were fourteen times
more likely to die of bladder cancer, six times more likely to die of
kidney cancer, and four times more likely to die of pancreatic cancer
than typical Ohio men. The overall cancer death rate was 23 percent
higher than the statewide rate.[21]

The evidence was irrefutable: The Cincinnati Chemical Works had
a major cancer problem. This time, the company could not ignore it.
In 1958, soon after Arthur Wendel finished his examinations, the

company halted manufacturing of benzidine in Cincinnati and belatedly instituted improved protections for workers handling the chemical. But the plant's Swiss owners, who had already stopped using benzidine in many of their European plants, did not stop using it in Cincinnati. It was too economically important to abandon. Instead, they began shipping in benzidine via railcar from Allied Chemical's huge factory in Buffalo, New York, where it had been made for more than forty years.

Soon after, in 1959, the Cincinnati Chemical Works closed down, and azo dye production moved to Toms River. A few of the benzidine workers, including two with bladder cancer, transferred to New Jersey. Most, however, stayed in Ohio. Benzidine itself made the transfer, of course. The powder-laden railcars from Buffalo were re-routed to Toms River, where a new generation of employees began working with the carcinogenic chemical. Most knew nothing about what had happened in Cincinnati. Medical researchers around the world did not know either, because Mitchell Zavon, the consultant hired by the Cincinnati Chemical Works, did not publish the stunning results of Wendel's examinations. After all, those embarrassing findings were mere case reports, not a massive, Richard Doll–style case-control study.[22]

Fifteen years later, Zavon had a change of heart and decided to update the case count among the twenty-five workers who had been exposed to benzidine in Cincinnati. He discovered four more bladder cancer cases, raising the total to thirteen out of twenty-five long-term benzidine workers. This time, Zavon and Richard Wendel, Arthur's son, wrote up the updated results and published two journal articles, in 1973 and 1974.[23] By then, though, it was largely a moot point. The freight cars from Buffalo had made their last trip to New Jersey in 1971, when Toms River Chemical finally stopped using benzidine. The company had little choice. Animal tests had at last definitively shown that benzidine was a carcinogen, and the newly formed Occupational Safety and Health Administration was cracking down hard on its use.[24] Allied Chemical, the last benzidine maker in the United States, ceased production in 1976, the same year it finally revealed the bladder cancer toll from six decades of production in Buffalo: one hundred and fifty one cases, and probably many more because record keeping in the early years was poor.[25]

What about at Toms River Chemical? Benzidine was just one of many known or suspected carcinogens—including epichlorohydrin, anthraquinone, trichloroethylene, and naphthalene—used in huge quantities at the plant at various times during the 1950s, 1960s, and 1970s. Were cancer cases piling up there as well? No one tried to find out. After the shocking study results in Cincinnati, Ciba was not eager to stir up more anxiety by bringing in another set of outsiders to study cancer in its Toms River workforce. There was little interest from outside researchers anyway, since most of them were now focusing on lifestyle risks like smoking and diet. In the absence of authoritative information, younger workers like George Woolley and the Talty brothers gradually developed their own ideas about possible cancer hotspots at Toms River Chemical, but their theories were based solely on unsubstantiated chatter around the plant.

It was as if the events in Cincinnati had never happened—no stunning cluster of bladder cancer cases, no investigations, no irrefutable scientific confirmation. The workers and residents of Toms River were still on their own.

The realization that there might be a cancer problem at the Toms River chemical plant was slow to take hold among the workforce, but once established it could not be shaken. The tipping point came in the summer of 1985, after more than a year of ceaseless pummeling in the newspapers over the toxic content of the waste that the factory was sending into the ground and ocean. The level of anxiety among employees who handled those chemicals every day grew so high that George Woolley and other union leaders felt that they had to ask Ciba-Geigy for a health study—even at the risk of generating more bad publicity that could threaten the company's future in Toms River. "We didn't necessarily *demand* a study, but we felt it had to be done, even if it couldn't prove anything one way or the other," Woolley said.

Ciba-Geigy agreed to take the first step, authorizing a search of the plant's personnel records to count the number of cancer cases. To interpret the results, the company hired Philip Cole, a predictable choice. The Harvard-trained Cole was the chairman of the epidemiology department of the University of Alabama at Birmingham, where he was building a long roster of chemical industry clients for whom he consulted on health controversies and testified in court cases. Cole was a forceful critic of what he considered overreaching by researchers like Eula Bingham and Thomas Mancuso who were close to labor unions and believed, as Wilhelm Hueper did, that workplace chemicals were important causes of cancer. Instead, Cole endorsed Doll's view that occupational exposures were responsible for just 4 percent of cancers. In fact, in an interview many years later, Cole said, "My estimate would be two or three percent."

Cole had equally strong opinions on the *kinds* of studies that were scientifically meaningful. Like many of his peers, he believed that case series studies, such as the one Arthur Wendel conducted in Cincinnati, were of little value in identifying potential causes of workplace cancer because they lacked control groups for comparison. And even case-control studies were not very helpful, he believed, unless they included rigorous efforts to account for the confounding influences of luck and lifestyle. Philip Cole was, in short, an occupational cancer epidemiologist who was highly skeptical of the tools of occupational cancer epidemiology. One of his frequent antagonists, Richard Clapp of Boston University, called him "the father of the negative study."

When Cole traveled to Toms River in the summer of 1985 to meet with the plant managers, he learned that the company had found a few anomalies in its search of employee medical records. While there was nothing as obvious or dramatic as the cluster of bladder tumors Wendel had found in Cincinnati in 1958, one number did leap out: There were five fatal cases of brain cancer—seven if you counted two men who worked at the Toms River factory but died in Europe. There also seemed to be a surprisingly high number of deaths from bladder cancer and lung cancer. Were those case totals truly high? It was not an easy question to answer. Lung cancers were not particularly rare, but brain and bladder cancers were. On the other hand, more than sixteen hundred men, and a few women, had worked in the production areas of the sprawling factory at one time or another since 1952. Among such a large group, perhaps those cancer totals were not so high after all. "I told them that based on the information they had given me, it was not possible to decide whether or not Ciba-Geigy had a problem with cancer in Toms River," Cole recalled years later. "The only way to find out would be to do a formal study." Company officials told him that they would think it over.

Ciba-Geigy and its employees had yet another reason to be jittery that summer about cancers at the factory: A federal investigator was in town to scrutinize the safety of the azo dye operation. Bruce Hills worked for the National Institute for Occupational Safety and Health, whose research was frequently used by the Occupational Safety and Health Administration to set workplace rules. Hills was there because of two suspect chemicals: ortho-tolidine and ortho-dianisidine. Both were closely related to benzidine, and Ciba-Geigy had depended on them since the mid-1970s, after OSHA's crackdown on benzidine. The problem was, the two benzidine derivatives appeared to be almost as carcinogenic as their parent compound, which is why OSHA in 1980 had urged dye manufacturers to stop using them.[26] In Toms River, that advice was not followed. By 1985, despite the overall decline of Ciba-Geigy's dye business, the factory was still on track to make more than thirty million pounds of dyes, many of them black, blue, or deep red azos made from ortho-tolidine or ortho-dianisidine.

Bruce Hills was not shocked by what he saw on his first visit to Ciba-Geigy. The production areas were cleaner than some other dye plants he had visited, where the air was so thick with colored particles that "you would think you were wearing tinted glasses." He told company managers that he would return in a few weeks to test the factory air for ortho-tolidine and ortho-dianisidine, and he made a request: He wanted to collect urine samples from azo workers to test for dye chemicals. "I asked, and the company told me they'd talk to the union about it, and that it wouldn't be a problem," Hills recalled.

Word that a federal cancer researcher was visiting spread quickly around the factory, generating more anxiety for workers who were already upset about all the bad news of the previous months. Benzidine had already been taken away as a dye ingredient because of previous cancer scares. If they lost ortho-tolidine and ortho-dianisidine, too, there would be no way left to make azo dyes. Ciba-Geigy's most important remaining product—and all the jobs that went along with it—would disappear. Even more chilling than the economic threat was the sense that perhaps Greenpeace and the plant's other critics were right. Maybe there really *was* a cancer problem at Ciba-Geigy.

Three days after Bruce Hills completed his first visit and left town, plant manager Victor Baker gathered the factory's employees together and announced that Ciba-Geigy would cooperate with Hills and would also launch its own cancer study. Philip Cole's protégé at the University of Alabama, Elizabeth Delzell, would conduct a case-control epidemiological study of the company's past and present workers in Toms River, with a special focus on brain cancer, and the results would be published in a medical journal. Ciba-Geigy's new official spokesman, Thomas Chizmadia, told reporters that the company's actions were spurred solely by a desire to assure its workers and neighbors that the plant was safe. The company had not found "anything of an emergency nature," he said.[27]

The low-key news coverage of the decision belied its significance. For a company that had suffered through so many unwanted "firsts" since the discovery of the pipeline leak sixteen months earlier— protests, sabotage, a record fine, a criminal investigation, slashing news coverage, and public scorn—here was another one. For the first time in its thirty-two-year history in Toms River, Ciba-Geigy was acknowledging the legitimacy of the longstanding worries of its workers and neighbors. No one had ever done a health study in Toms River; talk of cancer had always been confined to anxious exchanges over back fences and lunchroom tables. Now, at last, modern epidemiology—with its blazing controversies, tantalizing ambiguities, and frightening implications—was coming to town.

There were limits to the company's new openness, however. When Bruce Hills returned at the end of September to take air samples and collect urine, George Woolley told him that the union had instructed its members not to participate. "I was shocked and taken aback that the union said no," Hills remembered. "I mean, if you were a worker, why wouldn't you want this?" Woolley did not want it because the company had convinced him that the study design was flawed. Hills wanted to find out whether dye chemicals in the workers' urine were mutagens capable of altering DNA—an indication that they might also be carcinogenic. But company officials argued that if the urine tested positive for mutagenicity, it could be due to a chemical exposure from one of those "lifestyle" exposures Richard Doll and Philip Cole considered so important: smoking, alcohol, or even diet soda or barbecued meat. Since there would be no control group for comparison, there was no way to know if dyes were truly at fault.

Hills thought the objection was spurious—if the workers' urine turned out to be mutagenic, then he could do a more extensive study to assess other potential causes. If it were not mutagenic, then the workers would have one fewer thing to worry about. But to Woolley, who knew how tense the workers were and how endangered their jobs were, the company's objections made sense. "If we had been a part of that study, you'd wind up telling people they had mutagens in their urine but you wouldn't be able to tell them why," he recalled. "I just couldn't see putting people through that, considering how much stress everyone was already under." There was another reason, too, he acknowledged. "We didn't want to have any more bad press and make the situation worse for the company and for the workers, especially for a study that wouldn't prove anything."

Hills went ahead with his investigation, producing an internal report that described what he saw inside the azo production buildings.[28] Even though dye workers generally wore respirators and other protective gear during the dustiest steps of the manufacturing process (at least while Hills was watching), his tests of the cassette filters the workers wore outside their respirators revealed airborne concentrations of dye particles as high as 1.64 milligrams per square meter— a result Hills later called "quite high, for something that's a likely carcinogen." But he dropped plans to publish his findings in the scientific literature because without urine testing there was no way to know how much dust the workers were actually inhaling. He tried to find azo workers elsewhere to test, but Ciba-Geigy was one of the last large factories in the United States that still used ortho-tolidine and ortho-dianisidine. Without the participation of the azo dye men from Toms River, Hills could not find enough workers to test to yield credible results, especially because so many were heavy smokers and thus already had mutagenic compounds from cigarettes in their urine. Thanks to the success of Richard Doll's huge population-based smoking studies, a factory-based study that was too small to take smoking habits into account seemed hopelessly unreliable by comparison.

Years later, Hills would look back and conclude that his abortive research effort had come too late. By the 1980s, production of ortho-tolidine and ortho-dianisidine was shifting to factories in Asia, just as benzidine had shifted overseas in the 1970s and BNA in the 1950s. Soon, Ciba-Geigy would give up on azos, too, in Toms River—and also epichlorohydrin, nitrobenzene, and all the rest. Then there would be nothing left to do but to clean up the messes, count the cases, and try to figure out who to blame.

CHAPTER TWELVE

≈

Acceptable Risks

In early 1986, Michael Gillick, still alive and beating the odds at age seven, was undergoing experimental treatments at Memorial Sloan-Kettering Cancer Center in Manhattan, perhaps the most famous cancer hospital on the planet. Families from all over the world brought their children to its pediatric oncology ward, yet Linda Gillick seemed to meet someone from the Toms River region every time she was in the family room. During one remarkable week in 1986, she counted *seven* Ocean County kids on the ward, including Randy Lynnworth, whose medulloblastoma had recently returned. Now that there was so much publicity about Ciba-Geigy, including the pipeline breach, which had occurred only a few blocks from her home, Gillick began to wonder whether her son's illness was really the result of nothing more than "bad luck or bad genes," as someone in town had told her. She wondered whether it was, in fact, part of a larger pattern.

The experimental drug did not curb Michael Gillick's neuro-blastoma; his stay at Sloan-Kettering ended with no improvement. The Gillicks went home to their precarious life in Toms River, with one addition. Linda Gillick taped an Ocean County map to an interior door at the house. Whenever she heard of another local child with cancer—she knew of dozens already—she inserted a red pushpin at the location where the child lived. In her eyes, the pins seemed to cluster in Toms River, in an area that included her home.

Gillick mostly kept her observations to herself, but her caution hardly mattered because by the spring of 1986, Toms River was roiling. After months of public pressure, the county health department began testing backyard wells near the route of Ciba-Geigy's leaky waste pipeline. In the Shelter Cove neighborhood, where the pipeline entered Barnegat Bay, they found eighteen wells tainted by perchloroethylene, a probable carcinogen that was a component of Ciba-Geigy's oceangoing waste. The more likely source was a nearby dry cleaner, but the factory's critics blamed Ciba-Geigy and noted that there had been several small pipeline leaks in addition to the large one in 1984. Frank Livelli, a canny retiree who headed the newly formed Save Our Ocean Committee and was fast becoming the company's chief local antagonist, announced that he knew of two brain cancer cases among children in Shelter Cove. He suggested that there could be a connection between those two cases and the recent disclosure that five current or former Ciba-Geigy workers had brain cancer.[1]

The job of trying to answer those accusations fell to Chuck Kauffman, who was unprepared for the task. Lithe, balding, and gregarious, Kauffman was a former egg farmer who had lived in Ocean County for all but a few years of his life (he was fifty-seven in 1986) and now ran the county health department, where his title was health coordinator. He had witnessed, firsthand, the shift in local environmental attitudes. Back in the mid-1970s, there was no controversy over the feckless efforts of his department, and the state, to prevent contamination from spreading off the Reich Farm property. Similarly, no one complained in 1974 when the county abandoned a halfhearted effort to survey the residents of Pleasant Plains about their health problems. Now, a dozen years later, the situation was very different. Ciba-Geigy's misdeeds were chronicled daily in the local papers, along with many reports of drinking water contamination and unexplained illness. Now residents had different expectations for their health department.

Kauffman's department could not meet those expectations. He supervised almost two hundred employees (up from just ten in 1973, when he took the job), but most were nurses who worked with the poor or homebound. The department's newly formed environmental section had a staff of fewer than a dozen, none of whom had any expertise in groundwater contamination. In the past, Kauffman had dealt with that deficiency by asking the chemists at Ciba-Geigy for assistance, but that was impossible now that the company was seen as a polluter instead of a protector. "There wasn't a lot we could do or say," he recalled years later. "We didn't have the environmental tools and the expertise."

Kauffman watched with dismay as more people every day became convinced that his county was a cancer hotspot and that Ciba-Geigy was the cause. He found that hard to believe—it seemed too simplistic to blame one company for cancer cases all over the county. Still, he could not shake the feeling that there did seem to be an unusually high number of local children with cancer, especially brain tumors and leukemia. He even called several hospitals to try to find out how many children from Toms River had been admitted for cancer treatment. It was a fool's errand, he discovered, because the hospital records were inaccessible, inconsistent, or hopelessly out of date.

Things came to a head at a packed meeting of the Ocean County Board of Health. For the first time in memory, a camera crew was there—from WNET-Thirteen, the biggest public television station in the New York City market. The station was working on a documentary about Ciba-Geigy and cancer in Toms River. As the camera rolled, an obviously uncomfortable representative of the Environmental Protection Agency—her name was Maria Pavlova—tried to explain why Ciba-Geigy would not have to clean up all of the groundwater it had contaminated at the factory site and nearby neighborhoods. A partial cleanup would be "acceptable" to the EPA as long as the resulting cancer risk to the surrounding population was no greater than one additional case per million people, she said. Frank Livelli, who was in the audience, pounced. "I really think what you're trying to do, Dr. Pavlova, is to make the unacceptable acceptable," he interrupted. "What you're trying to say to the people is: Don't worry about it. There will be one case of cancer in a million, and that person who gets it is someone you don't know. He's not a neighbor. He doesn't have a family. He doesn't have friends. He's just an isolated incident and that's the guy that's going to get poisoned and going to get cancer. But you don't look around the audience and say it could happen to you or to your child."[2]

Chuck Kauffman, who was also in the audience, watched the faces of others in the crowd as Livelli spoke. Young mothers, retired couples . . . they all lived in Ocean County, *his* county, and they were racked with anxiety. His own efforts to keep track of cancer and other diseases around the county were haphazard, relying entirely on his own network of friends and colleagues. But he knew of an alternative system, housed in Trenton, that seemed more reliable. It was a cancer registry, a searchable list of every case in the state. So in August of 1986, Kauffman called the state health department and requested a registry-based investigation of childhood cancer incidence in Toms River. Was there a cluster, or not? He was hoping to settle the controversy, once and for all.

Ciba-Geigy had fought hard, and spent big, for the right to keep manufacturing dyes and plastics in Toms River. By mid-1986, the company had already spent millions to comply with the terms of the tough new ocean discharge permit the state had issued the previous year. Ciba-Geigy had improved its wastewater treatment system so much that when mysid shrimp were placed in a seawater mixture that was 50 percent effluent, more than half survived—a huge improvement over the results from five years earlier, when a mixture of just 4 percent effluent had killed half the mysids.[3] In fact, the company's wastewater was now less toxic than the residential effluent the Ocean County Utilities Authority was pumping into the Atlantic through its three outfall pipes. Ciba-Geigy executives did not come right out and say it, but the message was clear: If residents were worried about the health of the ocean, and not just interested in bashing Ciba-Geigy, they should clean up their own sewage, even if they had to pay higher sewer taxes as a result.

With the help of newly hired image consultants, Ciba-Geigy by 1986 was getting smarter about influencing public opinion. At the beginning of June, as local environmentalists prepared to hold another

"Save Our Ocean Day" rally on the Lavallette boardwalk, Ciba-Geigy plant manager Victor Baker sent a "Dear Neighbor" letter to thirty-eight thousand Toms River homes touting the mysid shrimp results.[4] The company even sent canvassers into the nearby Oak Ridge and Pine Lake Park developments—east and north of the factory—to try to rebuild trust with the neighbors.

Larger forces were conspiring against Ciba-Geigy, however. The price competition from Asian factories was brutal and unrelenting, and the wage differential was vast. On June 2, 1986, the same day many town residents got Victor Baker's letter in their mailboxes, the company made an announcement that was, if not a complete surprise, nonetheless staggering: Ciba-Geigy was ending almost all chemical manufacturing in Toms River. After thirty-four years in which it had produced about three billion pounds of dyes and plastics—along with perhaps forty billon gallons of wastewater and two hundred thousand drums of toxic waste—the company was pulling out. Production of resins and other plastics was moving south to Ciba-Geigy's huge pesticide plant in Alabama, where wages were much lower and there was far less environmental oversight. Manufacture of dye ingredients was shifting to Louisiana and to Asia, where wages and oversight were lower still. Only a small dye finishing operation and the research labs would stay in Toms River. Fewer than three hundred jobs would be left after the three-year transition period, down from the peak of more than thirteen hundred in 1968. Victor Baker and the other executives did not publicly blame their critics for the downsizing, but in private they said that the hostile environment in town was a factor.[5]

The workers were shocked and angry. No one really believed that chemical manufacturing had a long-term future in New Jersey—many large plants in the state had already closed—but Ciba-Geigy's employees did not expect that the end would come so soon, and so abruptly, in Toms River. The union was not ready to give up completely on Ciba-Geigy, however, because Ciba-Geigy was not giving up completely on Toms River. As soon it was done shedding all those jobs, the company hoped to start adding them again—at a new pharmaceutical plant on the factory property. Compared to dyes, drug manufacturing was relatively clean and highly automated and thus less vulnerable to changing environmental standards and competition from cheaper labor elsewhere. Only eighty people would work there, but there might be many more jobs later.

The would-be renaissance came with a catch, however: Ciba-Geigy would build the pharmaceutical plant only if it could continue to use the ocean pipeline—a condition the company's opponents vowed they would never accept. The pharmaceutical plant would produce far less wastewater than the dye and plastics operations—about one and a half million gallons per day, down from five million gallons—but the company's critics thought that even one gallon would be too many. Ciba-Geigy's ocean discharge permit expired in 1990, and the environmentalists resolved to prevent its renewal under any circumstances. "We had accomplished a lot already, and we just told ourselves that whatever it took, we were going to shut down that pipe," remembered Nancy Menke Scott, a cofounder of Save Our Ocean. The company and its union, meanwhile, were just as resolved to the contrary. Ciba-Geigy had already shown that it was willing to spend millions to preserve its investment in Toms River, and its workers were ready to supply the muscle to go along with the millions.

The stage was set for a final struggle over the fate of Ciba-Geigy in Toms River. Resolving it would take three more years, during which time the local newspapers would be filled with several hundred anxiety-inducing stories about cancer and pollution. After being ignored for so long, both had become inescapable in Toms River.

Chuck Kauffman's request for a study of childhood cancer incidence in Toms River ended up on the desk of Michael Berry, who was new to the state health department and excited about his work. That would change later. For now, though, he was fired up. He was supposed to spend 80 percent of his time working on air pollution issues, with the rest of his time occupied by his other assigned task: responding to calls about possible cancer clusters around the state.

After Kauffman's call in August of 1986, Berry checked the department's records and saw that his colleagues had logged three other calls since 1982 regarding childhood cancer in Toms River. One was from a neuro-oncologist in New York who had treated several local children. Yet none of the calls had led anywhere; there had not been any investigation—not even a cursory one—of childhood cancer in Toms River. In fact, the health department had investigated very few residential clusters anywhere in New Jersey, in part because the state cancer registry was still relatively new and did not yet have much data. While tips about possible occupational clusters were taken seriously— there was a major ongoing investigation of a cluster of the rare lung cancer mesothelioma at an asbestos factory, for example—calls alleging neighborhood hotspots often were not.

There were good reasons for the disparity. Ever since Paracelsus, sharp-eyed scientists had identified suspected cancer-causing agents by looking for aggregations of disease in groups of highly exposed workers such as chimney sweeps and cobalt miners. In an occupational setting, it was easier to find workers who were exposed to very high levels of a few specific chemicals over many years. With an employer's cooperation, an investigator could visit a factory to collect environmental samples and interview workers and consult a centralized repository of health and employment records to confirm diagnoses and estimate how long workers had been exposed to specific chemicals. Wilhelm Hueper, among others, had proved the worthiness of that approach.

All of those tasks were immensely more complex in a residential study. Medical records were spread out over dozens or even hundreds of physicians' offices and hospitals, each with its own method of classifying illnesses. There was no way to track who had moved into or out of the area, or when. Measuring a neighborhood's current exposure to pollution was extremely difficult, and an investigator could only guess at past exposures. In a factory, a researcher could consult old records showing which chemicals were used in which years. But there were no such records for past chemical exposures in neighborhoods.

The upshot was that it was nearly impossible to get clear answers about the causes of residential cancer clusters. If the history of occupational cancer epidemiology was like a good detective story, with plot twists, multiple suspects, and—in some cases, at least—a satisfying conclusion in which the miscreants got their comeuppance, residential cluster investigations were closer to existentialist drama. Important, provocative questions were asked; nothing was ever resolved.

Most of the cluster calls that came in to Michael Berry's office were about neighborhoods, not workplaces. Since 1980, the state health department had something more to offer those callers than just a calm voice on the other end of the line. That was the year the New Jersey State Cancer Registry came on line. The concept was a simple one: The registry was designed to be the central clearinghouse of information about the approximately thirty-two thousand (at the time) cases of cancer diagnosed yearly in New Jersey.[6] That confidential data included the home address, age, and occupation of each afflicted person as well as the type and stage of cancer and the type of treatment. The health department could analyze it to answer all sorts of questions, including whether some parts of the state were getting substandard medical care. The registry's primary purpose, however, was to identify clusters and provide "more complete and more precise statistical data . . . necessary to determine the correlations between cancer incidence and possible environmental factors," in the words of the 1977 legislation that established the registry.[7] In other words, it was designed to answer the very questions that were being asked in Toms River during the summer of 1986.

Cancer registries were an old idea, though not a very effective one. In England, the Factory Act of 1895 required employers to report all cancers in workers who handled coal tar products, including dyes. Compliance by factory owners was poor, but the idea proved popular—a natural extension of the infectious disease registries that had started centuries earlier as an early warning system for plagues. In theory, a cancer registry could serve the same function as an observant physician like Percivall Pott or Walther Hesse, but on a larger scale. Instead of being limited to Pott's patients or Hesse's miners, a population-based registry could search for patterns in much larger groups—all the dye workers in England, for instance, or every resident of Connecticut, which in 1935 became the first American state to

have a cancer registry. By 1986, 181 city, state, or national cancer registries were operating in sixty countries.[8]

The registries proved to be much easier to create than to use, however. Even if a government agency managed to get all of the hospitals,
clinics, and physicians in its jurisdiction to report the cancer information as required, analyzing those statistics was a nightmarish process.
Diagnoses were frequently ambiguous, varying greatly from clinic to
clinic. Metastatic cancers were often improperly double-counted
when tumors showed up in a second bodily organ, and so were recur-
rences. The same legislators who were eager to take the credit for establishing cancer registries were far less enthusiastic about staffing
them adequately, so it was a Sisyphean task to keep up with the torrent of reported data that poured into the cancer registry offices.

New Jersey's registry was especially overwhelmed. In 1975, the
National Cancer Institute had singled out New Jersey for having the
highest cancer death rate in the United States. For men, the annual
toll was twenty-nine deaths per ten thousand; for women, eighteen
deaths per ten thousand.[9] Those rates were about 10 percent higher
than the national average. Speculation about the cause of the excess
focused on pollution, but death rates were not a very good way to
identify possible causes of cancer. The key statistic for causation research was incidence—the frequency of diagnosed cases, regardless
of whether the patients died. So the state legislature responded in
1977 by creating a cancer registry that collected incidence data, for
the first time in New Jersey.

By the time Chuck Kauffman asked for an incidence study in Toms
River in 1986, however, it was clear that the registry was a failure. The
office administering it was so understaffed that there was already a
backlog of more than two years, which meant that in 1986 the most
recent complete set of data was from 1983. (By the mid-1990s, the
backlog would rise to three years.) By law, doctors and hospitals were
supposed to report cases within six months of diagnosis, but they almost never did. The worst offenders were the huge cancer hospitals in
New York City and Philadelphia, which treated thousands of New
Jersey residents every year. Children like Michael Gillick or Randy

Lynnworth might not be included in the registry at all if they were treated out-of-state at Memorial Sloan-Kettering, New York Hospital, or the Children's Hospital of Philadelphia—three of the largest pediatric cancer centers in the world.

Even though he knew about the cancer registry's limitations, Michael Berry wanted to be as responsive as he could to Kauffman's request. He decided to conduct what the state health department called an "incidence analysis" of childhood cancer in Dover Township, the official name of Toms River. The process could not have been simpler. First, Berry called the registry office and asked for a computer printout listing all of the cases of cancer diagnosed between 1979 and 1983 in children under age twenty who lived in the township. He found fifteen cases. Then, he determined the total number of cases that "should" have occurred in the town if pediatric cancer in Toms River was as prevalent as it was in the state as a whole. Based on the township's population, he came up with an "expected" rate of 14.7 cases. In other words, the actual number of childhood cases was almost exactly the same as the expected number. Childhood cancer incidence in Toms River was neither unexpectedly high nor low. It was average. As far as Berry could tell from his crude analysis, there was no childhood cancer cluster in Toms River.

Berry reported his results to Kauffman, but neither was satisfied. For one thing, Berry's analysis did not include totals for specific types of cancer, and even if it had, the numbers would have been so small that it would have been hard to draw any significance from them. In addition, the final year of the analysis was 1983. Through his informal network at the county health department, Kauffman had heard of some recent cases, diagnosed since 1984, but they were not included because the registry was running two years behind. Finally, Berry's analysis did not include at least some local children who were treated solely in New York and Philadelphia. If it turned out that an unusually high number of Toms River children with cancer were being treated out of state—as Linda Gillick's recent experience at Memorial Sloan-Kettering strongly suggested—then the registry-based analysis was misleading.

"There's no indication according to records that cancer in the township is elevated," Berry told the *Asbury Park Press* when word of the study leaked out in November.[10] But the article, which carried the unnerving headline "Toxic Waste, Cancer Incidence Raising Unanswered Questions," was anything but calming. It noted that patients treated out of state were not included, quoting a lawyer and a neurosurgeon who each said that they knew three Toms River children recently diagnosed with brain cancer. Just as unsettlingly, the same Sunday newspaper included a small notice in the television guide for the upcoming week: The WNET-Thirteen documentary *Chemical Town USA* would air in prime time in three days, on November 19, 1986.[11]

The executives at Ciba-Geigy were on high alert; they were expecting to be pounded in the program. The writer and co-producer of *Chemical Town USA*, Michael Rosenblum, had attended college with factory spokesman Thomas Chizmadia and had convinced him to make plant manager Victor Baker available for questioning and to lead a tour of the factory. ("I told my cameraman, just shoot every drippy pipe you can find," Rosenblum recalled years later.) By the time the tour and interview were over, it was clear that the documentary would be critical of the company, especially because Rosenblum had asked about a possible brain cancer cluster in town. Ciba-Geigy's lawyers dashed off a letter to WNET president John Jay Iselin claiming that Rosenblum had misled Chizmadia about the content of the story. Iselin had special reason to pay attention, since Ciba-Geigy's pharmaceutical division had just bankrolled a major science documentary series for WNET (ironically, it was called *The Brain*). The station's lawyers insisted on reviewing a transcript of an early version of the documentary. Somehow, Ciba-Geigy also got a copy of the same transcript. That prompted another letter from the company's lawyers, this time demanding script changes.

In the end, Rosenbaum made only a few of the changes the company demanded. But he and Kathleen Hughes, the associate producer who had done most of the original reporting, both felt that the pressure from Ciba-Geigy's lawyers changed the overall tone of the documentary by the time the final cut was approved by station executives. They had hoped to feature Randy Lynnworth and emphasize the cluster allegations, but the final version instead stressed the tradeoff between jobs and pollution. Randy and his family did not appear until the tenth minute, long after the story of a Ciba-Geigy worker who was facing a layoff as a result of the planned downsizing. "I thought that it would be a much tougher story than it actually ended up being," Hughes remembered. "All the major points were still hit upon, but overall it was softer."

Softened or not, the documentary disturbed many in town when it finally aired. They saw an obviously ill Randy Lynnworth—he would live only a few more months—paging through a family photo album, supported by his mother's embrace. The Lynnworths could have been anyone's family, and Randy anyone's son. "Science today still cannot prove whether the tumor was caused by their proximity to the plant," the narrator intoned as Ray Lynnworth lifted his teenage son out of his wheelchair and placed him gently into bed. "But for the Lynnworths, the fear that cancer might have been environmentally induced haunts them continually." In case anyone had missed the point, Shelley Lynnworth added: "You really have to be aware of what goes on around you, because we never ever dreamed that our bubble would burst and our lives would change so drastically. And my God, I hope it doesn't happen to any of you but you really can't sit back and say it's the person across the street, because it's not. It's you and it's me."

By the end of 1986, it seemed like almost everyone in town was inclined to agree. It was obvious that the cancer issue was not going to go away anytime soon.

CHAPTER THIRTEEN

~~

Friends and Neighbors

Even after the noisy protests of the previous three summers, no one on the New Jersey shore had ever seen anything like August of 1987. This time, the instigator was Mother Nature, not Greenpeace or the Save Our Ocean Committee. Each incoming tide seemed to deposit a new horror on the beaches of Ocean County. On August 13, just in time to ruin a peak weekend, a massive slick of medical waste—probably dumped by a passing barge—washed ashore on a seventy-five-mile stretch of beachfront. The debris included syringes and needles, blood bags, and bandages.[1] On Long Beach Island, a boy stepped on a hypodermic needle; so did a mother in Point Pleasant, north of Toms River. The county beaches stayed closed for three days, at the height of the tourist season. The following week, dead crabs washed ashore at Barnegat Light, and high bacteria counts from seagull droppings closed the beaches in Seaside Heights—right next to the Ciba-Geigy pipeline terminus. Most disturbing of all were reports of dead bottlenose dolphins covered with bloody sores. Eighty carcasses were found that summer on New Jersey beaches, apparent victims of viruses and a toxic algae bloom of unknown origin.

None of those plagues was new to the region—there were actually fewer beach closings in 1987 than in the previous summer—but their cumulative impact in a single month was shocking. Tourism was Ocean County's largest industry by far, contributing more than a billion dollars a year to local businesses. For merchants, the summer of 1987 was an unmitigated disaster. *The New York Times* described children roaming nearly deserted beaches, collecting tampon applicators to use as sand castle turrets.[2] Shore tourism was not only New Jersey's second-largest source of income (after petrochemicals), it was at the core of New Jerseyans' self-identification—the best retort to outsiders who derided the state as a putrid industrial wasteland. The economic damage from the pollution was bad enough, but the ridicule was worse. It prompted outrage that was widespread and deeply felt. At two o'clock on the Sunday of Labor Day weekend, at the close of the nightmarish summer season of 1987, thousands of beachgoers joined hands in an unprecedented protest that stretched for miles on Long Beach Island.

For Ciba-Geigy, the timing was miserable. In April, the company had applied for six state permits it would need to build the pharmaceutical plant and stay in business in Toms River. For the plant's critics, on the other hand, the hellish August was a boon. On Labor Day weekend, while beachgoers were grasping hands down on Long Beach Island, one thousand people gathered at Ortley Beach for a protest rally organized by Frank Livelli's group, Save Our Ocean. Speaking through a bullhorn, Livelli urged the crowd to boycott Ciba products and tried to link the company's wastewater discharges to the washups of floatable waste. "What the hell is the difference if they're dumping it off a barge or through a pipeline under the water?" he asked.[3]

Actually, there was no connection at all between Ciba-Geigy's discharges and the dolphin deaths, beach closures, or medical waste washups. The company had been dumping wastewater into the Atlantic for more than twenty years; its effluent had never been cleaner than it was in 1987. In September, the factory's new manager, John Simas, sent out a "Dear Neighbor" letter to the entire town touting the results of a million-dollar study funded by the company and conducted by the state Department of Environmental Protection. The agency had found no traces of industrial chemicals in water samples taken from the beaches of Lavallette, Ortley Beach, and Seaside

Heights. In fact, the only places the chemicals were detected were within three feet of a pipeline discharge vent, a half-mile from the closest beach.

None of that seemed to matter. There was only one company in New Jersey allowed to dump treated industrial waste into the Atlantic through its own pipeline: Ciba-Geigy. So many of the ocean's problems seemed utterly beyond control: toxic algae, disease-carrying bacteria, and slicks of medical waste came and went at their own inscrutable whims. But Ciba-Geigy's discharge was entirely controllable. The factory's future depended solely on the decisions of public officials who had to answer to a roused electorate. "By the end of that summer, people were just fed up," remembered John Paul Doyle, the town's state assemblyman at the time. "The cancer fears were definitely out there, but what made the difference was what was happening with the ocean. After that summer, it was all downhill for Ciba-Geigy." Over the next three months, thirty-three bills aimed at curbing ocean pollution were introduced in the state legislature, including several that would ban discharges like Ciba-Geigy's.

In search of political muscle, Ciba-Geigy hired a local real estate lawyer who had rapidly become one of the most powerful men in New Jersey: Lawrence E. Bathgate II. If the acquisition of wealth via rising property values was the secular religion of Ocean County, then Larry Bathgate was its chief apostle. He owned a Rolls-Royce, an airplane, a small airport, and five homes, including a beachfront mansion in Bay Head and an estate in Rumson near Bruce Springsteen's. Bathgate was on intimate terms with Governor Tom Kean and soon-to-be-president George H. W. Bush, who appointed him finance chairman of the Republican Party in 1987 after Bathgate raised $600,000 for him at a single dinner. Not bad for a carpenter's son who still practiced law in his hometown of Lakewood, just north of Toms River. Bathgate had made his fortune buying up farms and turning them into subdivisions. Now, at forty-eight, he was at the peak of his power as New Jersey's preeminent fundraiser and one of the leading political moneymen in the United States.[4] While lesser fundraisers toiled for candidates in the hopes of being rewarded with government contracts, Bathgate moved in much more rarefied circles. His talent was forging relationships with public officials and using those contacts to assist his friends, who were often also his clients and investment partners. With Governor Kean not scheduled to leave office until 1990, Bathgate was perfectly positioned to assist Ciba-Geigy in securing the permits it would need to stay in business.

But Larry Bathgate was not the only power player in Ocean County. Democrats controlled the State Senate, and its president was Toms River's own John F. Russo, a longtime friend of the chemical plant whose support was wavering as public opposition grew. Russo had unexpectedly won his seat in 1973 (a horrible year for Republicans because of the Watergate scandal) and had never felt entirely secure in office, even after becoming one of the most powerful Democrats in the state. "I paid a lot of attention to my district. They're conservative in Ocean County, but that doesn't mean they wanted to have a health risk in their community," Russo remembered. "Once people became aware that there was a leak in the pipeline, and possible toxic emissions that may cause cancer, it wasn't hard to go from there to saying let's just shut the whole thing down."

Besides the newly hired Bathgate, the only local counterweight to this army of newly stirred voters was the badly outnumbered factory workforce, and many of those dispirited workers were already getting layoff notices. By 1987, there were only about five hundred union workers left, and in another year the total would shrink to three hundred and fifty. Ray Talty by now was working at Ciba-Geigy's newly upgraded sewage-treatment plant, where he would occasionally take a sip of the treated wastewater, just to satisfy himself that it was as clean as the company claimed. "The stuff was very clean, it looked just like our drinking water," he remembered. But his brother John, the union vice president, knew the improvements had come too late to save the company's reputation. "After a while, we figured out that it wouldn't have mattered if you had nothing but distilled water or milk in the pipe, the pipeline was going down," said John Talty. "It had become a symbol."

The workers were increasingly focused on their own health worries. In 1988, Ciba-Geigy had to post danger warnings on sixty-two water fountains at the factory due to contamination from the plant's aging lead pipes.[5] More importantly, employees were hearing the initial results of the worker cancer study being conducted by Philip Cole and Elizabeth Delzell of the University of Alabama at Birmingham. "Management called the union leadership up to the meeting room to hear the results of Doctor Cole's study and told us, 'This is great news—we don't have a problem,'" recalled George Woolley. "Then they gave us information packets about what we were supposed to say about the study out in the community." The message Ciba-Geigy hoped to communicate to its employees and to the public was that the plant's workers were healthier than the general population. Based on an analysis of medical records of 2,642 current or former male employees, the Alabama researchers concluded that the death rate among male workers was almost 20 percent lower than the overall rate for white American men of the same age. Deaths from cancer, meanwhile, were only slightly higher than expected: 106, compared to 97 expected. Many of the cancer victims were Cincinnati veterans who might have been exposed there, not Toms River.[6]

What the company did not say was that industrial workers, even at chemical plants, were almost *always* healthier than the general population because of a well-established concept called the "healthy worker effect." Simply put, factory workers are less likely to die prematurely because they are healthy enough to be hired in the first place and to hold on to their jobs. That is why the death rate among long-term factory employees is only 60 to 90 percent of the rate for the general population, which has more sick and disabled people.[7] Despite this well-known bias, many factory-based epidemiological studies made the same questionable comparison because obtaining mortality statistics for the general population was much easier than getting them for a more appropriate comparison group, such as workers at a different factory. For manufacturers, there was another advantage to this shortcut: Comparisons to the general population tended to make their factories seem safer than they really were.

The union leaders at Ciba knew nothing about the healthy worker effect, but when they took a close look at the study results, they were disturbed nonetheless. For one thing, many now-dead workers who the leaders knew had been treated for cancer were not included. The study was of deaths, not cancer incidence, so if cancer was not listed on a worker's death certificate, he was not classified as a cancer case. "We'd say, 'What about Frank Scarpone? What about John Jaczkowski?' We knew a lot of names, and not all of them were listed as dying of cancer," John Talty remembered. He and Woolley were also frustrated that the Alabama researchers, in their initial presentation to the union, did not subdivide the overall death rate into specific buildings, which meant that there was no way to gauge the risk faced by those workers who handled the most dangerous chemicals.

What worried them most were brain and central nervous system cancers. The Ciba-Geigy worker study had confirmed five such cases, higher than the expected three. But at the briefing, Woolley was told that the case totals were low enough that the apparent excess was probably a statistical fluke. This was exactly the type of excuse that Wilhelm Hueper had anticipated a generation earlier: Case-control studies gave reliable results when huge populations and common diseases were studied, but for a rare cancer at a single factory—even one as large as the Toms River plant—unsettling findings could almost always be explained away as random flukes because the case numbers were so small.

Not until the Alabama study was published in the *Journal of Occupational Medicine* in 1989—after the factory's fate had been sealed—did a more disturbing picture emerge. In that article, Elizabeth Delzell briefly mentioned that she had subsequently found at least *eleven* brain cancer cases—including four nonfatal cases and two deaths that were not included in the original study because they had occurred in Toms River dye workers who had moved to Europe. "This analysis confirmed that rates were relatively high among men employed in the azo dye and [plastics and resins] areas, and suggested, in addition, a relatively high rate among men employed in the laboratories," she wrote.[8] Delzell also noted that there were unusually high numbers of cancer deaths in several job categories, including azo production (sixteen cases instead of the expected eight) and maintenance (thirty-seven cases instead of the expected twenty-five).

Later, George Woolley and others in the union would be bitter about not getting the full story in 1987. For now, however, they had only their suspicions—and the company's assurance that there was nothing to worry about. "We didn't trust the study, but we didn't know the facts," Woolley said. The worries over the Alabama study further corroded morale at the factory, but the union leaders continued to support the company publicly as they geared up for a climactic showdown with Ciba-Geigy's critics in the spring of 1988. After months of delay, the state announced plans for three public hearings on the company's application for the permits it would need to build the pharmaceutical plant. Hundreds of people were expected to attend. Both sides promised a lively show. "We're not looking for a confrontation," explained the union's blustery president, James McManus, at a press conference. "If we need to be reasonable people, we'll be reasonable people. If we need to be wild men, we'll be wild men."[9]

Despite all the *sturm und drang* it was generating, the Ciba-Geigy factory was not the most serious water pollution threat in Ocean County. The larger risk in the 1980s came from the plume of contaminated groundwater from the old illegal dump at Reich Farm, which had been all but forgotten amid the tumult over the vastly larger chemical plant. While controversy raged over the theoretical possibility that Ciba-Geigy's waste might make people sick if they swam near the ocean outfall or drank from a contaminated irrigation well, tens of thousands of unknowing Toms River residents were almost certainly drinking tap water tinged with industrial waste, some of it carcinogenic. The waste had leaked from the five thousand Union Carbide drums that Nick Fernicola, back in 1971, had dumped on two acres in the rear of the now-abandoned egg farm. Still owned by Samuel and Bertha Reich, the Pleasant Plains property was now being used for storage by a stone-crushing business.

The farms of Pleasant Plains gave way to subdivisions at a leisurely pace during the 1970s and 1980s, even as other parts of town were growing feverishly. Underneath the fields, however, there was plenty of activity. Struggling as usual to keep up with demand, the Toms River Water Company was still very dependent on its Parkway well field, which was about a mile south of Reich Farm. There were six Parkway wells operating in mid-1987, providing one-third of the town's overall supply. The thick layer of saturated sand beneath the Parkway well field was a rich source of groundwater, and it was continually replenished by a natural north-to-south flow. Groundwater seeped southward at a brisk pace of about sixteen inches per day. Then, when it entered the huge "capture zone" of the Parkway field, the water would start moving much faster—zipping along at more than nine feet per day as it was pulled toward the intake screen of one of the six wells. The Parkway field was essentially a perpetual water source, a bottomless cup—the perfect resource to match Toms River's grow-now-worry-later ethos. The water company made the most of it. Back in 1974, when the Parkway field was new, Toms River Water extracted two million gallons per day. By 1987, the daily draw averaged nearly three million gallons—closer to four million on hot summer days.

A mile away at Reich Farm, meanwhile, there was also action underground. The chemicals Nick Fernicola had dumped in 1971 were on the move. There had been three attempted cleanups at the dumpsite—two in 1972 and one in 1974—involving the removal of about five thousand waste drums and eleven hundred cubic yards of soil. But the remediation was anything but thorough. Instead of testing the soil to look for chemicals, Union Carbide contractors carted dirt away only if it looked or smelled contaminated. If the dirt looked okay—whatever that meant—they left it alone. But the soil was not okay at all. Almost as soon as Fernicola dumped the leaky waste drums, globules of benzene, trichloroethylene, styrene, and other wastes began trickling down through ten feet of sandy soil and into the aquifer zone, where they hitched a ride with the groundwater seeping south. Within a year, the edge of the chemical plume had already moved beyond the Reichs' land. It kept going south, tainting dozens of private wells as it spread like a fat, uncurling finger—pointing right at the Parkway field and its six huge and very thirsty public-supply wells.

No one will ever know when the fingertip of that toxic plume first reached the intake screen of a Parkway well. The timing of that fateful event would eventually be a matter of great debate in Toms River.

Whether it happened as early as 1978 or as late as 1986, from that moment onward, thousands of customers of the Toms River Water Company were drinking low levels of toxic chemicals from Reich Farm. There had been signs of possible contamination as early as 1974, but the water company declined to install filters or order advanced tests on the Parkway wells. Instead, Toms River Water worked the wells harder than ever during the 1980s. The passage of the federal Superfund law finally broke the cycle of neglect. Along with the Ciba-Geigy factory, Reich Farm was placed on the original National Priorities List in 1983, and by 1986 contractors for the U.S. Environmental Protection Agency were busy at both Superfund sites conducting "remedial investigations" to determine how extensive the groundwater contamination was and what kind of health risks it posed.

An EPA contractor started testing the Parkway wells in May of 1986 and immediately found something that did not belong there: trichloroethylene. TCE, as it is known, was and still is a jack-of-all-trades for the chemical industry: degreaser, solvent, and key ingredient in hundreds of products, including Ciba-Geigy's resins and Union Carbide's plastics. The TCE concentrations in the three affected Parkway wells were relatively low—between three and fourteen parts per billion. But by the mid-1980s, evidence of TCE's hazards was building. Unlike benzene or toluene, TCE often could not be smelled or tasted in water unless levels were high, yet even low concentrations of its vapors could cause liver, kidney, lung, and heart damage—as well as cancer, at least in rats. As a result, the EPA had started enforcing a limit of five parts per billion of TCE in water, and New Jersey was about to set a limit of just one part per billion that would take effect in 1988.[10] The Toms River Water Company definitely was not ready to comply.

Was Reich Farm the source of the contamination that had struck the Parkway wells? There was still no way to know for sure. TCE had been dumped at the farm, but it also could have come from a machine shop, garage, or some other illegal dump. It could even have come from Ciba-Geigy—though that was extremely unlikely because even though the chemical plant was only a mile west of the Parkway wells, the river was a natural barrier and the groundwater flowed south, not east. On the other hand, there were many reasons to suspect that the source was Fernicola's illegal dump, since the EPA contractor had found not just TCE but also more than a dozen other industrial chemicals in the groundwater beneath Pleasant Plains, including several used in plastics manufacture.

The Toms River Water Company reacted to all of this information with a characteristic combination of lethargy and secrecy. Sixteen months after the EPA contractor found TCE in three Parkway wells, the water company finally closed down the well where the contamination was highest. But as soon as pumping stopped in one well, TCE levels began rising in two others. (This problem, which resembled the arcade game Whac-A-Mole, would haunt the water company in the 1990s too.) Unwilling to close all three wells, the water company—with the state's permission—decided to reopen the most heavily tainted well and mix its water with water from the five other Parkway wells before distributing it around town. The dilution reduced the overall TCE concentration in the blended Parkway water to two parts per billion, still slightly above the new limit of one part per billion that would take effect statewide in 1988.

The water company's customers knew nothing about any of this—just as they never knew about the contamination of the Holly Street wells in the mid-1960s. Back in 1974, water customers had learned about problems at the Parkway well field only because of a belated story in the *Asbury Park Press*. Something similar happened this time, in late 1987, when a copy of a well-testing report was sent to the county board of health, apparently by mistake.[11] The *Observer* jumped all over the story, especially after the water company tested water fountains at two schools served by the tainted Parkway wells and detected TCE at concentrations of three and two parts per billion, respectively. Facing more public outrage in a town that was already anxious about pollution, the water company finally shut down the three most affected wells. By then, it was November; the extra water would not be needed until next summer, when it would be needed very badly.

It was 1965 all over again. Back then, the Toms River Water Company had been rescued from a similar squeeze by the completion of the chemical plant's ocean pipeline, which diverted chemical waste away from the river and the Holly Street public wells. Now, twenty-two years later, the water company's Parkway wells needed a similar rescue—and it would have to come before the following summer, when the people of Toms River would be demanding more water than ever. They always did.

In early 1988, as the dates of three climactic public hearings approached, Ciba-Geigy's terrible luck continued. In mid-February, at a construction site across the street from the Ocean County Mall, a bulldozer ripped a four-foot gash in the company's waste pipeline. This time, about two hundred thousand gallons of wastewater spilled, an event the *Observer* marked with the headline, "Oops!" The next day, just in case anyone had forgotten that Ciba-Geigy was still in criminal jeopardy, a grand jury issued a new set of indictments on the familiar charges of conspiracy, illegal dumping, filing false reports, and misconduct. (The original charges had been dismissed on procedural grounds and then restored on appeal.) The new indictments targeted only two executives—William Bobsein and James McPherson—plus the company. Ominously for Ciba-Geigy, there were reports that the other two previously indicted executives, David Ellis and Robert Fesen, who were now out of legal jeopardy, were cooperating with prosecutors.[12] Ciba-Geigy was trying to look to the future, but events kept dredging up the ugly past.

The long-awaited public hearings were the circus everyone had expected. At Toms River High School North, the eighteen hundred students were sent home at noon; by then, hundreds of adults were already milling outside in an unruly mass. There was some shoving among the warring parties: union workers in white baseball caps, placard-waving environmentalists, and a dozen police officers trying to preserve order as state officials collected the names of the people who wanted to speak. By the time the hearing started, at one o'clock in the auditorium, all but a handful of the twelve hundred seats were full. Sentiment was divided—several hundred white baseball caps were visible in the crowd—but the environmentalists were louder.

They booed and heckled plant manager John Simas and Larry Bathgate, the first speakers, but fell silent when a woman most of them
had never seen before approached the microphone and began talking.

She was Linda Gillick, and she had carefully choreographed her
five minutes of allotted time for maximum effect. First, she motioned
for a group of children to walk to the front of the room; each carried
a red rose with a black ribbon tied to the stem. "I represent the families of Ocean County children with cancer. Some are with me today,"
she began. Ciba-Geigy, she said, was not the only polluter in town,
but the company should be forced to clean up its waste before producing any more. As she spoke, her message became increasingly dramatic, her voice louder and more insistent: "Ciba-Geigy helps sell
flowers—daffodils, to be exact—for the American Cancer Society to
raise money for research. Keep your daffodils; most of our children
are pushing them up from their graves, or will be. You may think I
have no facts and figures to substantiate the high incidence of cancer
in Ocean County. I do."[13]

Gillick then told the hushed crowd that she would recite the ages,
hometowns, and cancer types—but not names—of forty-one Ocean
County children diagnosed since 1983, including ten now dead. The
children at the front of the room would hand out their roses as she
read. "Watch your rose as its beauty fails slowly, silently and continuously, because you are watching my child and all the others around
you slowly, but not silently or painlessly, die," she said, as the children
began giving the black-ribboned flowers to the state officials on the
stage and to members of the audience, including several very uncomfortable Ciba-Geigy employees. "I leave their destiny and the destiny
of each and every child here, and those still to be born, in your hands."

And then, as Linda Gillick prepared to read her list, the small boy
beside her—he was nine, but looked closer to five—asked to speak.
He had not filled out a request card, but it made no difference. No one
would have dared to tell Michael Gillick he could not speak. As he
began, the television cameras scrambled for a clear shot of the boy
whose face seemed the very personification of cancer's torments. "If
you have a child, picture him with cancer because of this water," he
said. "Think of what it could do to him. He could die at any second,

any minute, so please stop!" As he spoke, there was no other sound in the huge auditorium but the clicking of cameras. Michael Gillick's voice broke, and he began to cry. "What Ciba-Geigy is doing is really wrong, but you guys keep going on and on doing your stupid job and making people sicker," he continued. "Please stop!" And with that, the boy ran up the aisle and out of the auditorium. His mother ran after him, after first remembering to hand the list of sick children to another parent and instructing her to finish reading it.

It lasted just thirty seconds, but Michael Gillick's speech was long enough to sear the memories of the hundreds who heard it. Amazingly, it would be reprised eight years later at an even more emotional public meeting in the same high school auditorium. For those who heard it, Michael Gillick's 1988 speech was a *Rashomon* moment, open to many interpretations. Michael remembered it with a child's simplicity: "I gave them the rose, I gave them the lecture, and I ran out of there." Ciba-Geigy's opponents remembered it as deeply moving and immensely powerful—the ultimate expression of what was at stake in their crusade. "It was amazing. I can still picture it," said Sheila McVeigh, who still lived on Cardinal Drive. "The room was packed, and the Ciba people just kept quiet. Everyone cheered for the little children when they finished. It was spellbinding, really, and very sad." Many of the factory employees had a different interpretation. They could not deny young Michael Gillick's sincerity or ignore his pain. But they thought his mother's tactics were terribly unfair. John Talty, who had been handed a black-ribboned rose, knew the long war was lost. "When Linda Gillick's son gave me that rose, I just looked up and said to someone, 'How do you beat that?'"

Having lost what little was left of its support in the community, Ciba-Geigy had one last hope: that Larry Bathgate could convince his good friend, Governor Tom Kean, to issue the permits the company needed, no matter what the people of Ocean County wanted. For a while, it seemed possible that Bathgate might succeed, especially after Christopher Daggett came to town. A former top aide to Kean who still played tennis with the governor, Daggett was the regional director of the U.S. Environmental Protection Agency. He came to Toms River in June to announce that the EPA did *not* want Ciba-Geigy's ocean pipeline closed. Instead, the agency wanted to use the pipeline to assist in the long-awaited cleanup of the site. There were billions of gallons of contaminated groundwater beneath the factory property, and the EPA wanted the company to pump it all up (a process that could take thirty years), treat it at the factory's newly upgraded wastewater plant and then send it through the pipeline into the ocean. The only feasible alternative, Daggett added, would be to discharge the treated groundwater into the river. Either choice, he said, was a "very low risk" to public health—and fully compatible with the planned pharmaceutical plant. That last point was crucial because he would soon have the authority to decide whether the pharmaceutical plant would be built; Governor Kean had just announced that Daggett would be leaving the EPA to become commissioner of the New Jersey Department of Environmental Protection.

Daggett's plan was a tremendous lift to Ciba-Geigy. Not only would it clear the way for the pharmaceutical plant, it would also be a financial windfall for the company. Using the ocean pipeline to get rid of the treated groundwater would be about $90 million cheaper than building a new piping system to discharge it into the river or, as the local environmentalists preferred, injecting it deep underground.[14] But Ciba-Geigy's reprieve was only temporary. A few weeks later, the town council—in a final break after decades of slavish deference to Ciba-Geigy—voted to rezone the factory site (and Reich Farm, too) to bar new construction, despite a threatened lawsuit by Bathgate.

The final blow, fittingly, was self-inflicted. Larry Bathgate's most famous friend, the Republican presidential nominee, George H. W. Bush, came to Ocean County on July 22, 1988, for a fundraiser at Bathgate's mansion on the beach in Bay Head. Bathgate gave Chris Daggett and his wife two free tickets, which would have cost $5,000 if Daggett had paid for them. That turned out to be a huge mistake. When word got out, Democrats up and down the Jersey shore attacked Daggett as a pawn of Ciba-Geigy. Frank Livelli and the Save Our Ocean Committee successfully demanded an ethics investigation by the EPA inspector general. The investigation eventually cleared Daggett, but the imbroglio held up his confirmation as state environmental commissioner for months. By then, his plan to keep the pipe-line open was dead.

After the free tickets fiasco, the last dominoes fell in rapid succession. Daggett's successor at EPA quickly dropped plans to use Ciba-Geigy's pipeline to discharge the treated groundwater into the ocean. Then, in October of 1988, the state DEP denied Ciba-Geigy's application for a permit to build the pharmaceutical plant on the grounds that "too many unresolved concerns about the existing site remain to consider adding a new industrial facility."[15] Two weeks later, despite a last-minute advertising blitz by Ciba-Geigy that included a final "Dear Neighbor" letter from plant manager John Simas, eleven Ocean County towns voted overwhelmingly in favor of closing the pipeline—including even Toms River, the plant's dearest neighbor of all. In 1952, Toms River had welcomed the Swiss with brass bands and flowery speeches; now voters were, in effect, telling them to leave, by a two-to-one margin. The vote was nonbinding, but the company was finally ready to surrender. In December, Simas announced that Ciba-Geigy would shut down the pipeline by the end of 1991 and focus instead on cleaning up the mess left behind from almost forty years of chemical manufacturing.

The Save Our Ocean Committee celebrated its victory with more rallies on the Lavallette boardwalk, where a plaque in memory of Frank Livelli would be placed after his death in 1995. He lived long enough to see the ocean portion of Ciba-Geigy's pipeline cut into pieces, towed to deep water, and dropped to the sea bottom—at a spot six miles east of the beachfront mansion where Larry Bathgate's fundraiser had doomed any hope of keeping it open. As for Bathgate, he never again flew quite so high as he did in the late 1980s. His real estate empire had been partially financed on zero-collateral loans from the First National Bank of Toms River, but the bank—which had grown at breakneck speed to fifty-five branches from its original Main Street location—abruptly failed in 1991. It was the state's largest-ever bank failure, triggered by more than $300 million in delinquent loans. A few days before it was declared insolvent by federal regulators, the bank sued Bathgate, who was late on more than

$16 million in loan repayments. He would eventually bounce back, but the lawsuit was an embarrassment for President Bush. Bathgate cut back his involvement in Republican politics, resigning as national finance chair in 1992.

After thirty-six years of uninterrupted production, the Toms River factory produced its last batch of dye chemicals at the end of 1988, its last resins in 1990. By then, fewer than three hundred people were working at the vast complex. Some were engaged in tearing down buildings or assisting with cleanup operations, while the rest mixed and tested dyes manufactured elsewhere. Whenever local business leaders would entreat plant manager John Simas to try again to win approval for the pharmaceutical plant, he would curtly refuse. "We've had a lot of disappointments," he told reporters in 1991. "We're here. We'll continue to be here. We have to remediate. But I don't think we'll operate in a community where we're not wanted."[16]

In the spring of 1988, a cylindrical tower appeared at the edge of the Parkway well field, a flash of dull white barely visible through the trees to drivers whipping past on the Garden State Parkway. About twelve feet in diameter and three stories high, the tower was stuffed inside with plastic packing material. It also carried the fervent hopes of the Toms River Water Company, which was counting on the contraption to help it survive the summer without another contamination crisis. It was called a packed tower air stripper, and it cost $350,000.

The technology was simple but effective, which was why similar towers were sprouting at hundreds of other polluted sites around the country. Contaminated water was pumped to the top of the tower and sprayed down onto the plastic packing material. As the droplets dribbled down through the nooks and crannies of the plastic stuffing, a fan at the bottom of the tower blew heated air upward through the material. When the rising hot air met the falling water, highly volatile chemicals in the water evaporated and were expelled into the atmosphere through a vent at the top of the tower. The cleaner water was then pumped out of the bottom of the tower and distributed to customers.

The Toms River Water Company could have dealt with its contamination problem in a much more comprehensive way, instead of relying solely on the air stripper, but that would have been much costlier. It could have abandoned the contaminated Parkway wells and drilled new ones elsewhere, or it could have finally added carbon filters to those wells, a move state and county health officials had first recommended—but had never insisted upon—in the mid-1970s. But the water company desperately needed its Parkway wells, and adding a carbon filtration system would have raised the total cost by several hundred thousand dollars. Besides, the contraption worked. In May, just in time for the start of the peak tourism and lawn-watering seasons, the water company restarted the three tainted wells and pumped their water through the new air stripper before distributing it to customers around town. The process lowered TCE levels from fourteen parts per billion to just one part per billion, New Jersey's new safety standard. Breathing easier, the water company's managers resumed their frenzied pumping of the affected wells, pulling more than a million gallons per day out of them that summer.

As far as the EPA was concerned, though, the air stripper was only a temporary fix. If they were maintained properly, air strippers worked well for TCE and other volatile compounds that vaporized easily. But TCE was just one of several dozen chemicals detected in the Reich Farm plume, and some of them did not vaporize so easily. They were what chemists called semi-volatile organic compounds, and air stripping could not remove them from groundwater—only carbon filtration could. Later, that fact would become very important. So the EPA, which in September completed its long-awaited Superfund cleanup plan for Reich Farm, ordered Union Carbide to remove volatile *and* semi-volatile compounds from the groundwater beneath Pleasant Plains and to do so by intercepting the contamination plume before it reached the Parkway wells. The water company's air stripper, which treated the water only after it had reached the wells and was capable only of removing volatile chemicals, was insufficient, the EPA concluded. A proper groundwater cleanup in Pleasant Plains, the agency estimated, would take eleven years and cost Union Carbide about $6 million—a modest sum compared to the massive cleanup planned for Ciba-Geigy, which was supposed to take almost three times as long and cost more than twenty-five times as much.

After almost forty years of unchecked dumping, Toms River was finally about to become a place where people pulled toxic waste out of the ground instead of putting it in. The pollution wars were over. The chemical plant was winding down, the cleanups were starting up, and the initial cancer surveys of the factory and the town were at least somewhat reassuring. The worst seemed to be over for Toms River, as long as everyone had finally absorbed the hard lessons of the previous four decades.

PART III

COUNTING

CHAPTER FOURTEEN

≋

Two Wards, Two Hits

It is no small challenge to spend long days and longer nights in a place where children die, but Lisa Boornazian had the knack. She began working at The Children's Hospital of Philadelphia in 1991, during the summer of her senior year of nursing school at Villanova University. The following year she found a home in the cancer ward at CHOP. Back then she was Lisa Davenport, and not much older than some of her patients. "I loved working in oncology. I saw plenty of nurses who came to work at the unit and it wasn't what they expected. They just couldn't stay. But I loved it," she remembered. The ward was a surprisingly lively place, where little kids dashed down the wide hallways with their wheeled intravenous stands clattering beside them. The older children, though, were much more difficult to deal with. "The teenagers had a grasp of death, and what the diagnosis meant," Boornazian said. "The younger kids mostly had no idea." Those with brain or bone cancers faced long odds. Survival rates were better for children with blood cancers, principally leukemia and lymphoma, but their treatments took many months and were brutal: chemotherapy, often followed by radiation and bone marrow transplants.

The work shifts on the oncology ward were organized in a way that made it impossible for the nurses to keep an emotional distance from their assigned patients, since the same three or four nurses would take care of a child for months on end. Their relationships with parents were equally intense. Many parents practically lived in the ward and went home only to shower and change clothes before rushing back. The nurses worked under the unforgiving gaze of mothers and fathers driven half-mad by lack of sleep and the sight of their children enduring a pitiless cycle of excruciating needle sticks, nausea attacks, and dressing changes. Parents would frequently take out their anger on the nurses, and the nurses who lasted on the ward learned to respond without rancor or condescension. The long shifts, especially the sleepless overnights, created an intimacy among nurse, parent, and child that no one else could share—certainly not the doctors and social workers, who were mere transients by comparison. The nurses were family. And when it was time for a funeral, the nurses did what family members do: They showed up, and they mourned.

Funerals were part of the ritual of life on the oncology ward, and Lisa Boornazian (she married in 1993) attended her share, in towns all over New Jersey, Pennsylvania, and Delaware. One in particular stuck in her mind. It was for a vivacious young woman named Carrie-Anne Carter who was diagnosed with Ewing's sarcoma, a rare bone cancer, during her senior year of high school. She was in treatment for more than a year, and Boornazian grew very close to her family. When the teenager died on February 6, 1995, Boornazian decided to make the two-hour drive from Philadelphia to attend the funeral in a town she had never visited before: Toms River, New Jersey.

Although Boornazian had never been to Toms River, she knew the name well. By 1995, everyone who worked on the oncology ward at Children's Hospital knew about Toms River. Many years later, she explained why. "We began to notice that we were getting a lot of kids from the Toms River area" in 1993 and 1994, Boornazian recalled. "It wasn't just me. All of the nurses noticed it." CHOP drew its young cancer patients from a vast geographic area of more than ten million people, and some families would travel even farther—from as far away as South America and the Middle East, if they could afford it. Yet when Boornazian and the other nurses would look at the charts of the two-dozen or so patients admitted to the oncology ward in a typical month, there always seemed to be at least one or two from a town with a year-round population of just eighty thousand people, or from communities nearby. "There was a certain point where, for a couple of months, we would get a new admission from Toms River every week," Boornazian said. It became a source of dark humor on the ward, with the nurses regularly asking one another: "I wonder when our next patient from Toms River is coming in?"

The doctors ignored the chatter. Boornazian raised the issue a few times with physicians she knew well. "I would ask them, 'Have you noticed this? Do you think there's something going on in Toms River?'" The doctors all told her the same thing: no. They said it in various ways, and with inflections that ranged from respectful to patronizing, but the core message never varied. "The general sentiment from the doctors was that it was just a coincidence, and that we shouldn't worry about it," Boornazian recalled. She did not take offense. The physicians ruled the hospital, and CHOP was not just any hospital. It was the oldest and most prestigious children's hospital in the United States, regularly finishing at the top of national rankings for pediatric care. How could the doctors be expected to pay attention to something as trivial as a home address? They were far too busy to get to know the families the way the nurses did. "The doctors rotate in and out, and I don't think they really realized how many patients there were who were from the Toms River area," Boornazian said. "It's the nurses who are there day in and day out, year after year. To us, it looked like a very unusual number of patients."

When Boornazian told other nurses on the ward that she was planning to drive to Toms River for Carrie-Anne Carter's funeral, they responded with wisecracks that were only half in jest. "People said, 'Oh, don't drink the water when you're there, and don't breathe too deeply,'" Boornazian remembered. From conversations with the Carters and other Toms River families, she had heard a little bit about the Ciba-Geigy chemical plant and its history of pollution. So she was a bit unnerved when, while driving to the funeral, she looked to her left at a stoplight on Route 37 and saw the fence and security gate of the sprawling factory complex. "I remember driving by and thinking, 'That's the plant that everyone talks about.'" For days afterward, she could not stop thinking about that big factory in the woods and about all the local families she had met in the oncology ward.

The animal experiments and population studies of researchers like Wilhelm Hueper and Richard Doll were phenomenally successful at identifying an ever-lengthening list of carcinogenic compounds, but how and why those chemicals triggered malignant tumors was beyond the scope of their assays and surveys. Toxicology and epidemiology alone could not provide a convincing explanation of the inner workings of a cell as it transitioned from normal to malignant.

One of the very first scientists to begin to lift the curtain on the hidden mechanisms of carcinogenesis was an underappreciated giant of biology named Theodor Heinrich Boveri.[1] Born in the Bavarian city of Bamberg in 1862, Boveri was a brilliant student, winning a fellowship in 1885 that allowed him to pursue any scientific question he wanted for five years. Boveri chose heredity. He wanted to understand how traits are passed from one generation to the next through fertilization, or from cell to cell through division. Working first with the eggs of roundworms and then of sea urchins, Boveri focused his observations on the squiggly, rod-shaped objects he saw inside the nuclei of cells as they divided. The pliable rods were one of the few parts of the nucleus that could be seen through the microscopes of the day—they stood out when stained, which is why they were called chromosomes, or "colored bodies"—and their behavior fascinated Boveri. During cell division, they shifted and danced like a squad of synchronized swimmers before receding into invisibility in the two new daughter cells.

Unlike many of his fellow biologists, Boveri believed that chromosomes were the structures of inheritance, the engines that made Darwinian evolution possible by transmitting an individual's variable traits to the next generation. He set about trying to buttress his belief by systematically observing the movement of chromosomes in animal cells. What looked a freeform dance of the chromosomes during cell division, he discovered, was actually a precise and highly organized gavotte in which the rods divided individually and then into separate but identical clusters, each going its separate way as the cell split.

Boveri also noticed that chromosomes did not dissolve after the division was complete, as many biologists thought at the time, but instead lived on in compacted form inside the nuclei of the daughter cells until they unspooled again during the next round of cell division. He spent the rest of his life trying to convince skeptics that chromosomes were the carriers of hereditary information and that variations in individual chromosomes were responsible for the inherited variations in all living things.

Cancer was a key reason for Boveri's theories about chromosomes. He chose to study roundworms because they had just two chromosomes, both of which were large enough to see through a microscope. He then moved on to sea urchins because a single urchin egg could be fertilized by two sperm. Whenever that occurred, the resulting zygote had an unbalanced number of maternal and paternal chromosomes. If the egg was then able to divide, the anomaly was passed on to the daughter cells. Those "unbalanced" urchins, Boveri realized, often had developmental problems, including cancer.[2] In an astonishingly farsighted book he wrote in 1914, a year before his death, Boveri asserted—correctly, as it turned out—that vulnerability to cancer could be inherited via mismatched or unstable chromosomes.[3] The rapid cell growth of tumors, he correctly theorized, could be triggered by growth-promoting chromosomes (he did not know about genes, the subunits within chromosomes that actually carry hereditary information) or by the absence of growth-inhibiting chromosomes or by chromosomal abnormalities that appear only after many generations of cell division.[4] The same year his book was published, Katsusaburo Yamagiwa managed to induce malignant tumors by painting rabbits' ears with coal tar. His success, combined with Boveri's observations, strongly suggested that carcinogens did their mischief by triggering mutations within chromosomes. In fact, Boveri was sure that pollution-induced mutations—and the resulting cancers—would be found in industrial workers who handled coal tar.[5] He was right about that, too.

What about young children with cancer? Could an environmentally induced mutation really be passed down from one generation to the next, just like hair or eye color? Hermann Joseph Muller thought so. Another in a long line of brilliant but abrasive biologists, Muller had been thinking about that question for practically his whole life. When he was a young boy in New York City, in 1899, his father had taken him to the American Museum of Natural History and explained, with the help of a display of horses' hooves, how organisms evolved via natural selection acting on accidental variation. His father died soon afterward, but the lesson stayed with his eight-year-old son. If evolution could occur naturally, Muller thought, surely the deliberate actions of humans could affect the future evolution of his own species—for good or ill.[6]

Small and bellicose—he compensated for his size by standing on tiptoe while arguing—Muller was a social misfit who, like Boveri, was prone to depression.[7] Despite his unhappiness, or perhaps driven by it, Muller did brilliant work as a young researcher at the University of Texas. He conducted a series of experiments in which he irradiated fruit flies with X-rays, which were already widely known to trigger cancerous tumors in animals and humans alike.[8] In 1927, Muller was ready to divulge his results, which caused an immediate sensation. By exposing the sperm cells of fruit flies to X-rays, Muller reported, he was able to generate heritable mutations—white eyes in male flies instead of red, for example—almost at will. Many of those mutations were lethal, triggering a wide range of abnormalities, including sterility.[9]

Carcinogens could mutate genes; Muller had validated Boveri's prediction. Not all mutagens triggered cancer, and not all carcinogens did their work via mutation, but there was a strong correlation between the two—as the people of Toms River would learn when they found out what was in their drinking water. More broadly, Muller helped to solidify the view that genes, with their unique ability to copy themselves and transmit information to daughter cells, must control cell development and therefore the growth of the entire organism. Genes were the wellspring of life—and also of death, depending on the particular the message they carried. Muller had answered the question that first occurred to him as a young boy at the museum: Yes, human beings could manipulate their genes in ways that would affect not only their survival as individuals but the long-term viability of their species. Muller was awarded a Nobel Prize for his genetics re-search in 1946, just after the atomic bombings of Hiroshima and Na-gasaki. He spent the next twenty years studying fruit fly mutations and campaigning against the use of radiation in weaponry, industry, and medicine—"time bombing our descendants," he called it.[10] The sins of one generation, Muller suggested, would be visited on the next through genetic mutation and cancer.

One of the Toms River children whom Linda Boornazian met on the oncology ward at The Children's Hospital of Philadelphia in the early 1990s was Michael Anderson. He was ten years old when he was first admitted to CHOP in September of 1991, the culmination of a sum-mer of intensifying agonies that began with a sore thumb on May 8. Mike's birthday was May 11, but by then he was already too sick to enjoy it—his ankles and hands ached, and he was running a fever. A month-long course of antibiotics did not help. By July, he was in a wheelchair, his feet and hands bent like an old man's and too painful to move. On an initial visit to Philadelphia, the doctors diagnosed juvenile rheumatoid arthritis, but the medications they prescribed proved as useless as the antibiotics. His mother, Melanie, had to feed and wash her son as if he were a baby. When Mike went back to school, his mother went with him, carrying him up and down the stairs. His father, Bruce, who worked as a reactor operator at the Oys-ter Creek nuclear power plant, began to worry about the family's health insurance. Mike's three brothers—ages seven, thirteen, and sixteen—watched, terrified, as the unchecked and unidentified dis-ease ravaged his small body. Circle-shaped bruises appeared all over him. They were not the impact bruises Mike used to get from playing soccer or riding his bike—the bruises all rambunctious young boys get—but something deeper, stranger.

His parents suspected leukemia, but the doctors said that was im-possible; the blood tests were negative. Even so, Bruce and Melanie Anderson insisted on another examination in Philadelphia in Septem-ber of 1991. This time, a new pediatric rheumatologist saw Mike. His name was Steven Goodman, and he was only a few years out of med-ical school. He looked at Mike's case in a new way—asking more questions, conducting a more thorough physical exam. Examining Mike's fingers, Goodman noticed that Mike's bones, not arthritic joints, were the source of his pain. He ordered another set of blood tests, and this time the results were different. Mike's platelets were low, and his white blood cell count was high—the same lethal combination Rudolf Virchow had dubbed *leukemia* after observing it under his microscope in 1845. The Andersons got the catastrophic news on Friday the Thirteenth: Mike had a rare form of leukemia that was essentially a combination of the two most common types, acute lymphocytic leukemia, known as ALL, and acute myelogenous leukemia, or AML. The prognosis was dire; the required double dose of extended chemotherapy would be vicious.

The next three years were "a roller coaster through hell," as Bruce Anderson would remember them. After six weeks of intensive chemotherapy at the hospital, Mike went home to Toms River, returning at least once a month for another debilitating round of chemotherapy drugs, administered through a catheter surgically attached to his chest. There were many other trips to the hospital, too—harrowing, late-night runs across New Jersey whenever Mike had a high fever. The regimen of chemotherapy and steroids was gradually killing the cancer cells, but it was also wrecking Mike's immune system, leaving him open to an array of opportunistic infections, any of which could have killed him. Melanie learned to keep an overnight bag packed and ready at all times in case they had to rush back to Philadelphia. She and Bruce also learned to be assertive with the nurses and doctors. They asked questions, checked medication dosages, and insisted that all of Mike's visitors first wash their hands. The Andersons connected with Linda Gillick via Ocean of Love, a support group she had formed in 1988. Raising money in the community for families struck by childhood cancer, Gillick delivered Easter baskets, Thanksgiving turkeys, and Christmas gifts to the Andersons and other families. She organized summer picnics and trips to Broadway shows. And for families who needed it, she would provide direct financial support—including for funeral expenses. She was driven, and she was tireless. The Andersons were amazed by Gillick's devotion to other sick children, even as her own son's condition remained perilous.

Very gradually, as Mike Anderson's health improved and the sense of perpetual crisis that had enveloped his family began to ease, his parents allowed themselves to ponder the deeper questions raised by their son's illness. Bruce Anderson knew about Linda Gillick's map of local childhood cancer cases, and he came to share her conviction that a chilling pattern was emerging—a cluster.

Robert Gialanella thought so too. He had met Linda Gillick in 1989 in a store checkout line at the Ocean County Mall, while she was waiting to pay for a huge basketful of holiday toys for the children. He was a local physician, and within a few weeks he was also a board member of Ocean of Love, where he would remain for thirteen years. Gialanella's oldest son was born one day before Michael Gillick, and he came to think of the Ocean of Love children as extensions of his own family. "Linda was so committed to these kids," he remembered. "I just felt I wanted to be involved." Gialanella was a gastroenterologist, not an oncologist, and he treated adults, not children. But he knew a bit about epidemiology, and in 1991 he and Gillick became alarmed as the number of pushpins on her map proliferated across Ocean County and especially in Toms River. If something in the environment really was causing those cancer cases, it seemed to be gaining potency every year. Gialanella decided to try to find someone at the state health department who might be willing to investigate. It took a while to locate the right person, but in 1991 Gialanella spoke to Michael Berry, the same health department employee who five years earlier had responded to a very similar request from Chuck Kauffman of the Ocean County Health Department.

That earlier study had found no statistically significant elevation in the local childhood cancer rate, but Berry agreed to Gialanella's request that he take another look because this time a few more years of registry data could be included in the analysis, increasing the overall number of cases and slightly reducing the effects of chance. In fact, there were enough cases by now that Berry could try to look at some specific categories of diseases instead of lumping all childhood cancers together. Also, there had been some improvements at the cancer registry in collecting data from out-of-state hospitals, so it was more likely that at least some of the Toms River children treated in Philadelphia and New York would be included this time. One problem, however, had gotten worse: The registry was now four years behind schedule; the most recent full year of data was 1987. The cases Chuck Kauffman had heard about in 1984 and 1985 would show up in Berry's second study, but all of the cases that Linda Gillick had heard about in 1989, 1990, and 1991 would not. The state cancer registry was like a surveillance camera whose film took four years to develop. By the time the pictures were available, the world had moved on.

Berry ran the numbers, and this time the results were ambiguous—the number of cases was higher than expected, but not by much. Gialanella was disappointed, but not surprised. "We knew there were more kids than were documented in the state reports," Gialanella remembered. "We were looking at real-time data, all of the kids we knew about. I think we kind of said to ourselves that if the state had more current data it might show something more significant." Berry's inability to confirm the existence of a cluster was extremely frustrating to Linda Gillick and Bruce Anderson, who by now were convinced the cluster was real and getting worse. Ocean of Love stopped asking the state to investigate—that route now seemed hopeless—but Gillick kept adding pins to her map, and Anderson started reading everything he could find about environmental causes of cancer.

Anderson's personality was as forceful and stubborn as Gillick's, but his intensity smoldered, while hers often burst into flames. He had an engineer's mind; he liked to pull things apart and see how they worked. He also liked to build with his own hands, including the house on Malcolm Street where the family had lived since 1984. Now he applied that same meticulousness to the process of learning everything he could about the causes of childhood leukemia, including the possible role of environmental pollutants in combination with inherited vulnerabilities. One of his son's doctors had mentioned something called the "two hit" theory of carcinogenesis, developed by a researcher who was now in Philadelphia. The idea intrigued Anderson; he wanted to know more.

More than forty years earlier, in another hospital ward in another city, the father of the two-hit hypothesis, Alfred Knudson, first encountered the mystery and pathos of childhood cancer. He would go on to become one of the world's most influential cancer geneticists, but in March of 1949 he was a twenty-six-year-old medical resident on a one-month rotation in the pediatric oncology ward at Memorial Hospital in New York City. Most of the patients in the twenty-bed ward had acute lymphocytic leukemia, and for the first time there was some hope of curing them. A new class of chemotherapy drugs seemed to slow tumor growth by impeding the function of folic acid. "Before the antifolates, there was nothing you could do except just watch the leukemia kids die. It would usually take three or four months," Knudson remembered many years later. "I was with those kids all of the time for that one month on the ward. It was total immersion, and it was exciting because there were remissions for the first time in leukemia."

Knudson's rotation in the ward soon ended, and so did the remissions. Most of the children with leukemia relapsed within a few months, with little to show from the experimental treatment except a few extra months of life and an awful array of debilitating side effects. Antifolate drugs would later become a useful part of the standard chemotherapy regimen, but their early failure was a wrenching experience for the young doctor-in-training. "When you see twenty kids all together on the ward," he said, "it just suddenly hits you: Why, and how, do these kids get cancer?"

His memories of those helpless young patients, many of whom would never go home, stuck with Knudson—so much so that he later decided to give up clinical medicine for a research career focused on trying to understand the causes of childhood cancer. In 1964, he was at the City of Hope Medical Center in California when Hermann Muller came through as a visiting researcher. The diminutive leftist was now seventy-five years old and was still working with fruit flies; briefly hospitalized for a heart ailment while in California, he insisted that an assistant bring his flies to his bedside so that he could inspect them. He was also still buzzing energetically over the unsolved problems of mutation and cancer. Why did it often take many years—or several generations, in the case of fruit flies—for radiation exposure to trigger cancerous tumors? The answer, Muller believed, was that several mutation "events" were required to transform a healthy cell into a malignant one.[11] Cancer, in other words, was more like a long-distance relay than a solo sprint.

Alfred Knudson agreed with his eminent visitor. To Knudson, a multistage sequence of carcinogenesis fit the evidence. He and Muller were not the only scientists who thought so. The idea had been embraced by cancer researchers as diverse as Katsusaburo Yamagiwa, Ernest Kennaway, and Wilhelm Hueper, each of whom knew from firsthand experience that inducing tumors in rabbits, mice, dogs, or other experimental animals required many toxic doses over long periods of time. They also knew that some individuals were especially vulnerable to carcinogenic exposures and that most victims were elderly. Those observations all suggested that there were multiple steps to malignancy and that the journey could take many years. Richard Doll theorized that carcinogenesis required six or seven mutations.[12] But that was just an educated guess, an estimation based on Doll's observation that the cancer death rate for the elderly was more than seven times higher than for young adults.

Knudson wondered whether he could push the multistage hypothesis beyond guesswork. He had read Theodor Boveri's groundbreaking 1914 book on mutation and cancer and was excited by a burst of recent findings that seemed to confirm many of Boveri's ideas. Genetics was in the midst of a revolution in the 1960s. James Watson and Francis Crick's success in elucidating the double helix structure of DNA in 1953 at last revealed exactly how hereditary information is stored in the chromosomes of living organisms, as Boveri had predicted sixty years earlier. Now researchers could peer inside the twenty-three pairs of chromosomes in every human cell and identify DNA segments, or genes, associated with specific traits and bodily functions—and malfunctions, too.

Many of the breakthroughs that followed concerned cancer. The most famous was the 1960 identification of a mutation known as the Philadelphia chromosome, named for the city where it was discovered. The bone marrow cells of 95 percent of adults with chronic myelogenous leukemia carried the telltale genetic defect. It was the first direct evidence that chromosome alterations preceded cancer and thus was another confirmation of Boveri's ideas. The Philadelphia mutation was a scramble: A chunk of DNA from Chromosome 9 swapped places with a chunk from Chromosome 22. What was puzzling was that not everyone whose cells had this chromosomal translocation got leukemia; many did not. To Knudson, that was additional evidence that more than one mutational event was required. The identification of the Philadelphia chromosome and other translocations associated with cancers would eventually lead to the discovery of oncogenes, the rogue mutations predicted by Boveri that promote the rapid cell division of cancer. But even the presence of an oncogene in a cell was not a surefire indicator that malignancy would result—one mutation was often not enough.

The multiple-hit theory of carcinogenesis made sense to Knudson, but his experience treating childhood cancer patients led him to doubt Richard Doll's suggestion that six or seven successive mutations were needed to trigger a malignancy. After all, some of Knudson's young patients were born with neuroblastoma; others developed leukemia as infants. Mutations were rare events; surely a baby had not lived long enough to take a half-dozen genetic hits. On the other hand, it did not seem to make sense that only one mutation was necessary. The human body produced one hundred billion white blood cells daily, yet leukemia was still a relatively rare condition. If just one mutation was required, why didn't everyone have it? With those questions in his head, Knudson began looking for a childhood cancer he could study to test his ideas about multiple hits.

He found retinoblastoma. Like Michael Gillick's neuroblastoma and Randy Lynnworth's medulloblastoma, retinoblastoma begins with the malignant transformation of precursor stem cells, in this case, in the eye. These retinoblasts produce the specialized light-sensitive cells that line the retina, making vision possible. A very rare disease, afflicting one in fifteen thousand children, retinoblastoma comes in two varieties: hereditary and sporadic. Hereditary retinoblastomas, about 40 percent of all cases, afflict children who have a family history of the disease, while sporadic cases do not. What caught Knudson's attention were three quirks that apply only to hereditary retinoblastomas. First, they sometimes skip a generation: A grandparent is afflicted, a parent is not, and then the disease returns in a grandchild. Second, kids with a family history of retinoblastoma usually develop the disease very early in childhood—even during infancy. And finally, children with hereditary retinoblastoma very often develop more than one tumor in one or both eyes.

Knudson developed a theory that could explain all of retinoblastoma's quirks. He called it the "two hit" hypothesis, and it was based on the idea that a retinoblast cell required two mutations to become malignant. Children with hereditary retinoblastoma, Knudson believed, were born with a mutation they carried in all of their cells and thus required just one more mutation "hit" in any retinoblast cell to develop a tumor. With about one hundred million cells per eye, there were many opportunities for that second mutation. In fact, it could easily occur in several cells in one or both eyes, triggering a separate tumor each time. The timing was interesting, too, because cell divisions, opportunities for mutation, occurred most often during the last months of fetal growth and the first few months after birth, when the eyes were developing most quickly.

The two-hit hypothesis thus explained why children with hereditary retinoblastoma tended to be younger and more likely to have multiple tumors. Children with sporadic retinoblastoma, on the other hand, got the disease only after an individual retinoblast cell suffered two mutational events—akin to lightning striking twice in the same place. Knudson's theory even explained why hereditary cases sometimes skipped generations, since the parent in the cancer-free middle generation carried the inherited mutation but luckily avoided the second hit. "It all fit together," Knudson recalled. "With these retinoblastoma kids, it just seemed obvious that inheriting this gene isn't enough to make the cancer. Something else has to happen. There has to be a second hit."

By the time he was ready to test his theory, in 1970, Knudson was working in Texas and had access to the medical records of forty-eight children treated for retinoblastoma. Twenty-three of them had tumors in both eyes, which meant that they were almost certainly hereditary cases. The next step was to set up a statistical test: If two-eyed cases required just one mutation while one-eyed cases needed two mutations, how many of each kind would be expected in a typical group of forty-eight children with retinoblastoma, and how old would each child be at diagnosis? Knudson calculated an expected distribution and then looked to see if they matched the actual results from the forty-eight cases. He was thrilled to discover that he got a hand-in-glove fit for all of his most important predictions. He had predicted, for example, that the average number of tumors among the two-eyed cases would be three—a prediction that fit the hospital data almost perfectly.[13] Knudson was right: Retinoblastoma was a two-hit cancer—the first multistep cancer with a fully plotted pathway to malignancy.

It is almost impossible to overstate the impact Knudson's discovery had on research into carcinogenesis, including the role of pollutants in places like Toms River. He would move on to Fox Chase Cancer Center in Philadelphia in 1976, continuing his cancer work and happily watching researchers around the world expand his findings in sometimes surprising ways.[14] The biggest surprise came in 1983, when the inherited "retinoblastoma gene" in the human genome was finally identified. The gene was on Chromosome 13, which is where Knudson had predicted it would be found back in 1976. The surprise was that it turned out not to be a villainous oncogene like the leukemia mutation created by the Philadelphia translocation. Instead, the newly named "Rb" gene was a hero—a new type of gene whose existence had yet again been foretold by Boveri seventy years earlier. It was a tumor-suppressor gene, and it was ultimately found to protect against not only retinoblastoma but also bone cancer and other malignancies. The problem for children with hereditary retinoblastoma was that their cells contained only one functional copy of the Rb gene instead of the normal complement of two, one from each parent. If a second mutation knocked out the remaining copy, then the affected cell would lose its ability to regulate its rate of replication. It would become a cancer cell.

The multi-hit model Knudson laid out for retinoblastoma became the dominant paradigm for carcinogenesis, the default assumption of how cancer begins. A few cancers required just one "hit," but most were now assumed to be the products of complex sequential processes that began with inherited mutations and continued with additional genetic "hits." If those subsequent mutations occurred at critical times, especially during fetal development or early childhood when so many cells were dividing, the results could be calamitous. That lesson would not be lost on the families of Toms River.

There was another, even more powerful message for Toms River in Knudson's work: Most of the chemicals that loomed large in the town's unhappy environmental history were mutagens—they were capable of altering DNA. As he developed his two-hit theory, Knudson did not speculate on the possible causes of the non-inherited mutations that were essential steps on the road to carcinogenesis. His mathematical model for retinoblastoma assumed that these mutations occurred randomly—due to collisions with cosmic rays, for example, or because of DNA copying errors during cell division. But Knudson and everyone else in the field knew that there were other sources of mutations. Hermann Muller had proved that X-rays were mutagens, and by the late 1970s, the California biochemist Bruce Ames—using the genes of bacteria, which were even easier to work with than fruit flies—had proved that dozens of known carcinogens, including many industrial chemicals, were also capable of mutating DNA.[15]

Just about any stray sample of soil, groundwater, or air from Ciba-Geigy or Reich Farm was likely to contain at least one mutagenic compound.[16] Many coal tar derivatives were mutagens, including benzene, benzo(a)pyrene, and benzidine. Epichlorohydrin, the resin feedstock that had so terrified Ciba-Geigy workers in the 1960s, was mutagenic too. Over at Reich Farm, meanwhile, the underground pollution plume seeping south toward the Parkway wells included several of the same mutagens, plus a few others.

For a town where mutagenic chemicals were now part of the landscape, the implications of Alfred Knudson's research were clear: If a substance was capable of delivering a "hit" to DNA, it was a cancer risk, and no one was more likely to take a hit—or two or three—in a short period of time than a fast-developing fetus or a young child. When Bruce Anderson and other Ocean of Love parents heard about that, they wondered how many hits their children had endured, and who had delivered the blows. Before long, the State of New Jersey and the United States of America would be spending many millions of dollars to try to find out.

Lisa Boornazian, the oncology nurse at The Children's Hospital of Philadelphia, was not a boat-rocker. In 1995, she was a twenty-four-year-old who loved her job and respected the hierarchy of the hospital. The doctors had told her it was just a coincidence that so many children on the ward were from Toms River, and she was inclined to take their word for it. They were very good doctors, and they gave their patients excellent medical care, even if they were too busy to get to know them. But she could not shake her uneasy feeling that the doctors might be wrong about Toms River.

She and her husband, Adam, came from large families and often got together on weekends with their siblings, many of whom still lived in the Philadelphia area. They were especially close to Adam's sister, Laura Janson, and her husband, Eric. A few weeks after Carrie-Anne Carter's funeral in Toms River in February of 1995, the two couples were having dinner on a Friday night, and somehow the conversation turned to the funeral and to Boornazian's worries about Toms River. The discussion was not something that she had planned. "It just sort of happened by accident," she would remember much later.

In their extended family, Laura Janson was an authority figure on environmental matters. She worked in the Philadelphia regional office of the U.S. Environmental Protection Agency, where she specialized in technical assessments of hazardous waste sites, though she had never worked on any in Toms River. After eleven years at the EPA, Janson was a bit jaded about information that came in from the public because it usually turned out to be confused, poorly documented, or otherwise unreliable. But this was different. Her sister-in-law was not an alarmist; she was a medical professional at a major hospital where thousands of children were treated every year. At dinner, when Boornazian started talking about all the sick children she had treated from Toms River, Janson listened. When Boornazian asked her to check to see if anyone at the EPA was looking into the issue, she agreed. Janson would later explain her decision this way: "When you work at EPA, people are always saying something's wrong with their water, but if it's your sister-in-law talking, and she's a nurse at CHOP who has made actual observations of cancer in children, you figure you'd better follow up."

Janson checked and learned that there were two Superfund sites in Toms River and that neither had been the subject of an EPA health study. Was the agency considering launching such a study in Toms River? No, it was not, she was told. So Janson decided to call another federal agency, one that few members of the public had ever heard of. The Agency for Toxic Substances and Disease Registry was, and still is, a backwater in the federal environmental bureaucracy. Congress created it in 1980, when anxiety about hazardous waste was at its zenith, just seven months after the evacuation of Love Canal. The idea was that the EPA would oversee dumpsite cleanups, while the ATSDR would advise the EPA on the health hazards posed by each waste site. But the ATSDR had very little money to do its job, especially as the number of Superfund sites ballooned in the late 1980s. By 1996, the EPA was spending well over one billion dollars per year (about 20 percent of its budget) on Superfund; the entire ATSDR budget, meanwhile, was just $60 million.[17]

Steven Jones came to the ATSDR in 1992 from the EPA, where he had worked on Superfund cleanups in the Midwest. His new title, deputy regional director, suggested sweeping authority; the reality was that the ATSDR's regional office, which occupied a small corner of the EPA's space in Manhattan, consisted of just two managers: Jones and his boss. Their main responsibility was to make sure that state and local environmental health agencies were doing the work the short-staffed ATSDR could not do itself. Only rarely did Jones field calls from the public about supposed cancer clusters; when he did, they usually ended in mutual frustration. Like Laura Janson, he had been around long enough to know that ordinary citizens rarely understood what constituted a true cancer cluster.

So when Steve Jones's office phone rang one morning in March of 1995 and the woman on the end of the line started talking about a possible cluster in Toms River, New Jersey, there was no reason to think it would be anything more than another dead-end conversation.

As Jones listened, though, he heard some things that caught his attention. For starters, the caller was not a typical anxious citizen. Laura Janson introduced herself as an EPA employee, explaining that she was calling unofficially to relay the concerns of an oncology nurse, her sister-in-law, at The Children's Hospital of Philadelphia. Unlike so many callers, Janson was not passing along a vague report in which all of the most common types of adult cancer—lung, breast, skin, and all the rest—were lumped together. She spoke instead about much rarer children's cancers, especially brain tumors. "What she was expressing to me was very specific," Jones remembered. "I felt like I wouldn't be doing my job if I didn't follow up. Someone needed to look at it."

Like Lisa Boornazian and Laura Janson, Jones knew almost nothing about the sordid history of chemical pollution in Toms River. Also like the two sisters-in-law, he was neither a rebel nor a starry-eyed idealist, and he was anything but a naive civilian. Jones, Boornazian, and Janson all worked inside sprawling, risk-averse bureaucracies. They had been around long enough to be a tad cynical about the outsiders, the amateurs who were always demanding investigations without knowing the facts. And yet, in this case, all three felt that they could not ignore what they were hearing. The numbers were too high to dismiss as coincidence, and each case represented a face, a child, and a family.

Under the circumstances, Steve Jones felt that he could not say no to the EPA employee who called him, just as Laura Janson felt that she could not say no to her sister-in-law, the oncology nurse. As for Lisa Boornazian, she, too, felt that she had no real choice. She had to speak up for the memory of Carrie-Anne Carter and all the other Toms River children who had come through the cancer ward. They deserved an answer.

So on March 13, 1995, Steve Jones called the New Jersey Department of Health and formally requested an investigation of childhood cancer in Toms River.

They sparked everything that came afterward in Toms River, yet Lisa Boornazian and Laura Janson quickly faded from view after Janson's phone call to Steve Jones in 1995. More accurately, the two women were never in view at all. Until the publication of this book, their identities were unknown to everyone connected with the Toms River saga except Jones and a few of Boornazian's fellow nurses at CHOP.[18] The ATSDR had a strict policy regarding calls from citizens seeking investigations: Callers had the right to remain anonymous. At the time, the two sisters-in-law did not want to be named, though they changed their minds many years later. Janson felt that she had nothing more to contribute, and Boornazian did not want to make waves at the hospital. "I felt like I had done my part, and now it was up to the experts," Boornazian explained later.

Over the years, the anonymous "CHOP nurse," as she was widely called, would acquire a sort of mythical standing in Toms River as the person who finally got the state's full attention. Yet no one ever knew who she was—not even Linda Gillick. In the absence of solid information, the nameless nurse was a blank slate upon which the various players in the Toms River drama drew the characteristics that suited them. Officials at the ATSDR asserted that the nurse had gotten interested in the Toms River cluster after attending an agency workshop on environmentally induced cancer held at CHOP. Some of the Toms River families, meanwhile, had heard that she became involved because she had relatives in town and was a mother herself. There was even a rumor that she had grown up in Toms River. None of it was true. Boornazian was a Philadelphia native with no children of her own yet. She had not attended any special workshops and had no connection to Toms River except through the families she met on the ward.

In the years that followed Janson's 1995 call to the ATSDR, she and Boornazian occasionally saw or heard snippets of news about the Toms River investigations: a stray newspaper headline here, a few seconds on the six o'clock news there. But life rolled on; they were too busy to pay attention to the extraordinary events they had initiated in Toms River. In November of 1995, Laura Janson gave birth to a son, Kevin. Lisa Boornazian's daughter, Amanda, was born the following January. Boornazian took a leave from the hospital, and then another one in 1997 when her second child was born. She quit for good in

1999. Boornazian had never stopped loving the job, but balancing the demands of motherhood and the oncology ward was too difficult—especially the emotional demands.

On her last day on the ward, she watched a mother climb into bed with her dying daughter. "She just held her little girl and said, over and over, 'What am I going to do without you? What am I going to do without you?' For me, that was it. Now that I had my own kids, it all just felt too real," Boornazian remembered. "It was time to leave."

Laura Janson, meanwhile, stayed at the EPA, working faithfully on Superfund and water pollution cases as they moved slowly through the bureaucracy. Thirteen years after her brief call to Steve Jones, she ruminated on its unintended significance: "I've been with the EPA for twenty-three years, and that one phone call may have been the biggest contribution I've made to the protection of public health and the environment," she said. "That was my moment, I guess."

CHAPTER FIFTEEN

≈

Cluster Busting

By the spring of 1995, when Steve Jones called to ask him to look into a possible cluster of childhood cancer in Toms River, Michael Berry had been New Jersey's chief cluster investigator for almost nine years. It was still just a part-time responsibility—Berry spent most of his time on other tasks at the state health department—but it was now the least enjoyable part of his job. One of the first "incidence analyses" Berry ever attempted was the 1986 study of childhood cancer in Toms River. Its ambiguous results turned out to be a harbinger of dozens of similarly unsatisfying cluster studies he undertook around the state—including another one about Toms River kids in 1991. "After a while, it got frustrating," he recalled many years later. "I mean, what were we accomplishing?"

A cluster study in New Jersey was like one of those old-fashioned Hollywood movie backdrops that looked fairly impressive until you leaned on it and it toppled over. Berry's frustrations ran much deeper than just the usual problems with the state cancer registry, which in 1995 was still in poor condition, its records incomplete and arriving three years late or longer. As he accumulated years on the job, Berry came to realize that even if the registry had been up-to-date and reliable, he would not have been able to tell callers what they really wanted to know: whether an environmental problem was causing cancer in their neighborhood. For reasons that were not easy to explain in a phone call, it was a question Berry could not answer. In fact, Berry spent almost as much time explaining the limitations of cluster studies as he did conducting them. Sometimes Berry felt his job was more about being a therapist to anxious callers—he got about thirty cluster calls per year—than investigating what they told him. "It was hard to believe I was really addressing anybody's concerns," he recalled.

Most of the people who called him had at least three major misunderstandings about cancer patterns, each of which led them to assume that all clusters had a hidden cause and that Michael Berry could unearth it—if only he tried hard enough. The first misunderstanding was the nature of clustering itself. As Berry knew, *everything* clustered to a degree, often for no reason other than chance. Nothing that was subject to the complexities of the natural world, whether birds in a flock or sick people in a city, was distributed evenly in space and time. Some clumping was inevitable. In cancer incidence studies, the challenge was not to find the clumps—that was usually pretty easy, thanks to the registry—but to identify which were likely to have an underlying cause other than randomness.

The second misunderstanding was about the ubiquity of cancer. In adults, it was a much more common condition than most people recognized. In the mid-1990s, there was one new case per year for every 230 New Jerseyans. A more striking way to think about that was that an American man faced a 44 percent chance of getting cancer at some point during his life; for women, the lifetime risk was 38 percent. With so many cases, it was inevitable that some neighborhoods would have surprisingly high concentrations of cancer—again, for no reason other than bad luck. "People just didn't realize how much cancer there is all over," Berry would later explain.

Finally, many of the people who called Berry to report a possible cluster assumed that cancer was a single disease instead of a catchall term applied to more than 150 distinct conditions. All cancers involved uncontrolled cell division triggered by genetic damage, but many had little else in common. Cervical cancer, for instance, was predominantly spread via sexual contact; including cervical cases in a residential cancer cluster study made little sense. On the other hand, focusing only on cancer types that had been plausibly linked to industrial chemicals—brain and blood cancers, for example—reduced the total number of cases in a cluster study and thus made it even harder to confidently identify nonrandom clusters. For a rare type of cancer, just one extra case in a neighborhood—raising the total from one to two cases, or from two to three—would be enough to make the neighborhood look like a hotspot, even though that one additional case could easily be coincidental.

By the time Berry finished clearing up those misconceptions about cancer and then moved on to the deficiencies of the state registry, with its out-of-date and incomplete records, many callers were so discouraged that they dropped their request for a cluster investigation. About half of the time, however, Berry's explanations did not satisfy a caller. In those cases—perhaps fifteen times a year—Berry would take the next step and conduct an incidence analysis, using registry data. The 1986 and 1991 analyses he conducted on childhood cancer in Toms River were typical. The analyses were simple comparisons between the number of known cases in a community and the number that "should" have occurred there based on the average incidence rate for all of New Jersey.

What frustrated Berry about those analyses was that their only real scientific value was as a first pass, a preliminary screening tool. They were a way to identify communities worthy of more sophisticated investigations that might include air, water, and soil tests as well as interviews to determine residents' past exposure to carcinogens. Yet his supervisors in the health department never authorized any follow-up work in neighborhoods, no matter what Berry had found initially. Identifying true pollution-induced clusters amid the sea of unlucky flukes, Berry discovered, was beyond the resources, expertise, and inclination of the State of New Jersey. If a community really did have significantly more cancer than expected—and over the years, Berry had found several communities that seemed to—he would confer with his supervisors, send a letter explaining his findings to the person who had asked for the study, and then . . . nothing. There was no next step, no follow-up. Just the letter explaining the anxiety-inducing results and reiterating what Berry had already told the caller in their first conversation: The cluster was probably due solely to bad luck—but no one could say for sure.

New Jersey was hardly alone in its resistance to conducting anything more than the shallowest cluster studies. Only a few states or countries with cancer registries did more; most did far less. As Toms River would soon find out, full-blown environmental studies were very expensive and highly controversial—and in the end usually failed to deliver clear-cut results. The Hollywood movie set approach to cluster investigation, on the other hand, had some distinct advantages for the state health department. That Michael Berry was doing something—*anything*, no matter how futile—allowed the state health commissioner to assure the governor and legislature that his department was responding to every cluster call it received. That was more than many health departments could claim, and it was a political imperative in New Jersey, where industrial pollution was a perennial campaign issue. Incidence analyses were inexpensive, and their ambiguous results did not create conflicts with the chemical industry and other powerful interests. They were a clever political solution—and a scientifically illegitimate one.

"That was the conundrum of doing these analyses," recalled Berry, who retired in 2010, after twenty-four years with the department. "The incidence analyses were something we could do, and the registry was a resource we could use, and arguably *should* use, to respond to people who have concerns. But what happens next? Even if I thought there really was a problem in a particular place, to do a follow-up study would be very expensive, and who knows if we would find anything? Besides, it wasn't like I was working in an environment where I could say, 'Look at this, maybe we should do something.' It became clear that what the organization preferred was that we just respond, because people complain if you don't respond. But that's it. Nothing else."

So when Michael Berry's phone rang on March 13, 1995, and Steve

Jones from the ATSDR was on the line requesting another investigation of childhood cancer in Toms River, Berry felt the familiar wave of frustration building.

There was a depressing logic behind New Jersey's faux approach to cluster investigation. It had its roots in a century-long quest to verify neighborhood cancer clusters scientifically—a quixotic effort that tantalized and ultimately frustrated everyone who attempted it, including some of the greatest statisticians of the twentieth century.

Back when most cancers were believed to be infectious—the triumphs of Louis Pasteur and Robert Koch in the late nineteenth century convinced many Europeans and Americans of that era that *all* diseases were transmissible—the existence of "cancer houses" plagued by high numbers of cases was taken for granted. Just like Linda Gillick a century later, Victorians of all social strata, from aristocratic reformers to the working poor, looked around their communities, saw that cancer was not evenly distributed, and assumed that a hidden cause must be at work. With rare exceptions, cancer was treated like a shameful plague; many people thought it was related to venereal disease, and others believed that its victims should be barred from hospitals as risks to public health.[1]

A few physicians tried to apply scientific scrutiny to the notion of "cancer houses." One of the most dedicated was an otherwise obscure Englishman named Thomas Law Webb.[2] He compiled the addresses of 377 people who died of cancer between 1837 and 1910 in the industrial town of Madeley. In 1911, Webb gave his data to the person in Britain best qualified to analyze it, a hot-tempered polymath named Karl Pearson. A man of breathtakingly broad interests— he was a philosopher, poet, songwriter, and novelist in his spare time—Pearson's greatest passion was the development of mathematical statistics as a full-fledged academic discipline and as a tool for solving social problems.[3] In 1911, the same year he analyzed Thomas Law Webb's cancer records from Madeley, Pearson founded the world's first academic department of applied statistics at University

College London, which became the global incubator of the nascent discipline of biostatistics.

There is no indication in the historical record of how Pearson found out about Webb's remarkable collection of cancer data, but it is easy to see why he would be eager to analyze it: Webb's records were a way for Pearson to use his new methods of statistical analysis to test the widely held belief in "cancer houses." He was especially interested in what came to be known as significance testing, or statistical significance.[4] The concept is simple: Any apparent pattern within a group of numbers, or apparent correlation between two or more groups of numbers, should be tested to determine how likely it is that the pattern or correlation is due to chance and not to some other cause.

The 377 Madeley residents who had died of cancer between 1837 and 1910 lived in 354 houses, according to Webb's records. To determine whether cases were clustering for reasons other than chance, Pearson first needed to estimate how many homes would have multiple cases if those fatal cancer cases were distributed at random. Using the statistical methods he had developed, Pearson calculated that if cancer were distributed randomly among the nearly three thousand residences in Madeley, there would be about 331 houses with one cancer death, twenty-two with two deaths, and one with three. But Webb's records showed there were actually 315 with one death, twenty homes with two, six homes with three, and one unfortunate home in which four residents died of cancer over the seventy-three-year period.

To a non-statistician, the two sets of numbers might not have looked very different. But to Pearson they were night and day. Could it just be a fluke? Not according to Pearson. "The probability that such a distribution could arise from random sampling is only one in many, many millions," he concluded after conducting a series of probability experiments.[5] Pearson thought that his provocative findings merited a comprehensive follow-up investigation, including comparisons to nonindustrial towns and a detailed breakdown of cases by age and occupation.[6] But there was no follow-up study. In that sense,

Pearson was the first of a long line of cluster hunters whose tantalizing tentative findings failed to attract the interest and resources needed
to confirm or refute them. He never published again on the topic in
his long career, which ended with his death in 1936.

The cluster studies that came afterward were similar cautionary
tales.[7] One of Pearson's protégés, the reclusive Percy Stocks, published
a more sophisticated analysis in 1935, two years after he was appointed chief statistician in the British General Registry Office, the
same position William Farr had held almost a century earlier. Stocks
picked up where Pearson left off by studying approximately 3,500 cancer deaths in the cities of Bristol and Worcester. He identified
ninety-four houses in which there were two deaths, when only
forty-four would be expected if cancer were distributed randomly. He
also counted an unexpectedly high number of cases in adjacent
houses, which seemed to support the notion that cancer was infectious or environmentally induced, or both.

But then Stocks took two more steps that would create so much
uncertainty in subsequent cluster investigations, including in Toms
River. Knowing that most cancer victims were over fifty-five, he wondered whether what appeared to be cancer clusters were actually just
clusters of older people. What he discovered was that the distribution
of people between age fifty-five and seventy-five in the two cities
showed just as much clustering as cancer cases did.[8] Then he analyzed
the data by type of tumor, reasoning that similar kinds of cancer
should cluster if they were infectious or environmentally triggered.
Dividing the Bristol and Worcester cancer deaths into seven groupings, based on the organs in which the primary tumors occurred, he
found that none of the groupings showed a tendency to cluster, even
if cancer cases overall seemed to do so. What looked like evidence of
carcinogenic infection or pollution almost certainly was not, Stocks
concluded. He would remain a skeptic of cluster studies for the rest of
his career, and his doubts would strongly influence his close colleagues
Richard Doll and Austin Bradford Hill as they launched their seminal
studies on cigarette smoking and lung cancer.

After Percy Stocks, every respectable cluster investigator, including
in Toms River, had to account for the influence of age and had to look

at specific types of cancer. The latter requirement was especially oner-ous for researchers studying clusters of rare diseases because it meant that they had even fewer cases to work with and therefore less confi-dence that the clustering was not caused by chance. For example, if just two cases of cancer of the larynx were expected in a community over a ten-year period, did the appearance of four cases really consti-tute an alarming excess of 100 percent? If an investigator saw thirty cases instead of the expected fifteen it almost certainly did, but what about four cases instead of two? A small cluster like that could easily be a random blip, but there was also a chance that it was a sign of something much more serious. A cluster analysis, however sophisti-cated, could not differentiate between the two.

By the 1960s, biostatisticians like Percy Stocks had turned cancer clus-ter analysis into a well-structured scientific endeavor that almost no one wanted to pursue. Their insistence on statistical significance tests and age-adjusted data and their refusal to lump all cancers together in a single analysis legitimized the study of residential clusters while si-multaneously making it seem like a waste of time. Cluster studies now produced results that were scientifically credible but hopelessly ambiguous. At a time when Richard Doll was showing the exciting potential of population-wide studies in identifying risk factors for common cancers, the study of small, location-specific clusters in-creasingly looked like a dead-end pursuit, a statistical trap from which it was impossible to emerge with useful information.

There was still one major branch of epidemiology that remained very interested in cluster studies, and it was the same branch that had pioneered them one hundred years earlier. Infectious disease spe-cialists conducted cluster studies all the time, looking at incidence patterns for tuberculosis, measles, and other highly transmissible dis-eases. In the United States, the leading institution for such studies was, and still is, the agency now known as the Centers for Disease Control and Prevention, based near Atlanta, Georgia.[9] The CDC had made its reputation studying malaria, but health departments would send it reports of other disease clusters, too, in hopes of enticing the agency to investigate.

One of the most compelling of those reports originated with Sister Mary Viva, the principal of the Saint John Brebeuf elementary school in Niles, northwest of Chicago. During the first three months of 1961, four young girls from Niles—which had a population of just twenty thousand—died of leukemia. Two were students at Saint John Brebeuf; a third was a preschool-age sibling of a student. The CDC had never conducted a cancer cluster study before, but by 1961 evidence was accumulating that leukemia might be infectious: Researchers had identified viruses associated with leukemia in cats and dogs. A young CDC investigator, Clark W. Heath Jr., was dispatched to look into the reported cluster. He discovered it was twice as large as Sister Mary had realized: There were eight childhood leukemia cases diagnosed in Niles between September of 1957 and August of 1960— a rate almost five times higher than expected.[10] Seven of the eight were either students at Saint John Brebeuf or their younger siblings, which meant the incidence rate for families at the school was more than eight times higher than for other Niles families.

Heath and his collaborators spent more than a year working on the investigation. They mapped the home addresses of the victims, measured radiation levels, looked for local pollution sources, interviewed almost five hundred families, collected statistics for neighboring towns, and even looked to see if there was an unusual number of local feline or canine leukemia cases. Nothing clicked. They found no common factors among the cases other than their association with the school, though there were some intriguing hints about viral infection in Niles, which was growing very quickly and absorbing wave after wave of newcomers, just as Toms River was.[11] Heath would go on to investigate fifty clusters of childhood leukemia and lymphoma, finding seven others besides Niles in which there were indications of infection but no corroborating evidence.[12] "We were not able to find the decisive evidence we were looking for," he remembered many years later.

The federal government's first foray into cancer cluster investigation had accomplished very little, yet by the late 1970s the CDC was conducting more cluster studies than ever and was increasingly focusing on chemical pollutants, not viruses—all because of demands from citizens and politicians. Publicity over Love Canal and other environmental disasters had sparked a boom in requests for cluster investigations, especially in states that had cancer registries.[13] State health departments were fielding about fifteen hundred such requests per year.[14] The most worrisome of those requests—the ones with a plausible suspect cause and rates high enough to make random variation an unlikely explanation—were passed on to the CDC, which by the 1980s was conducting an average of five or six cluster investigations each year. Hundreds more were at least crudely investigated by state health departments, as Michael Berry was doing in New Jersey.

The complaint-driven genesis of almost all of those cluster investigations was turning out to be a profound weakness, and not just because anxious members of the public often reported cancer patterns that turned out to be unexceptional. There was a deeper issue that could not be solved by the clever use of incidence comparisons and statistical significance tests. This was the problem of hidden multiple comparisons.[15] The case-control studies popularized by Richard Doll in the 1950s were scientifically elegant not only because they were large enough to reduce statistical uncertainty but also because they began with a hypothesis. Doll wanted to test the proposition that smoking was a risk factor for lung cancer, so he assembled a large group of cases and compared them to a similar but cancer-free control group. Most cluster studies, by contrast, turned deductive science on its head. Instead of starting with a testable cause-and-effect hypothesis, they began with someone cherry-picking a suspicious cluster of cases out of a much larger population.[16]

For example, when Lisa Boornazian in Philadelphia confided to her sister-in-law that she had noticed an unusual number of sick children from Toms River on the oncology ward, she was making an unstated comparison to the hundreds of other communities that sent patients each year to the ward. Within such a large comparison group, sheer chance could easily explain why several towns—including Toms River—were overrepresented in the ward's patient population. Similarly, when Sister Mary Viva became concerned about a three-month period in Niles, Illinois, when leukemia was diagnosed in four local girls, she was making an unspoken comparison to dozens of other three-month periods, and dozens of other diseases, during her years as school principal.

In other words, out of the total universe of cancer cases—distributed throughout an almost infinite number of communities and time periods and encompassing more than 150 types of cancers—only an extremely small percentage of case aggregations were ever reported to a government agency as a cluster. With so much pre-screening, it was not surprising that many of the reported clusters were able to pass a test of nonrandom statistical significance, suggesting that they had some hidden cause. But did they really? Perhaps many of these clusters—maybe even *all* of them—were actually random, since only the flukiest of case aggregations were referred to investigators in the first place.

Investigators of workplace clusters could console themselves with the knowledge that even if they could never know for certain whether an apparent cluster was "real" or random, there was a good chance that their efforts would not be wasted. Chemical exposures in factories were high enough, and specific enough, that a cancer incidence study could help identify possible chemical suspects whose risks could then be confirmed—or refuted—through tests on lab animals and in case-control studies. Indeed, many industrial carcinogens—including asbestos, benzidine, and vinyl chloride—were all first identified as potential cancer-causers in occupational cluster studies. Residential cluster studies, on the other hand, had a spotless record: Not a single one had ever led to the identification of a new carcinogen.[17] Between 1961 and 1983, the CDC completed 108 residential cancer cluster investigations and failed to identify a likely cause in any of them.[18] Workplace cluster studies had a better record, but only slightly.[19]

Even when governments made extraordinary efforts to confirm a reported neighborhood cluster via environmental testing, the results were ambiguous. That was certainly true of what was the most famous and carefully documented residential cancer cluster of the era:

the twelve cases of childhood leukemia in Woburn, Massachusetts, where just five cases would have been expected based on the demographics of that blue-collar town north of Boston. Later, in the 1990s, the Woburn cluster would become famous (and very influential, in Toms River) because of the book and movie *A Civil Action* and because of a state study that found an association between childhood leukemia and mothers who drank contaminated water—an exceedingly rare cause-and-effect confirmation of a residential cluster. But in the 1980s, two smaller studies in Woburn—one conducted by government scientists, the other by biostatisticians working with the affected families—looked at the leukemia–drinking water hypothesis and came to opposing conclusions.[20]

By the late 1980s, there was no avoiding the unsettling conclusion: Neighborhood cancer cluster studies appeared to be a fool's errand, a source of perpetual embarrassment to the agencies that conducted them and the politicians who had to defend their unsatisfying results. In fact, a rough consensus was emerging among cluster researchers in state health departments and the CDC: Governments should get out of the business of investigating residential cancer clusters, no matter how vociferously the public demanded them. To lay the groundwork for such a controversial policy change, they organized a meeting at the Hotel Intercontinental in Atlanta, near the CDC headquarters. The 1989 gathering was officially known as the National Conference on the Clustering of Health Events, but it quickly acquired a much catchier name: the cluster buster conference.

To deliver the opening address, the organizers selected a paragon of the epidemiology establishment. Kenneth Rothman of Boston University had written two popular textbooks and was the founding editor of the journal *Epidemiology*. He got right to the point: "I am about to tell you that there is little scientific value in the study of disease clusters," he bluntly told the assembled scientists, some of whom—including Clark Heath—had spent their professional lives doing just that. "With very few exceptions, there is little scientific or public health purpose to investigate individual disease clusters at all."[21] Many of the researchers who followed Rothman at the podium agreed, especially for residential clusters. But they all acknowledged struggling with the consequences of ignoring requests for investigations. As one of the most experienced cluster investigators, Alan Bender of the Minnesota Department of Health, later told *The New Yorker* magazine: "Look, you can't just kiss people off."[22] Instead, he suggested a step-by-step response system that emphasized establishing a rapport with worried callers. Seventy-five percent of the time, he reported, "one or two telephone calls and a follow-up letter will satisfactorily answer the caller's concerns."[23]

The cluster buster conference had a powerful effect. Just months after it ended, all investigations of non-occupational cancer clusters in the United States had stopped, with very few exceptions. The CDC issued guidelines urging states to adopt Minnesota-style systems and ended its own cluster investigations, at least for a while.[24] "The state health departments didn't want to do these cluster investigations anyway, and now they could stop and say they were just doing what the CDC wants," remembered Daniel Wartenberg, a New Jersey epidemiologist who attended and argued in vain against the majority view. Instead, Minnesota's Bender carried the day with his categorical dismissal of cluster studies. "The reality," he told *The New Yorker,* "is that they're an absolute, total, and complete waste of taxpayer dollars."

There were other, less obvious ripples radiating from the conference. Perhaps the most far-reaching was the effect that it had on the attitudes of those who attended. New Jersey's Michael Berry was in the audience as Kenneth Rothman and then Alan Bender spoke. By the end of the conference, he knew what some of the biggest names in epidemiology were saying about cluster studies. Berry had been on the job for less than three years in 1989, and while he did not quite embrace Rothman and Bender's extreme position, their overall message resonated with him. It was consistent with his own frustrating experiences analyzing neighborhood clusters in New Jersey.

Michael Berry was not going to stop taking calls from citizens or occasionally conducting incidence analyses—his supervisors at the state health department would not let him stop even if he wanted to.

This was New Jersey, after all, the Superfund capital of the nation and a place where environmental health was a perennial political issue. Berry received more cluster calls every year than his counterparts in every state but New York and California, which were much more populous.[25] In New Jersey, callers reporting clusters could not just be ignored. But now some of the most prominent cluster-hunters in the world were confirming Berry's own doubts about what he was doing.

The request Michael Berry received on March 13, 1995, for another investigation of childhood cancer in Toms River sounded to Berry like another exercise in cluster-hunting futility: a vague complaint, a small community, very few cases of cancer and no obvious culprits—at least, as far as Berry knew at the time. Yet he did not try to talk Steve Jones into withdrawing his request. Jones was not an ordinary citizen. He worked at the ATSDR, and he was passing along a complaint from another authority figure, an oncology nurse in one of the most prestigious children's hospitals in the world. Just as importantly, Toms River was not just another community. By 1995, the logbook in Berry's office showed that the state health department had received five calls about childhood cancer in Toms River. The first three—in 1982, 1983, and 1984—were not followed up, but the 1986 request from Chuck Kauffman and the 1991 request from Robert Gialanella had each prompted Berry to undertake an incidence analysis, the second of which revealed that pediatric brain tumors and leukemias seemed to be on the rise during the late 1980s, even if the increase was not large enough to be statistically significant.

There was another worrisome factor, too. The state health department had just completed a study comparing childhood cancer incidence in New Jersey's twenty-one counties. The 1994 analysis found that from 1980 to 1988, the overall childhood cancer rate in Ocean County was well above the statewide average.[26] That troubled Berry, and it bothered him even more that the rates in Ocean seemed to be especially high for the category of cancers that Robert Gialanella and others had been most concerned about: brain tumors. Thirty-seven

Ocean County children under age fourteen had been diagnosed with brain and nervous system tumors between 1980 and 1988, when the overall rate for New Jersey suggested there should have been just twenty-two. In a county with eighty thousand children, that was 70 percent more than expected. And now Steve Jones was telling him that the Philadelphia nurse was especially concerned about brain tumors in Toms River kids.

Berry set aside his reservations and told Jones that he would look into it.

≈

CHAPTER SIXTEEN

Moving On

Almost no one in Toms River was paying attention as Michael Berry prepared to conduct the first comprehensive analysis of cancer in their community—four long decades after residents first voiced worries that chemical pollution was making them sick. The long fight over the future of the Ciba-Geigy factory and its ocean pipeline had been so all-consuming that its resolution seemed to affect the town like the breaking of a prolonged fever followed by a deep, exhausted sleep. The dye and resin jobs had fled south to Alabama and Louisiana or across the ocean to Asia. The workforce was down to only about three hundred people, and the trickle of wastewater the plant still produced—about 1 percent of its former total—was now flowing into the municipal sewer system. (Ironically, that trickle of treated wastewater still ended up in the Atlantic because the Ocean County Utilities Authority was still quietly operating its three ocean outfall pipes for treated domestic sewage, even though the factory's pipeline had been shut down.)

Ciba-Geigy was doing all it could to prolong the slumber. In 1992, the company reached a tidy resolution of the long-running criminal case against it. More than six years after they were originally indicted, the company and former executives James McPherson and William

Bobsein pleaded guilty to reduced misdemeanor charges of illegally dumping hazardous liquids and other banned wastes into the factory's lined-but-leaky landfill between 1981 and 1984. Avoiding jail time, the two men were fined $25,000 each, and the company agreed to pay $9 million in civil and criminal penalties, reimburse the state for more than $2 million in expenses, and donate $2.5 million for local environmental projects.[1] The plea deal allowed the company to keep claiming that its illegal actions were unintentional, because the state dropped the more serious charges that Ciba-Geigy had deceived regulators by filing false reports and altering records. "In settling with the state, we take responsibility for mistakes that were made at Toms River many years ago. We apologize for them," said Richard Barth, chairman of Ciba-Geigy's United States subsidiary, after the guilty pleas in Trenton. "Fortunately, no harm to health or the environment has resulted."[2]

Not everyone in town believed that. Linda Gillick kept hearing about more kids with cancer and inserting more pushpins into her map, which now hung at the Ocean of Love office. She still tangled occasionally with county officials—objecting to the use of asbestos in water pipes, for example—but Gillick and her board had more or less given up on asking for cancer studies. Besides, Ocean of Love was busier than ever providing services to families of stricken children. There were more families than ever to look after.

Gabrielle Pascarella was born on February 4, 1989, the year most people in Toms River thought their environmental troubles were over. Her parents, Kim, a lawyer, and Linda, a teacher, brought Gabrielle home and introduced her to her two big sisters, aged eight and four. Gabrielle was a beautiful baby but her pediatrician was mildly concerned about the large moles on her back, called nevi. The marks probably meant nothing but could be a symptom of a more serious condition. A surgeon removed the nevi, but Gabrielle cried a lot and seemed prone to infections. In early December, when she was ten months old, her parents felt a strange hardening of her fontanelle, the soft spot atop an infant's skull. The bulge indicated that Gabrielle's brain was under pressure. She was initially diagnosed with meningitis but large doses of antibiotics did not work, so the Pascarellas, like the

Andersons and many other Toms River families, made the long drive through the Pine Barrens to see specialists at The Children's Hospital of Philadelphia. On December 24, they got a diagnosis: malignant neurocutaneous melanosis, an exceedingly rare cancer of the meninges, which are the membranes that envelop the brain and spinal cord.

"The doctors told us there was nothing they could do other than make her comfortable," Kim Pascarella remembered. "They told us it was a terminal case." The Pascarellas had no illusions about the likelihood of success, but they wanted to keep trying. They found a doctor at Memorial Sloan-Kettering who was trying an experimental therapy, which Gabrielle started on her first birthday. Linda Gillick, who had made dozens of trips to Sloan-Kettering already, made another one to see the Pascarellas. They were grateful for her support, even as Gabrielle's condition worsened. Diagnosed on Christmas Eve, Gabrielle died on the day before Easter. She had lived fourteen months.

It was hard to believe that life could go on, but Ocean of Love and the fellowship of other families helped. The Pascarellas hosted the group's annual "family reunion" at their home, and some early fund-raising dinners were held at a restaurant they owned. "We decided to make it a family project to stay involved," Kim Pascarella remembered. "We saw what Linda was doing for these families, and we wanted to be a part of it." Soon the annual reunion was too large even for the Pascarellas' spacious house, and the annual dinner was too big for their restaurant. That was in part a tribute to Linda Gillick's fundraising acumen, but it was also because there were more cases every year. Cancer seemed to keep finding the families of Toms River.

The steady accumulation of cases bothered Kim Pascarella, just as it bothered Linda Gillick and Bruce Anderson. Each diagnosis was a deeply personal tragedy that could only truly be understood by the other families who had been through it. But the large number of cases also seemed to have a *collective* significance, Pascarella thought. "It was hard to put your finger on it," he remembered, "but there was just something in your gut that said this just wasn't right." Pascarella's law practice included part-time work for the town; he was used to addressing problems in the community. Childhood cancer in Toms River, he thought, was starting to look like more than just a series of indi-

vidual calamities; it was growing into a community problem that
needed a community response, though he had no idea what that response should be.

Having just agreed to conduct a study of childhood cancer incidence
in Toms River, Michael Berry needed to figure out where Toms River
was. That was a harder task than it seemed. There was no "Toms
River," strictly speaking. The Toms River region was a crazy quilt of
overlapping jurisdictions stitched together since colonial times. Ever
since 1768, when King George III issued a royal charter establishing
the town, it had been officially known as Dover Township, but no one
ever said that they were from Dover. The sprawling school district
took the Toms River name, and the post office gave a large swath of
the region, including parts of three adjacent townships, a Toms River
address.

This confused history was not a trivial issue. Before he could figure
out whether the rate of childhood cancer was unusually high in Toms
River, Berry needed to know how many people lived there. Because
Toms River was the best-known place-name in Ocean County, some
families who told nurse Lisa Boornazian they were from Toms River
may have actually lived in, say, Brick or Berkeley. Berry decided to deal
with this by focusing his study on children diagnosed with cancer
while living in three overlapping geographic areas. The first two were
straightforward enough: Ocean County and the township boundar-
ies. Then, to represent the town's "core" section, Berry selected four
census tracts covering about six square miles, coinciding roughly with
the area the U.S. Census Bureau called Toms River.[3]

Berry also had to decide which cancers to study. If he counted all
childhood cancers together, he would be perpetuating the fiction that
cancer was a single disease. On the other hand, counting each type of
cancer as its own category was unworkable because there would be
too few cases. Fewer than five thousand children lived in the Toms
River "core"; a community that small would typically have just one
case of childhood cancer—of any type—per year.[4] The incidence rate
for any specific cancer type would be much lower still. In the Toms
River core, even over many years, there might be just one or two cases

of any particular type of childhood cancer—so few that just one additional case would make a huge difference statistically, yet could easily be just a chance event.[5] Berry settled on an imperfect compromise between medical legitimacy and statistical validity: He would sort childhood cancer cases into fifteen groupings of similar diseases, plus an "all childhood cancers" group. Even though the groupings were large—perhaps too large to be medically defensible—most would generate only a handful of cases in the core zone. It was a compromise approach, one that Berry thought offered at least a chance of figuring out whether there really was something unusual going on in the town.

For all sixteen categories but one, Berry included all cases in children under age twenty, since Boornazian had said she was worried about young children and teenagers. But he decided to treat brain and central nervous system cancers differently by carving out a special subcategory for those tumors in children under age five. He did so because Boornazian had reported seeing a lot of brain tumor cases from Toms River and because a statewide study conducted the previous year had identified Ocean County as having a sky-high rate of childhood brain cancer: 70 percent higher than expected between 1980 and 1988. If pollution really were the cause, he reasoned, then very young children would probably be affected most, since there was solid scientific evidence that fetuses, infants, and toddlers were especially vulnerable to chemical exposures.

The last big issue Berry needed to face was time. Which years would he examine? On this question, he was entirely dependent on the flawed state cancer registry. There had been a few improvements, especially in reporting by out-of-state hospitals in New York City and Philadelphia, but by 1995 the registry was running four years behind, the most out of date it had ever been. The delay meant Berry's analysis could not include cases diagnosed after 1991 or before 1979, which was the first year of fairly complete registry data. (It was also the year of Michael Gillick's birth, which meant that he would be one of the first cases included in Berry's analysis.) The only way Berry's study could be both historically comprehensive and up to date would be to include years outside the 1979-to-1991 window of the registry—but that idea was too impractical to take seriously. Who would pay the huge cost of digging up records in hospitals and doctor's offices to find reliable data for 1975, or 1995, or any other year the registry did not cover?

The upshot was that if the deluge of industrial chemicals dumped and burned in Toms River during the 1950s and 1960s had triggered a cluster of childhood cancer in those years, Berry's analysis would not be able to discern it because he had no information about cases diagnosed before 1979. Even worse, because of the four-year time lag at the registry, Berry would not be able to address, even indirectly, the question that so alarmed Linda Gillick and Lisa Boornazian: Was there *still* a cancer cluster in Toms River?

Having set the parameters of his study, Berry was ready to begin. He used the cancer registry to identify the birth address of every child under age twenty who had been diagnosed with cancer between 1979 and 1991 while living in Ocean County. Then he consulted local maps to double-check all of the addresses, making sure that they were classified correctly. Finally, he added up the cases, categorized the cancers, and laid out the results in a table. The one for the town and the core zone looked like this:[6]

	Male	Female	Total
Township			
Bone and Joint	2	1	3
Brain/Central Nervous System	8	3	11
Endocrine	1	1	2
Eye	0	1	1
Hodgkin's Disease	2	1	3
Leukemia	8	8	16
Oralpharynx	1	0	1
Other digestive	0	1	1
Other respiratory	1	1	2
Ovary	—	2	2
Non-Hodgkin's Lymphoma	4	0	4

	Male	Female	Total
Township			
Prostate	1	—	1
Renal	0	2	2
Skin	1	1	2
Soft Tissue	1	3	4
Testis	1	—	1
Total:	31	25	56
Toms River core zone			
Brain/Central Nervous System	3	2	5
Endocrine	1	1	2
Eye	0	1	1
Hodgkin's Disease	1	0	1
Leukemia	2	2	4
Other respiratory	1	0	1
Total:	8	6	14

The data table confirmed what Berry already knew: There were precious few cases to work with. There was no point in even trying to analyze fourteen categories—everything but brain/central nervous system tumors, leukemias, and overall cancers, he decided. Even in those three relatively large categories, however, there were still so few cases that even if the totals turned out to be much higher than expected, he might not be able to rule out bad luck as a likely cause, especially if boys and girls were counted separately.

To find out if the local totals really were high, Berry calculated the number of pediatric brain cancers, leukemias, and overall cancers that would be expected in the county, town, and core zone if their rates were identical to the statewide average of all New Jersey children.[7] Then he updated his results table, calculating simple ratios that expressed the relationship between observed and expected cases in each category. (Any "incidence ratio" over 1.0 was higher than expected.) Finally, he added the special category he had decided to include for brain/nervous system cancers in children under five, as well as the countywide totals. The new table looked like this:

	Age	Observed	Expected	Incidence Ratio
COUNTY:				
All Cancers	0–19	230	215.5	1.07
Brain/CNS	0–19	54	38.8	1.39
	0–4	21	12.0	1.75
Leukemia	0–19	60	53.4	1.12
TOWNSHIP:				
All Cancers	0–19	56	42.7	1.31
Brain/CNS	0–19	11	7.6	1.45
	0–4	5	1.9	2.64
Leukemia	0–19	16	10.2	1.58
TOMS RIVER CORE ZONE:				
All Cancers	0–19	14	9.4	1.49
Brain/CNS	0–19	5	1.6	3.05
	0–4	3	0.4	7.14
Leukemia	0–19	4	2.2	1.80

All it took was a quick glance at the results table for Berry to see that there was nothing typical about these children. In every remaining category, they had more cancer than expected. Just as importantly, all the remaining categories showed the same bull's-eye pattern: Whatever mystery factor was affecting cancer rates seemed to be strongest in the heart of Toms River. For overall childhood cancer cases, for example, there were 7 percent more cases than expected in Ocean County, 31 percent more in the township, and 49 percent more in the core area. Most disturbingly, the biggest disparity was in the category thought to be the best indicator of a potential environmental problem: a sevenfold excess (three cases instead of the expected 0.4) for brain and nervous system cancers in children under age five in the Toms River core zone.

The message in the numbers seemed clear: Linda Gillick was right, and so was Lisa Boornazian. Something unusual really *was* happening in Toms River—or at least had happened between 1979 and 1991. For Berry, there was only one remaining question, and it was an exceedingly difficult one to answer: Could it all just be due to random variation, to a run of very bad luck? He had attended the cluster buster conference in 1989, and he knew how misleading apparent clusters could be. Now he would need to follow in the footsteps of Karl Pearson and the other biostatisticians who had confronted the same problem by performing tests of statistical significance. Berry needed to know how confident he could be that chance was not the cause of the cancer patterns he had identified in the county, the township, and especially in the heart of Toms River.

The significance test Berry employed was one of the most widely used in epidemiology: a 95 percent confidence interval, very similar to a margin of error in an opinion poll (though not quite the same). Pollsters employ margins of error because the fewer people they poll, the less confident they can be that the results accurately represent the sentiments of the larger population. To account for this uncertainty, statisticians apply a formula—its basics were first worked out by Siméon Poisson in the mid-nineteenth century—that assesses a poll's accuracy based on the number of people polled and the size of the larger population those people are supposed to represent. Instead of expressing the results as a single number ("55 percent of voters approve of the president's performance"), pollsters can apply the formula and express the results as both a number and a range ("55 percent approve, with a margin of error of plus or minus 3 percent"). The wider the margin of error, the less reliable the result. Usually, these ranges are based on a 95 percent confidence level, which means that if the poll were conducted the same way twenty times, the result would fall within the margin of error every time but once.

Cancer rates fluctuated by chance, just as opinion poll results did. Rates were especially wobbly in small communities and for rare diseases. For childhood brain and nervous system cancers in the Toms River core zone, for example, the incidence ratio was 3.05—three times higher than expected. But if there had been just three fewer cases over the thirteen-year study period—a variance that was quite possible for chance reasons alone—the ratio would have been only 1.25, barely an excess at all. On the other hand, just three more cases would have hiked the incidence ratio all the way up to a truly alarming 5.0. Those random fluctuations were the "noise" that made it so difficult to identify the "signal" of a nonrandom cancer cluster. By calculating 95 percent confidence intervals for each incidence ratio, Berry could assess how confident he could be about his results. The tighter the interval, the more confident he could be. And if the entire interval was above 1.0, then Berry could reasonably conclude that there really was more cancer than expected, for reasons other than random fluctuation. His result, in other words, would be statistically significant. The problem was, for a study of rare cancers in an area as small as the Toms River core, Berry would need to find a staggeringly high excess of cases to avoid an interval that was hopelessly wide and dipped below 1.0.

With all that in mind, Berry calculated 95 percent confidence intervals for each of his categories, taking special note every time that a confidence interval was entirely above 1.0. And then, one last time, he revised his results:

	Age	Observed	Expected	Incidence Ratio	95% C.I.
COUNTY:					
All Cancers	0–19	230	215.5	1.07	0.93–1.21
Brain/CNS	0–19	54	38.8	1.39	1.05–1.82
	0–4	21	12.0	1.75	1.08–2.68
Leukemia	0–19	60	53.4	1.12	0.86–1.45
TOWNSHIP:					
All Cancers	0–19	56	42.7	1.31	0.99–1.70
Brain/CNS	0–19	11	7.6	1.45	0.72–2.60
	0–4	5	1.9	2.64	0.85–6.17
Leukemia	0–19	16	10.2	1.58	0.90–2.56
TOMS RIVER CORE ZONE:					
All Cancers	0–19	14	9.4	1.49	0.81–2.50
Brain/CNS	0–19	5	1.6	3.05	0.98–7.11
	0–4	3	0.4	7.14	1.44–20.87
Leukemia	0–19	4	2.2	1.80	0.48–4.61

These results were precisely the kind that regularly drove cluster investigators nuts. Every number in the "incidence ratio" column carried the same message: Something was wrong in Toms River. But the numbers in the next column, the one that showed the 95 percent confidence intervals, muddled that message in every possible way. All twelve confidence intervals were wide, especially in the township and the core, where the case numbers were lower. This meant that luck could be having a large influence on those ratios, each of which could easily be much higher or lower than the ratio indicated. And in all but three categories, the lower bound of the confidence interval was below 1.0, which meant that there might not be a problem at all. Could it be, for example, that if the effects of chance were eliminated, the Toms River core might have 50 percent *fewer* childhood leukemia cases than the statewide average, instead of 80 percent more, as Berry had calculated? Yes, it could, since 0.50 lay within the 95 percent confidence interval. But there was also a plausible chance that the leukemia rate among Toms River children was actually almost five times higher than expected, since the upper bound of the confidence interval was 4.61. The best that Poisson's mathematics could confidently predict was that the true, nonrandom, actual-to-expected case ratio almost certainly lay somewhere within the gaping chasm of 0.48 to 4.61. (Actually, Poisson—and Berry—could do a little better than that, because not all values in between 0.48 and 4.61 were equally likely to be the true risk. When graphed, confidence intervals form a bell curve that peaks at the calculated ratio—in this case, 1.80. So, if forced to pick just one number, Berry's best guess would be that leukemia risk for Toms River children in the core zone was 80 percent higher than expected. But he could not be *confident* about that guess. He could confidently predict only that the true risk for Toms River children lay somewhere between 52 percent lower than expected and 461 percent higher than expected.)

It was a distinctly unhelpful prediction, as if a weather forecaster had studied the radar, measured the temperature, humidity, and wind, and then declared that tomorrow's weather would be either hot or snowy or—the best guess—something in between. There was no way to know whether to wear a sunhat or a parka.

The New Jersey Department of Health, like most other public health agencies, had already established a clear precedent on how to handle this kind of uncertain finding: Ignore it. A result was not credible, and therefore not worthy of further attention, if its 95 percent confidence interval crossed under 1.0, no matter how slightly. It did not matter that an interval of, say, 0.98 to 7.11 (the range for brain cancers in Toms River children) meant that the cancer rate was *far* more likely than not to be much higher than expected.

There were good reasons for New Jersey's conservative approach. For one thing, it helped to minimize the distorting effects of hidden multiple comparisons. There were almost two thousand census tracts in New Jersey, and therefore tens of thousands of groupings of four contiguous tracts that Berry could have studied. But Berry had counted cancer cases in only one such grouping, the four census tracts he designated as the Toms River core. If he had surveyed the entire state, the sample would have been so large that Berry would almost certainly have found many groupings of four contiguous census tracts that had high numbers of cases for no reason other than sheer chance. The 95 percent standard was a rigorous check against such a coincidence. It was a high hurdle, and almost all the confidence intervals Berry calculated failed to clear it.

But was the hurdle too high? The 95 percent significance test was designed for use when just one statistic was being analyzed in isolation from all others. So while there could easily have been other groupings of four census tracts in New Jersey that by chance could have had very high numbers of brain and central nervous system cancers, Berry's analysis showed that the Toms River core zone was unusual in other important ways, too. It had an unusually high number not only of brain cancer cases but also of leukemias and all childhood cancers combined—and so did the township and the county. In fact, Berry found elevated cancer rates in every single category—all twelve of them—in which there was any hope of a statistically valid result. He had never been involved in a cluster study in which every single category was elevated, and this one was in a town with a notorious history of chemical pollution, as Berry was learning. Finally, three of those categories—brain and nervous system cancers in county children under age twenty, county children under age five, and "core zone" children under five—had managed to clear even the very high hurdle established by the state's rules and were thus statistically significant. Could *all that* be just an extremely unlucky series of coincidences?

The only way to find out would be for Berry to go beyond the state's standard protocol for incidence analyses, something he had never done before for a residential cluster. He would have to *investigate,* not just calculate.

Over at the chemical factory, the few workers who remained were more worried than ever about cancer, as was the much larger group of retirees. Evidence was accumulating that their worries were well founded, though few employees knew it. For his doctoral dissertation, one of Elizabeth Delzell's students at the University of Alabama at Birmingham, Fabio Barbone, undertook a more detailed analysis of her 1987 survey of cancer among long-term employees. Barbone completed his work in 1989, concluding that "six or seven" of the eleven cases of malignant central nervous system cancers Delzell found were likely caused by exposures at the factory and that workers in the azo, vat dye, and epichlorohydrin production areas all faced elevated risks. He also concluded that about seventeen of the fifty-one lung cancer cases were probably due to exposure to chlorine, anthraquinone, and epichlorohydrin.[8] This time, employees were not briefed on the results of the new analysis, which Barbone did not publish until 1994. Still, relatives of workers who died of cancer knew enough to sue the company. There were three lawsuits in 1995 alone.[9] Ciba (the company dropped the "Geigy" in 1995) settled them all out of court for undisclosed sums, without the public airing that would come at a trial.

Ciba was being just as careful in how it handled what was now its most critical issue in Toms River: the massive Superfund cleanup it was about to begin. Everything about the cleanup was gigantic, including the price tag: $165 million and rising. The old dumpsites on the factory property were believed to hold more than fifty thousand intact or crushed drums of hazardous waste and at least one hundred and fifty thousand cubic yards of severely contaminated soil. That was enough chemical-soaked dirt and sand to fill the passenger compartments of one hundred and thirty 747 jumbo jets and enough drummed waste to fill four Olympic-sized swimming pools. But those were just guesses. The truth was, no one knew what had been buried back in the 1950s and 1960s. There were no reliable records, only the memories of longtime employees and the results of preliminary tests that had detected ninety-five industrial chemicals in the soil or groundwater—seventeen of them, including four known carcinogens, at concentrations higher than those permitted under state law.

It would take another decade to assess all the old dumps and figure out how to clean them up, but pumping up the contaminated groundwater was a more straightforward process. It was also much more urgent, since groundwater plumes were still spreading chemicals—and anxiety—beyond the factory grounds, across Cardinal Drive, and into Oak Ridge. The Lynnworths had moved away, but Sheila McVeigh and other residents wanted the cleanup started quickly. Life next to a Superfund site could be disconcerting. One day, McVeigh was sunbathing in her backyard when a truck appeared on the other side of her fence, on Ciba property, and a crew wearing full-body protective "moon suits" got out to check a groundwater well that was just a few dozen feet away from where McVeigh, clad in a swimsuit, was relaxing on a chaise longue. "Stuff like that happened all the time," she recalled. "It could be a little scary." By the spring of 1995, an interconnected system of forty-three recovery wells on the factory grounds and in Oak Ridge was sucking up about two million gallons of contaminated groundwater per day and sending it through three miles of piping to the company's wastewater treatment plant, where the tainted water was treated and then reinjected into the ground elsewhere on the factory property.[10]

Now that chemical manufacturing had ended, Toms River was a much happier place for Ciba. The company gave storage space in its empty buildings to the Boy Scouts, who used it for their canned food drives for the homeless. Ocean County Citizens for Clean Water, which had started the rebellion against Ciba eleven years earlier, was now its partner, using company funds to monitor the Superfund cleanup. Even the *Observer*, which had been so scathing in its coverage, now ran stories like "Ex-Pariah Ciba Gets Big Honor for Eco Policy" and editorials headlined "Ciba Success Benefits All."[11] No one in Toms River—Ciba least of all—wanted to think about any latent consequences from forty years of toxic emissions into the air, the sandy soil, and the fragile river.

There were no citizen's groups monitoring the cleanup at the other Superfund site in town and no newspaper articles chronicling every step of the process. As usual, Reich Farm was ignored, even though it posed a much more direct threat to many more people.

The winter and spring of 1995 were uncharacteristically busy at the old dump, where there had been so little activity the previous twenty-four years. In March, contractors finished digging up and treating more than fourteen thousand cubic yards of tainted soil, at a cost of $3 million to Union Carbide. It was not much in comparison to the planned cleanup at Ciba—enough soil to fill only twelve jumbo jets instead of one hundred and thirty—but it was far more elaborate than the slapdash cleanups of the 1970s at the old egg farm. This time, the EPA-supervised effort took six months and required dumping hundreds of truckloads of excavated dirt into a steel-walled device mounted on a trailer. The device, called a thermal desorber, looked like a giant spider with a smokestack rising from its belly. Dirt was poured in one end of a tubular kiln and came out the other end ten minutes later after being heated to 700 degrees Celsius—hot enough to vaporize chemicals like trichloroethylene and perchloroethylene. The vapors were captured, blown through a series of filters, and then sent up the smokestack.

The thermal desorber did its job but could not solve a fundamental problem: The worst of the pollution had long since left the Reich Farm property. Attempting a soil cleanup at the farm now, almost twenty-four years after the dumping, was like locking your doors after thieves had already taken everything you owned. Having finally mapped the huge swath of tainted groundwater seeping southward beneath Pleasant Plains, Union Carbide discovered that it stretched more than a mile, in a band four hundred feet wide and one hundred and fifty feet deep. At its southern end, the plume lurched eastward and made a beeline for the two closest Parkway wells. Now there could be no more ambiguity about what was happening: The wells, slurping up nearly two million gallons of groundwater every day, had altered the plume's direction and were sucking up its chemical constituents. If nothing changed, almost every drop of chemical waste in the plume would be drawn into the intake screens of those two wells and then—unless the air stripper removed them—distributed to the people of Toms River. A lot was riding on that solitary air stripper tower at the Parkway well field—too much, according to the EPA.

Unwilling to place its faith entirely in the air stripper, the EPA had initially insisted on a much more comprehensive cleanup. In addition to heat-treating tainted soil at the dumpsite, the agency told Union Carbide to drill extraction wells to intercept the groundwater plume before it reached the Parkway well field. The contaminated water would then be run through carbon filters (to remove semivolatile compounds) and at least one additional air stripper (to remove volatiles). Then the cleansed groundwater would be injected back into the ground, just as Ciba would be doing at its factory site. The EPA's cleanup plan for Pleasant Plains, the agency estimated, would cost Union Carbide another $3 million (in addition to the three million it was spending to heat-treat the soil) and would require eleven years of pumping, treating, and reinjecting.

Union Carbide had reluctantly accepted this EPA cleanup plan in 1988, but the matter was not closed as far as the company was concerned. Its new map of the plume, company officials argued, showed that the EPA plan was not the best approach. So much pollution had already moved so far south of Reich Farm, the company contended, that it was too late to try to intercept it with extraction wells. Thanks to the powerful suction of the Parkway wells, the center of the plume was now more than four thousand feet south of Reich Farm and just seven hundred feet from the closest public well. "The EPA wanted to cut off the plume, but we showed that the plume had already arrived. It was already at the well field," recalled Jon Sykes, a Canadian hydrologist and consultant to Union Carbide. "Even if you tried to cut it off, an awful lot was still going to slip through."

Union Carbide had a simple, if brazen, counterproposal. The better approach, the company's experts told the EPA, would be to do no additional groundwater cleanup at all. The air stripper at the Parkway field was already successfully removing trichloroethylene and other volatile chemicals from the two million gallons of water that were sucked up daily by the two tainted Parkway wells, so why change anything? Why not let drinking water wells continue to serve as pollutant-recovery wells too?

Amazingly, the Toms River Water Company agreed with Union Carbide. The whole point of the EPA plan was to protect the water company's wells, but Toms River Water was much more worried about running out of water than they were about chemical contamination. The Parkway field was still the mother lode, supplying more than one-third of the town's water. In fact, Toms River Water—shockingly, in light of past history—was planning to drill *more* wells at the Parkway field to double its capacity and keep pace with the town's explosive growth. The EPA's plan was a direct threat to this expansion. If Union Carbide drilled extraction wells farther north to intercept the plume and pump up millions of gallons per day of tainted groundwater, there would be much less water available for the Parkway wells. To ensure the water company's support for its counterproposal, Union Carbide promised to reimburse Toms River Water for all past and future expenses associated with operating the air stripper tower. How much the water company would be paid was not disclosed, but the EPA later estimated that Union Carbide's do-nothing approach would cost the chemical company just $1 million in groundwater-related expenses, instead of $3 million under the EPA plan.

All Union Carbide needed now was the approval of the EPA—the only party to the negotiation that was supposed to be looking out for the general public, not stockholders. In January of 1993, after a friendly preliminary meeting with Toms River Water and the EPA, Union Carbide's Craig Wilger made his formal proposal to the agency. The company's leave-it-alone plan "assures continued containment and recovery of Reich Farm contaminants while providing a vital need for safe municipal water," Wilger wrote.[12] Union Carbide was, in essence, asking the EPA to reward the company financially—to the tune of $2 million—for its extreme tardiness in cleaning up the mess at Reich Farm. The company's plan would allow a public well field to double as a cleanup system for a toxic spill, and would take sixteen years to complete instead of eleven, leaving the water supply vulnerable for an extra five years if the air stripper ever malfunctioned.

The irony was striking: Back in the 1970s or early 1980s, Union Carbide could have contained the plume by drilling extraction wells at Reich Farm and pumping up and treating the tainted groundwater. Instead, the company had convinced regulators that the contamination was not a proven risk to drinking water supplies. Now that the plume was huge and the risk to the town's drinking water was proven, Union Carbide had come up with an entirely different rationale for avoiding the expense of additional treatment. The best approach, the company now claimed, would be to keep relying on a single air stripper at a public well field—the same idea the EPA had rejected in 1988 as a potential "long-term risk to the community."[13]

By July of 1995, the EPA was ready with its answer to Union Carbide's audacious counterproposal: yes. The company would not have to do anything else to clean up the huge toxic spill beneath Pleasant Plains, the agency declared. "What happened is that Union Carbide came to us and showed us that they were the problem at the well field, so that was the right place to control the plume," remembered Jon Gorin, the EPA's project manager for Reich Farm. The agency was also worried, Gorin said, about triggering water shortages if it stuck to its earlier plan to force Union Carbide to intercept the groundwater before it reached the water company's wells. "We would have been competing with Toms River Water for available water, and we didn't want to do that," Gorin said. To the EPA, what mattered was that by the time the water was distributed to residents, it would meet all state and federal standards. Thanks to the air stripper, the "finished" Parkway water was apparently clean, so why build a new cleanup system on the assumption that something might go wrong with the current one?

Viewed in hindsight, it was "a mind-boggling decision," as a lawyer would later put it, to deliberately allow carcinogenic chemicals to reach public wells, when there was an alternative—already approved by the EPA—that would have intercepted at least part of the plume and cut the period of risk by five years. Instead, the EPA allowed drinking water wells to be used as pollutant recovery wells, putting its trust in a single air stripper without requiring more protection, including the carbon filters that had been discussed since the 1970s but never installed. For all its riskiness, however, the EPA's reversal was not controversial at the time. "We had done this before at other sites," recalled the agency's Carole Peterson, who oversees Superfund cleanups in New Jersey. "It just wasn't considered something strange to do." Reich Farm would be the last place such shortcuts were attempted in New Jersey, for reasons that would soon become ferociously obvious.

The final step to change any Superfund cleanup plan was a required public hearing, scheduled for August 16, 1995, at town hall. Though it was advertised in advance in the *Observer*, almost no one showed up. The only people who spoke for more than a minute or two were from the EPA, Union Carbide, and United Water Toms River, which was the water company's new name.[14] There were just five questions from the public; none expressed any skepticism about the revised cleanup plan.[15] Over the previous eleven years, Toms River had been the site of dozens of protests, acts of civil disobedience, and public hearings that had attracted hundreds of angry residents. But that was all in the past; the people of Toms River had moved on. It was summertime, and they wanted to go back to watering their lawns in peace.

Seeing no opposition, the EPA approved Union Carbide's plan.

Fifteen days after the perfunctory public hearing on the weakened cleanup plan for Reich Farm, on August 31, 1995, Michael Berry mailed out his final report on childhood cancer incidence in Toms River. The recipient was Steven Jones at the Agency for Toxic Substances and Disease Registry, who had asked Berry to conduct the incidence study at the request of Laura Janson and her sister-in-law Lisa Boornazian, the Philadelphia nurse. It had been an unusually difficult report to write. Berry had come up with results that could not easily be dismissed as a fluke. Childhood cancer rates were higher than expected—sometimes much higher—in every category he analyzed. The confidence intervals were wide and often dipped below 1.0, but even when Berry applied the health department's conservative approach to significance testing, he still found statistically significant clusters of childhood brain cancer in Ocean County and the Toms River core zone. He wondered if he should push for a deeper investigation instead of just dropping it, as had always happened in the past with residential clusters. There was at least a chance that in-depth interviews with the affected population and a comprehensive assessment of local environmental conditions might unearth a likely explanation for what was happening in Toms River—and maybe even help find a way to stop the suffering.

Berry talked it over with his colleague and friend Jerald Fagliano, an epidemiologist at the state health department who would soon assume a central role as events unfolded in Toms River. Together, they decided not to push for a follow-up investigation. The uncertainty of Berry's results made them uncomfortable. What if the apparent clusters really were just coincidences? Or what if a second investigation also failed to reach a clear conclusion? The two men also knew that they would get strong resistance from their bosses if they asked for a follow-up study. A deeper investigation would stir up the community and the press, which always made the top brass of the health department very uncomfortable. And besides, where would the money come from?

"Jerry and I had talked about what we should do next, if anything, and we decided not to try to scrape together the money for any more investigation," Berry recalled years later. "I guess that was, in part, my fault. We could have defused a lot of the stuff that came later if we had done more in 1995, but we didn't." In any case, Berry had no reason to believe that the health department's senior staff would have authorized a full-blown study even if he had asked for one. "It was clear that they would only support the least amount of effort, generally aimed at making the issue go away," said Berry. "Meaningful follow-up to any of these investigations was only going to happen due to outside pressure."

It was a paradox: Public pressure would come only if Berry's incidence study was publicized, but the state health department made no announcement about his findings. His final report was a public record under state law, but Berry, Fagliano, and their newly appointed supervisor—Elin Gursky, the senior assistant state health commissioner—decided not to take any affirmative steps to tell the people of Toms River about it, even though they knew that pollution and cancer had been high-profile issues in the town for years. "We were writing this to Steve Jones at ATSDR, so we kind of thought of it as an internal document, not something for the community," Fagliano explained years later. "In retrospect, should we have organized some sort of formal release? Absolutely, we should have. The fact that we didn't do that created an initial lack of trust, to put it mildly, between the state and the community once it all came out. We have learned a lot since then about the importance of involving the community."

Instead of making a public announcement, Berry simply called Steve Jones to explain the results and then drafted a two-page letter to Jones summarizing his conclusions. After a delay of a few weeks, Gursky approved the wording. The key sentence of Berry's letter appeared in the penultimate paragraph: "Because of the small number of cases included in this analysis, it is not possible to conduct studies to determine possible causes at the municipality or even county level."[16] In other words, investigation over. In addition to Jones, he also sent a copy to Herb Roeschke, who had replaced the retired Chuck Kauffman as the senior staffer at the Ocean County Health Department. As an afterthought, Berry made another copy and sent it to Robert Gialanella, the local physician who had asked for a similar analysis back in 1991. The doctor, Berry figured, might be interested in what the updated report had found.[17]

And that, Michael Berry assumed, was the end of the Toms River investigation.

~

CHAPTER SEVENTEEN

Invisible Trauma

All three men who received copies of Michael Berry's study in the mail during the first week of September of 1995 recognized that they were holding the epidemiological equivalent of a lit stick of dynamite. Yet they responded in three very different ways.

Steve Jones, whose request had prompted the study, talked to his supervisors at the Agency for Toxic Substances and Disease Registry and began thinking about how the agency might conduct a health study in Toms River in cooperation with the state health department. The perennially strapped ATSDR rarely got involved in residential cluster investigations. But if the state supported the idea, Jones thought, there could be a follow-up study in Toms River in a year or two.

At the Ocean County Health Department, meanwhile, Herb Roeschke was frustrated and more than a little worried when a copy of the cancer incidence report landed on his desk. Berry's letter, he remembered later, "was based on sketchy information from a very small sample, and we had no way to verify any of it." Roeschke raised the issue briefly with his senior staff but did not tell his bosses on the board of health about Berry's findings. Roeschke had worked for the county and town health departments for seventeen years and had been through the environmental wars of the 1970s and 1980s. He remembered Pleasant Plains, the pipeline, Greenpeace. Over the years, he had been the target of several verbal lashings from Linda Gillick over drinking water issues. How would she react if she found out about Berry's letter?

He would soon find out, because Linda Gillick had the letter too. She got it from her friend Bob Gialanella, the physician. Michael Berry had sent Gialanella a copy of his 1995 letter as a courtesy, because the doctor's request in 1991 had prompted one of Berry's earlier studies of childhood cancer in Toms River. Gialanella was on the board of Gillick's group, Ocean of Love, but that did not matter to Berry. His studies were public documents, and Gialanella had been interested back in 1991, so Berry sent him a copy. "When I looked at the new data, I could see there had been a change," Gialanella remembered. "All of a sudden it looked like something was really going on that hadn't shown up before." He called Gillick right away. "I don't think either one of us were surprised," he recalled, "but it confirmed what our gut feelings were."

For almost ten years, Linda Gillick had been convinced—"in my heart and in my mind," as she later put it—that there was a childhood cancer cluster in Toms River, but the authorities had always told her she was wrong. Now the State of New Jersey, via Michael Berry's letter, was confirming the cluster. The state's decision not to do a follow-up study only made Gillick more determined. It was a transformative moment in the Toms River story: Alarming and scientifically credible information was now in the hands of someone who was dead set on forcing the authorities to act on that information.

The only question now for Gillick and her allies was how best to catch the attention of the state's political power brokers. Here Bob Gialanella had an idea: He was a longtime friend of Tom Curran, who had just been hired as special projects editor at the largest paper in New Jersey, the *Star-Ledger* of Newark. Curran's wife had graduated from medical school with Gialanella, and the families got together for dinner occasionally. Linda Gillick had easy access to the *Observer* and the *Asbury Park Press* thanks to her years of cancer fund-raising and activism in Ocean County, but the *Star-Ledger* was

in a different league. It was required morning reading for Governor
Christine Todd Whitman and her commissioners, including the health
commissioner, Leonard Fishman. Linda Gillick knew that if she could
get the Toms River cancer story into the *Star-Ledger*, she would catch
the attention of the entire state.

So Bob Gialanella called his old friend Tom Curran, the editor. It
was the final link of a remarkable chain of highly improbable personal connections: The Toms River kids in the oncology ward in Philadelphia happened to have an exceptionally committed nurse, Lisa
Boornazian, who happened to have a sister-in-law who worked at
the EPA, Laura Janson, who happened to know how to reach the
ATSDR's Steve Jones, who happened to pass on her concerns to the
one man who could investigate them, Michael Berry, who happened
to send an unsolicited copy of his results to Robert Gialanella, who
happened to know both Linda Gillick and an editor at the state's top
newspaper, Tom Curran. If any of those links had been missing, it is
hard to imagine how the Toms River childhood cancer cluster would
ever have become such a big deal.

A few hours after Gialanella spoke to Curran at the *Star-Ledger*,
Linda Gillick got a call from the newspaper's medical reporter, Gale
Scott, who was both experienced and aggressive. She spent the next
few months working on the story, in between other assignments. Initially skeptical, she reviewed Berry's report and made calls to dozens
of people in Toms River. Scott called Roeschke and several current or
former Ciba workers, including George Woolley. She researched the
groundwater contamination at the chemical plant, reading old state
and federal studies detailing what was in the factory dumpsites. Over
time, her skepticism faded. The more people she spoke to, the more
convinced she became that the undisclosed cluster of childhood cancer in Toms River was an important story. Her main concern was how
the town would react. "The thing we wrestled with," Scott remembered, "was that we knew it was a good story and we knew everyone
would be very upset—panic is not too extreme a word, really."

Early on Sunday morning, March 10, 1996, a senior official at the
state health department awoke to a phone call from a press aide with

an urgent message about a story on the front page of that morning's *Star-Ledger.* "It was a horrible shock," the official recalled. The headline of Gale Scott's story was "Kids' Cancer Rate Alarms County." The article's more than twenty-five hundred words were carefully chosen, but their overall effect was incendiary. Beginning with an anecdote about the then-anonymous oncology nurse in Philadelphia ("Cancer nurses are hard to shock. . . ."), the article detailed the findings of Berry's incidence analysis and then shifted to a lengthy explanation of the soil and groundwater pollution discovered on or near the Ciba factory. There was no mention of Reich Farm and the Parkway wells, which were overlooked as usual. Steve Jones was quoted saying that the ATSDR wanted to do a health study because "there's something going on out there." His position was endorsed both by Linda Gillick—"finally, someone is listening," she said—and by a researcher at the Cancer Institute of New Jersey, Michael Gallo. "There is a significant increase there, no question," Gallo told the *Star-Ledger.* But there were no comments from the state health department, and Herb Roeschke, the county health official, was quoted skeptically. "Certainly we are concerned about all disease in Ocean County, but this letter is talking about three cases of children with brain cancer," Roeschke said, apparently referring to the three cases in children under age five in the Toms River core zone. "That's not many cases."[1]

The next day the state threw more gasoline on the fire. Instead of meeting with Gillick and other concerned residents, the health department called a press conference in Trenton. Elin Gursky, who was the top physician in state government as senior assistant commissioner of health, declared that the cluster "is statistically elevated, but not to the point where we are overly concerned."[2] Gursky had known about the Toms River cases since the previous summer, when she saw Berry's report and approved its wording. Now that the report was public knowledge, she said, the state health department would conduct another review of childhood cancer rates in Toms River but would not undertake a study of possible environmental causes. In Toms River, meanwhile, people were panicking. The Ocean County office of the American Cancer Society was flooded with calls from terrified residents, as were the radio talk shows. By Tuesday, two days after the initial *Star-Ledger* story, many stores in town had sold out of bottled water; replacements were unavailable because the wholesalers had sold out too. Scam artists filled the void, peddling bogus water filters door-to-door. The anxiety was feeding on itself. The more people heard about the cluster, the more upset they became.

Kim Pascarella, who had stayed active in Ocean of Love after his daughter's death in 1990, had a very different reaction to the *Star-Ledger* story. Like many of the affected families, the Pascarellas had long believed that the aggregation of childhood cancer cases in Toms River was not a coincidence, though they rarely talked about it with anyone outside of the group. "When I saw that story, I thought, 'Ah, it all kind of makes sense now. We're not crazy after all.' As soon as that story hit, it gave us some credibility, some validation," he remembered. The story also represented an opportunity, he thought. Years later, with the benefit of hindsight, he would express it this way: "Without the *Star-Ledger* article, there would have been no uproar. Without the uproar, there would have been no government involvement. Without the government involvement, there would have been no legal case. And without the case, there would have been no truth." No one was ready to predict any of those subsequent events on the day after the first newspaper story appeared. But Kim Pascarella, Linda Gillick, and Bob Gialanella, among others, did immediately recognize that they now had a golden opportunity to press for an investigation. They resolved to make the most of it.

The families' cause was aided by the seeming indifference of state officials, who were digging themselves into a deeper hole by the day. In a network television interview a few days after the initial story appeared, an uncomfortable-looking Elin Gursky, who had been ordered to do the interview by State Health Commissioner Fishman, tried again to explain why the health department would not conduct an environmental study: "To go on those kinds of fishing trips is very, very costly, and would probably yield nothing."[3]

Gursky and Roeschke were speaking in the language of probability: Local families almost certainly had nothing to worry about, and the payoff from an environmental study would almost certainly not be worth the expense. But Toms River had heard those kinds of arguments before. "We were ripe for this because Ocean County had been the dumping ground for a lot of environmental hazards, and we had this long history of wells that were tainted with chemicals," said Gary Casperson, a local banker who was the chair of the county board of health and Herb Roeschke's supervisor. Now there was evidence—*official* evidence, not rumor—that the children of Toms River really did face a higher risk of cancer. Even if the increased risk meant only a handful of extra cases per year, that did not matter to many residents. Like Randy Lynnworth before him, Michael Gillick was not an abstract statistic to anyone who had seen his tumor-ravaged face on television. He was a flesh-and-blood reminder of the torment of a cruel illness and the terror that anyone's child might be next. Yet the government experts, who were supposed to be the protectors of public health, were unwilling to do anything but justify their inaction by citing probabilities. They could not even say whether there was still a problem, since the last available incidence data was an inexcusable *five* years old. The people of Toms River were not only terrified, they were furious.

Linda Gillick quickly provided a way for them to channel their rage. She helped to organize a March 15 protest outside Roeschke's office at the county health department and invited the press to attend. When Casperson and Roeschke walked outside to address the crowd, they faced a row of television cameras and about a hundred angry residents. "There are many causes of cancer . . . ," Roeschke began, but was quickly interrupted by a woman who yelled: "Stop giving me the old story! I don't want to hear it!" When Roeschke tried to explain that the state and county were forming a task force, a man screamed, "We're tired of committees, you've done nothing!" And when Roeschke said that the residents of Toms River should feel safe because Berry's study had found only a handful of brain cancer cases—just five in the Toms River core zone, three times more than expected but still just one case per thousand children—another woman interrupted, shouting: "I don't feel safe at all, my kids could get cancer!" Another mother held up a photograph of her son, who had died the year before from brain cancer. "These kids don't have time to wait. I have two other children, and I'm scared to death," she said.[4]

What especially infuriated the crowd was that so much information had been kept from them. The *Star-Ledger* article had disclosed not only Michael Berry's 1995 Toms River study but also the 1994 statewide study that had identified Ocean County as having the highest childhood cancer rate in New Jersey. When Roeschke said that the reports were not released because they were "just statistics" whose meaning was unclear, he was interrupted by more shouting. "That's unacceptable!" yelled one woman. "Clear out your office!" shouted another. (It was a prescient jeer: The beleaguered Roeschke would soon resign under pressure and take a job in a different county.) Years later, a state official involved in the initial decision not to release Berry's report explained it this way. "We had done some good science, but we didn't know how to use it," said the official, who insisted on anonymity. "To use the information would mean sharing it, showing it, talking about it, and fixing it if there was something to fix. We weren't prepared to do any of that, so we just held on to the information. That was a mistake, certainly."

Roeschke, Gursky, and other officials seemed bewildered by the rage their comments provoked, but they should not have been. Similar storylines had already played out in other communities with toxic sites, including Love Canal. Sociologists studying the phenomenon came to regard fury as a predictable and even rational response to a uniquely nightmarish situation.[5] Facing a threat they could not see, residents had to rely on experts who did not seem to know much more than they did. If children were involved, the trauma of this loss of control was magnified; parents felt crushed by their inability to protect their children.

Research from other "hotspot" communities suggested that the attitude of the relevant government agencies was crucial in determining how people reacted. If the authorities tried to avert panic by seeming to understate the risk of an invisible threat—or, worse still, by withholding information—they would only increase the trauma they hoped to avoid. The best strategy was to acknowledge the fear, to provide as much information as possible (even if it was tentative), and most importantly, to give the community a substantive role in a genuine search for answers.[6] By that standard, the leaders of the New Jersey Department of Health had horribly mishandled the first stage of the Toms River crisis. They would spend the next five years trying to repair the damage.

The March 15 protest outside the county health department was only the beginning. The fear and fury kept building, reinforced by a stream of television and newspaper reports. "HIGH ANXIETY; Local Cancer Scare Spins Out of Control," was the screaming headline in the *Observer* on March 17. "Where have all the children gone?" asked a New York City television newscaster, introducing a story that opened with a shot of an empty Toms River playground. For the first time since Greenpeace and the "Blind Faith" Marshall murder case a dozen years earlier, Toms River was national news. Network camera crews came to town, and politicians were right behind them, heading straight to Linda Gillick's living room. At the front of the line was Robert Torricelli, who visited both the Gillicks and the Pascarellas. A North Jersey congressman known for his pugnacious personality (his nickname was "the Torch"), Torricelli was running for the United States Senate in 1996 and made environmental health a centerpiece of his campaign. He began pushing hard for a full-scale investigation in Toms River. President Bill Clinton was eager to assist a fellow Democrat, but Governor Whitman, a Republican, said nothing publicly, as her health department continued to resist doing anything except rechecking Berry's data. Gillick kept the pressure on, telling reporters that the governor's stance was prompting "frustration, anger and tears" at Ocean of Love.

Under heavy political pressure, State Health Commissioner Fishman did consent to hold an "information session" with the community on March 21, eleven long days after the initial *Star-Ledger* article. Expecting a huge crowd, the health department picked the biggest venue in town, the same auditorium at Toms River High School North that had been the site of the climactic hearing on the Ciba pipeline almost exactly eight years earlier. This time, the department made plans to bring in state police troopers for extra security, which turned out to be a wise decision.

The meeting itself proved to be less an information session than a demonstration of mob rule. The plan was for Fishman, Steve Jones, Elin Gursky, and a few other officials to explain what Michael Berry's study had found and what it meant. Gursky had come prepared with several hundred copies of Berry's letter and data tables, which were passed out to the crowd. Jones had some news too: The ATSDR would conduct an environmental study in cooperation with the state health department, which until then had refused to do so. The Clinton administration had come through with financial support, although Governor Whitman was still balking at the use of state funds. But the meeting did not go as planned. Instead of Fishman, the surprise first speaker was Michael Gillick, now seventeen but still only about four feet tall. He climbed on stage and took a microphone, reprising and updating his electrifying speech of eight years earlier. Once again, no one dared tell him to stop. He began by blasting Gursky, who was sitting just a few feet away, and the other state officials who had suggested that an environmental investigation would be a waste. "Is it a waste of time to save lives?" he said. "Is it a waste of time to save *children's* lives? I ask you to honestly think of the answer, not with your brains but with your hearts. I've battled this infestation of the body and soul for seventeen years. I know what it is like to live in pain and fear, not knowing when you are going to die."

The crowd of nearly thirteen hundred people roared its approval, but when Fishman began to speak he was shouted down. "Shut up!" a woman yelled. A man shouted, "We've got to do something right now! Not a year from now, not three months from now—now!"[7] More than ninety minutes later, Fishman was still trying to get through his opening statement as incensed residents excoriated him every time he tried to speak. The crowd took control of the meeting, lining up at the microphone to tell one wrenching story after another about cancer and pollution in Toms River. It was as if a padlocked door had been flung open and all the demons of the previous half-century were suddenly loosed. Retired Ciba workers spoke about colleagues felled by tumors, an adult cancer victim tore off her wig to reveal a scalp bald from chemotherapy, and parents of dead children wept as they described their ordeals. What they were seeking from the state was not always clear, but their chief demand seemed to be for a truly comprehensive investigation—starting immediately—of the water, soil, and air in Toms River and its possible role in causing cancer.

A slight, professorial lawyer who wore owlish glasses, Fishman was at a loss. His expertise was in healthcare finance and eldercare; he had little interest in environmental issues and no experience with unruly crowds. "He lost control of the meeting almost immediately, and he never got it back," recalled Michael Berry, who was present. "It was a pretty ugly proceeding, not one of the most stellar moments in health department history," remembered another longtime state official, James Blumenstock. "You had a group of almost thirteen hundred people who were truly at the end of their rope."

The chaos grew as people began shouting from their seats instead of waiting for a turn at the microphone. "Somebody open your mouth and tell all of us how can we explain it to our kids!" one woman screamed at the thirteen officials on the dais, who had given up trying to talk and instead sat in stunned silence. Some people in the audience walked out, repulsed by the shrieking, but others moved forward toward the stage as they sought to be heard over the din. The state troopers edged forward, too, forming a barrier between the officials and the increasingly aroused crowd. And then, just as the raw emotions inside the auditorium seemed about to explode, Linda Gillick stepped onto the stage, picked up a microphone, and—in her sternest schoolteacher voice—commanded the crowd to sit down and be quiet.

She was the only person in the auditorium who could have gotten away with such a demand, with the possible exception of a trooper brandishing a gun. Gillick had the personal credibility that came from having a son with cancer, and she had the respect of everyone in the room because of her years of charity work. "We all trusted her—all of the families, the whole community," recalled Melanie Anderson, who was at the meeting along with more than a dozen other parents of afflicted children. "People described her as a bulldog with a bone. She just could cut through all the garbage." Gillick had been omnipresent on television and in the newspapers as an advocate for children with cancer and their suffering families, and she had the no-nonsense, I'm-in-charge manner Fishman lacked. In a community that felt powerless and hopeless, she exuded authority and confidence. She was now the essential woman.

The officials on the dais watched, with a mixture of wonderment and relief, as Gillick quieted a crowd that five minutes earlier had seemed completely beyond anyone's control. She had defused, at least temporarily, a potentially violent confrontation. But her wizardry also carried an implied threat: I made them stop, and I can make them start again. She was now the most important person in the room, a startlingly adept leader who did not seem to care about budgets and probabilities and the limitations of cluster epidemiology and was instead demanding immediate action. Gillick was no longer a supplicant in the audience; she was on stage, in the power position. After calming the crowd, she turned to Fishman and asked a pointed question: You have heard from the people, she said, and you know what we want. What are you going to do now?

Fishman, who had seemed paralyzed throughout the meeting, suddenly came to life. Making an on-the-spot decision, he declared that the state would take its cues from the community, and specifically from Gillick, in designing a state-federal environmental study. (This was the same study that, until the day before the hearing, his department had opposed as impractical and still did not want to pay for.) Gillick would not only chair the study's citizens' advisory committee, Fishman announced, she would also choose its members. The state would also launch an emergency program to test drinking water in local schools, he said. By the time the meeting broke up shortly before midnight, after five excruciating hours, Gillick was setting the agenda. She declared that she would begin work the following morning, and she expected Fishman and his staff to be in Toms River tomorrow, too. The weary-looking commissioner promised he would.

In the frenetic days that followed, government officials struggled to figure out how to deliver on the promises they had just made under extreme duress. The community wanted fast action, but it would take months or years to satisfy most of Linda Gillick's demands. She wanted Michael Berry's incidence data to be checked and updated to reflect recently diagnosed cases, but getting the cancer registry up to date would be a massive chore requiring outreach to hundreds of doctors, clinics, and hospitals. She also wanted a comprehensive environmental study capable of unearthing the connections between cancer and pollution, but a scientifically meaningful study would need to start with one or more specific hypotheses—testable theories about what may have caused the cluster. Before they could even think about developing hypotheses and designing a study, the ATSDR and the state health department would have to do a sweeping investigation of past environmental conditions. Funding was also an issue. After the trauma of the public meeting, Governor Whitman relented and agreed to spend state money on a local investigation. The county legislature also set aside $250,000 to assist. But there was no doubt that most of the needed funds—probably millions of dollars—would need to come from Washington. Politicians as high up as Vice President Al Gore had promised to help, but there was no specific monetary pledge yet.

As they began work on those longer-term projects, the agencies were under pressure to do something—anything—to generate fast and preferably soothing results. Whitman, who would be up for re-election in nineteen months, had taken a political beating from Robert Torricelli and others over the initial failure to disclose Berry's cancer incidence data; she was now determined to show that her administration was moving quickly. On March 28, just seven days after the fiasco in the auditorium, a phalanx of state workers fanned out to twenty-one local schools, with water testing kits in hand. If people were frightened about their kids' health, what better way to begin regaining their trust than by testing at the schools? That same day, Whitman led a group of kids from Toms River on a tour of the state laboratory in West Trenton where the water samples would be analyzed. A large group of reporters and cameras captured every stage-managed minute.

There was an element of condescension to the water tests; their goal was clearly to placate the residents. Federal and state law already required water suppliers to test their wells regularly, so another round of tests seemed like a purely political exercise. "I was skeptical that it would tell us anything that we didn't already know," remembered Jerry Fagliano, the health department epidemiologist. A *Star-Ledger* columnist from Toms River, Paul Mulshine, berated the "howling mob" in the auditorium for its "aggressive ignorance" and said that the state officials should have refused to give in to its demands.[8] But the crisis was getting so much attention that appeasing Toms River had become a political imperative for Whitman. As the state lab began to analyze the water samples, she made plans to visit the town in person so that she could make the expected announcement that the water was safe. Any other result seemed too absurd to even consider.

The new water tests were not the usual ones, however. Gillick wanted the local drinking water tested "for everything," and she made it clear that if the state did not agree, she would complain to the newspapers. As she had promised, Gillick had plunged right into the work of the newly formed Citizen Action Committee on the Childhood Cancer Cluster. Even before she had settled on all twelve of its members (her first appointment was her son, Michael), Gillick's committee was huddling with Fishman and Gursky. "We decided early on that our function as a committee was to keep the governmental agencies to the grindstone," said Kim Pascarella, another early committee appointee, along with Bob Gialanella. In addition to their devotion to Ocean of Love, Pascarella and Gialanella shared one other important characteristic with the Gillicks: They were already convinced that pollution was the root cause of the cluster. A key role of the committee, as they saw it, was to push skeptical government officials to ferret out the supporting evidence needed to prove what the families were sure they already knew.

Their first push was for a water-testing program that was much more ambitious than had ever been tried in New Jersey. The EPA required that drinking water be tested for just eighty-three potential contaminants.[9] New Jersey's list was slightly longer and included some concentration limits that were stricter than the EPA's, but United Water and the state's other drinking water suppliers were still testing for only a tiny fraction of the total number of possible pollutants. The utilities were earning passing grades for not much testing effort;

Linda Gillick, the former teacher, wanted a much more thorough examination. The question was, what else should the state health department look for in Toms River's drinking water? There were more than fifty thousand chemicals in commerce; the state could not possibly test for them all.

Toms River did have one perverse advantage: It was the home of two thoroughly investigated Superfund sites. As a result, the EPA knew the names of dozens of arcane chemicals that had been dumped at Ciba and Reich Farm and were not on the standard checklist for drinking water testing. Some of them, in fact, were so obscure that there was no reliable test for them in drinking water. But even if regulators were not sure of the exact molecular structures of those arcane industrial chemicals, they could tell that almost all of them contained nitrogen or phosphorus. So Health Commissioner Fishman told Gillick's committee that the state health department, in addition to testing for the usual suspect chemicals, would also look for more than fifty other nitrogen- or phosphorus-containing compounds in Toms River's drinking water. In all, the state would be testing for about two hundred and fifty chemicals instead of the usual eighty-three. It was still only a few drops in the bucket, considering that there were thousands of other potential pollutants that also contained nitrogen or phosphorus, but it was a start.

The first results became available just two days after the water samples were collected from twenty-one schools and twenty public wells in Toms River. This was itself a huge change, since the results of drinking water tests usually were not available to the state for weeks or even months after sampling. The usual procedure in New Jersey was for water companies to hire private laboratories to do the testing and then, eventually, to send the results to the state Department of Environmental Protection. (The DEP did not have nearly enough staff to conduct the required quarterly tests on its own, since there were more than two thousand public water wells in the state.) The DEP normally did not insist on a quick turnaround, so samples often sat for weeks at the private labs before they were analyzed. Now, however, Toms River was getting special treatment. Governor Whitman had promised fast, trustworthy information about what was in the town's water, so the samples collected at local schools on March 28 were tested directly, and immediately, by the DEP's analytical laboratory in West Trenton.

The results were a shock. For starters, there was trichloroethylene in the drinking water at ten of the twenty-one schools. None of the samples exceeded the state's limit of one part per billion, but two schools equaled it. The findings suggested that the all-important air stripper at the Parkway well field was not doing its job perfectly. The bigger surprise was that water in thirteen schools was slightly radioactive. The levels were not an immediate hazard; the safety limit was fifteen picocuries of radiation per liter, while the highest school reading was twelve. But any reading over five was supposed to prompt extensive testing, and all thirteen schools were over it.[10] Four of the town's five public well fields were also over the limit; the two highest were twenty-six and twenty-nine picocuries.[11]

What that meant is that many of the ninety thousand people served by United Water Toms River were drinking water the government considered unacceptably radioactive and may have been doing so for years without anyone knowing. This alarming discovery deepened the sense of crisis among officials working on the Toms River testing. The radiation almost certainly came from naturally occurring radium in soil, not pollution, but there was no quick way to confirm that. Besides, jittery residents were sure to be terrified, no matter the origin. Even more than industrial chemicals, radiation was a source of psychological dread because of its long association with nuclear weapons and cancer. One of New Jersey's three nuclear power plants, Oyster Creek, was in southern Ocean County and had been a source of community anxiety ever since it opened in 1969.

"It was a real challenge when the radiation was found," remembered James Blumenstock, a senior state health official at the time. "With the contaminated sites you could blame corporations or midnight dumpers, but how do you deal with something that Mother Nature put there?" United Water was completely unprepared for the uproar, he added. "They weren't accustomed to dealing with this level of scientific scrutiny and pressure. They were out of their league."

What was especially puzzling was that a private lab had been testing local water for radiation for years and had never found such high levels. Later, the state would learn that naturally radioactive soil—and therefore water—was all over southern New Jersey, but the problem had been hidden by the slowness of the typical water-testing process, which Whitman had just ordered sped up for Toms River only.[12] As far as anyone knew in April of 1996, however, the radioactivity problem applied exclusively to the star-crossed town of Toms River, which now had yet another unwanted distinction: chemical town, cancer cluster, *and* home of radioactive water.

Governor Whitman had promised to keep the town fully informed of the investigation as it proceeded; now she was going to have to announce another scary surprise that defied an easy solution. Simply shutting down the affected wells was not an option, since almost every well field in town was affected. There had been a drought emergency the previous summer, and United Water was operating at the edge of capacity, as always. No governor wanted to risk the political consequences of depriving ninety thousand people of water. The imperative in Toms River had always been to keep the water flowing, no matter what. That was true in the 1960s, when Ciba's dyes invaded the Holly well field, and it was true in the 1980s, when the Reich Farm plume hit the Parkway wells. Now, in 1996, even in the midst of a cancer scare, it was still true. Federal law required corrective action only if radiation levels averaged more than fifteen picocuries per liter over four consecutive quarters; a single test was not enough to trigger a well closure. So the state told United Water to temporarily shut down only the two wells where readings were highest—closures the company could manage without disrupting service.

As she had promised, Whitman traveled to Toms River to deliver the bad news in person. She tried to put the best face on the discovery of a known carcinogen in the town's drinking water. "One thing I'd like to stress is that people can still drink the water," Whitman told reporters. She was obliged to add, however, that the two wells were being shut down immediately. There was no question about which storyline would dominate the television news that night. "Wells Shut over Radiation" was the front-page headline the next morning in the

Star-Ledger. The smaller print below added, "Officials assure Toms River water use is safe." The following day, the same newspaper reported that demand for bottled water in Toms River was "skyrocketing" and quoted a distributor whose sales had tripled overnight. "Right now, everyone is our customer," he said.[13]

It was hard to imagine that Toms River could become an even more anxious place, but the disclosure that the town's water was radioactive did the trick. Could the situation get worse still? Everyone involved in the state's emergency water testing program knew that it could. If the tests found industrial chemicals that were manmade and unique to Toms River, then life would become much more difficult for the public officials struggling to manage an already volatile situation. The EPA and the DEP had known about groundwater pollution at Reich Farm and Ciba for more than twenty years, yet as recently as 1995 agency officials had assured anyone who asked that there was nothing to worry about. If an enterprising scientist now managed to identify carcinogenic pollutants in the water supply and then traced those contaminants back to waste dumping at Reich Farm or Ciba, there would be recriminations all around—and probably lawsuits, too. In that case, the wrath of Toms River would fall upon not just the dumpers but also the politicians and government regulators who had let them get away with it for so long.

CHAPTER EIGHTEEN

≈

A Cork in the Ocean

Floyd Genicola was an outsider within the dense bureaucracy of the New Jersey Department of Environmental Protection. A solemn-faced man who favored crisp white shirts and solid ties, Genicola was a chemist by training and a dissenter by temperament, if not by dress. He had started his career with the state as a forensic analyst with the New Jersey State Police, where his job was to identify seized narcotics by analyzing their chemical structure. The scientific detective work played to Genicola's strengths, which included precision, patience, and a deep-seated aversion to taking anything on faith. He would sometimes spend weeks on a case, subjecting a sample to dozens of tests to make certain that his identification was correct. He left for a job at a pharmaceutical company, but when a position opened up at the DEP in 1982, he jumped at it. By then, Genicola had a master's degree in mass spectrometry and had learned how to identify compounds based on the masses of their component atoms. The DEP was getting its first spectrometer, and Genicola wanted to use his new knowledge for a good cause. Catching polluters, he figured, would be at least as interesting as catching drug dealers had been.

But the DEP was a very different kind of agency than the state police. Its job was not to investigate or prosecute but to interpret and

apply complicated rules governing everything from fishing and for-
estry to recycling and radioactive waste. The police solved crimes; the
DEP negotiated compliance. It was a process-oriented agency, and
Genicola was now immersed, uncomfortably, in the minutiae of environmental regulation. He was eventually put in the office of quality
assurance, where his job was to check the accuracy of test results generated by the DEP or by the businesses it regulated, including water
suppliers. Too often for his supervisors' comfort, Genicola thought
that the data was not up to snuff and was not shy about saying so.

In April of 1996, a few weeks after the near-riot at the high school,
Genicola's bosses invited him to sit in on meetings of the state task
force that had been hurriedly assembled to manage the emergency
water-testing program in Toms River. Prickly or not, Genicola got the
invitation because he was one of the department's in-house experts
on identifying pollutants in water. Genicola was happy to be asked;
he thought the project was interesting and a bit exciting, too. It reminded him of his days working for the state police. The health department's focus was on making sure that no one in Toms River was
being harmed right now, and the DEP was trying to make sure that the
water company was following its rules. Both missions were worthy,
Genicola thought, but he also saw Toms River as a *forensic* challenge.
An unusually large number of children had gotten cancer, and now he
would get a chance to try to figure out why, and who was responsible.

There was a huge amount of data to review. In their initial checks
of the local drinking water, Genicola and the state team found little
out of the ordinary besides the radioactivity and very low levels of
trichloroethylene.[1] After the trauma of the radiation announcement,
the otherwise routine results were an immense relief. Whatever had
caused the childhood cancer cluster—if anything had caused it—did
not seem to be in the water. On May 7, Health Commissioner Fishman made a triumphant return to Toms River to declare that after
"the most comprehensive, the most intensive, the most in-depth study
of a public water system ever undertaken in New Jersey, the water is
safe to drink, bathe in, and cook with."[2]

But that was not quite the whole story, and at least some members
of the state's water-testing team knew it. Buried in the reams of data

generated by the emergency testing program was a subtle signal that something unusual was in Toms River's drinking water, possibly something important. Months would pass before Floyd Genicola figured out what that signal meant, and years before others grasped its full significance.

The discovery Floyd Genicola and his colleagues eventually made in a droplet of Toms River water had its antecedents in the work of ancient Egyptian alchemists and all those who shared their obsessive quest to pull substances apart and identify their constituent parts. The Egyptians' successors, first in the Arab world and then in Europe, drew inspiration from texts such as the Emerald Tablet of Hermes Trismegistus, said to be authored by the Egyptian god Thoth. "Separate thou ye earth from ye fire, ye subtle from the gross sweetly with great industry" is a line from an early English translation of the Emerald Tablet. The translator was a secretive alchemist of the seventeenth century who made a name for himself in other pursuits: Isaac Newton. His ideas about light and gravity formed the basis of modern physics, but Newton spent much more time trying to transmute lead into gold, recording his experiments in secret code in his private journals.

Alchemists knew that if they burned, boiled, dissolved, distilled, or otherwise disturbed a seemingly stable compound, they would often end up with two or more constituents. But it was a laborious process of trial and error, and it did not always work because many compounds were impervious to conventional chemistry. What was needed was an instrument that could fulfill the charge of Hermes Trismegistus by separating *anything* into its most fundamental components. Two Germans showed the way. The first, Robert Bunsen, did not invent the gas burner that bears his name, but he perfected it and used it to study the colors of the flames he generated by vaporizing various metals. In 1860, his friend Gustav Kirchhoff came up with the Newtonian idea of differentiating the flames more precisely by viewing them through a prism and studying the spectral lines produced by each hot gas Bunsen tested. When a beam of incandescent light was passed through a cooler version of the gas and then viewed through the prism, it produced the identical set of lines except that they were dark instead of light. Kirchhoff was the first to read nature's own bar code, with his discovery that each molecule emits a specific set of spectral lines when hot and absorbs light at the identical wavelengths when cool. He and Bunsen did not know *why* every compound has a telltale spectral pattern (that answer would not come for another half-century), but they quickly grasped the usefulness of their "spectroscope" in identifying unknown substances.[3]

Before long, scientists all over Europe were finding new ways to compel molecular mixtures to reveal their secrets. In 1913, the British physicist J. J. Thomson ran a charged mixture of gases through an electromagnetic field in a vacuum tube and discovered that each gas was deflected at an angle determined by its mass and charge. By recording those angles of deflection on a photographic plate, Thomson could determine the masses of simple elements and molecules. Six years later, his student Francis Aston developed a much more accurate device he called a mass spectrograph, and later a mass spectrometer. Finally, in the late 1950s, two Dow Chemical scientists paired a spectrometer with another powerful tool, the gas-liquid chromatograph. In a chromatograph, the components of a gasified sample could be identified by the speed at which they passed through a liquid or polymer. When used in tandem, chromatographs and spectrometers could determine the composition of almost anything. By the end of the twentieth century, they were being used to establish the age of fossilized teeth, test athletes for illegal steroid use, search for undersea oil deposits, and monitor the breath of patients under anesthesia, among many other tasks.

With persistence and luck, a skilled investigator might even manage to use the two instruments to find and identify a vanishingly small amount of an unknown pollutant in a droplet of water. That would be the molecular equivalent of spotting a cork bobbing on the surface of the ocean and then figuring out which bottle of wine it came from.

The chemical signal in Toms River water was a tiny cork indeed. It was so small and strange that most of the people at the state DEP who had heard about it were convinced it was not there at all and was merely a measurement error.[4] Even so, the data kept bobbing to the surface. It was still there when the lab retested the samples, which came not from one well but two: Well 26 and Well 28 of the Parkway field, which happened to be the two most pollution-plagued wells in Toms River. As the DEP already knew, those two wells had been sucking up TCE and perhaps other contaminants from the mile-long Reich Farm plume since at least the 1980s and possibly earlier.

The only clue about the mystery compound's identity was a diagram generated by the DEP's gas-liquid chromatograph. It looked like a child's representation of a mountain range, with four peaks of varying heights drawn with a single unbroken line. Each peak corresponded to an unknown nitrogen- or phosphorus-containing compound detected in the water sample. The peaks were low, which meant that the concentrations were very low: six parts per billion, or 0.0000006 percent of the sample. Normally, the DEP would ignore such a faint detection, especially of an unknown compound. But Toms River was an opportunity to do more, to solve a mystery instead of merely satisfying the minimum regulatory requirements. So Floyd Genicola pushed his colleagues to cobble together a sample large enough to test in a mass spectrometer. After a lot of work, they succeeded.

The resulting spectrograms looked something like combs with broken teeth. There were discernible spikes of various heights at the molecular masses of 129, 156, 115, and occasionally 210 daltons. (A dalton is a unit of relative mass, named after the English chemist John Dalton.)[5] It was now clear that the trip through the spectrometer was breaking up the mystery compound into the same three or four large pieces, plus many tiny pieces that were almost too small to measure. But what chemical was it? No one Genicola knew had ever heard of a mass spectrum that corresponded to those four numbers. Genicola consulted the federal government's largest spectral library but found no matches with the combination of 129, 156, 115, and occasionally 210 daltons.

Very few people in state government shared Genicola's curiosity. To his bosses at the DEP, there were only two characteristics of the mystery chemical that mattered: It was not on any list of pollutants that they had to worry about, and it was only present at six parts per billion. The only reason anyone spotted it in the first place was that Governor Whitman had been pressured by Linda Gillick into ordering an extraordinarily thorough investigation. That was why the DEP had attached a nitrogen and phosphorus detector to its chromatograph, a detector capable of finding "non-target" industrial pollutants instead of only the ones on the EPA's checklist. Unexpectedly, that detector had found something. Unsurprisingly, Floyd Genicola wanted to find out what it was.

His chief antagonist inside the DEP on the issue was Gerald Nicholls, the director of environmental safety and analysis, who considered it a waste of time and money to try to identify a chemical that could barely even be detected in Parkway water.[6] "Normally, you're looking at fifty parts per billion before anyone got interested. Why would we make an exception for this one?" Nicholls wondered. He was not the only manager upset over Genicola's advocacy. The people in charge of assessing the safety of old dumpsites were apoplectic. Union Carbide and the EPA had assured them that Reich Farm was not a threat to Toms River's drinking water. What if it were implicated as the source of the mystery compound in the Parkway wells? Would the DEP and the EPA then have to revise their assessments of *all* of New Jersey's dumpsites—of which there were hundreds—to consider the impact of unknown compounds at very low levels? "They were frustrated," Nicholls recalled. "We were all frustrated."

Floyd Genicola thought the issue was clear cut, too: Kids were getting cancer, there was something strange in the water, and it was the state's responsibility to try to find out what it was and who put it there. He pointed out, to anyone who would listen, that the mystery compound had been found not just in the two Parkway wells. It had also been found, albeit at just one part per billion, in the tap water of a grade school, St. Joseph's, right in the middle of town. Besides, Linda Gillick and her advisory committee already knew that the DEP had detected the mystery compound. If the state did nothing, the next headline in the *Star-Ledger* or the *Asbury Park Press* might read: "Toms River Schoolchildren Drinking Unknown Chemical; State Refuses to Investigate." There was no real choice; they would have to let

Genicola keep chasing the mystery chemical. "We had this long debate and we ultimately came to the conclusion that even if we didn't want to look for this compound, the public was going to *make* us look for it because our process was very open and the public knew what was happening," Nicholls remembered.

Reluctantly, Genicola's supervisors told him to keep going.

Health Commissioner Fishman's announcement on May 7 that the water was safe had the desired effect: Toms River calmed down, slightly. Panic was replaced by a quieter but steady anxiety. Amateur epidemiology, meanwhile, was breaking out all over town.[7] Calls about possible clusters were pouring into Ocean of Love and a special state hotline; dozens of residents were making their own lists of cancer cases in neighborhoods and offices. Employees at the downtown post office demanded an environmental study, as did workers at the state office building across the street. The postal workers' union was worried because thirty-two of the three hundred employees at the post office had gotten cancer over the previous twenty-five years. But in 1995, New Jersey's incidence rate was 1.3 cases per three hundred people, so thirty-two cases over a quarter-century was almost exactly the expected rate.[8] That fact was irrelevant in the hothouse environment of Toms River, however. The postal union got its investigation thanks to the ever-helpful Robert Torricelli, the Senate candidate who practically lived in Toms River that spring, when the press coverage was at its peak. After exhaustive, costly tests of the air, soil, and water, federal investigators found no toxic hazards at the post office.[9]

If anyone had a strong case for an epidemiological study, it was the workers at Ciba. Union president John Talty kept hearing about retirees who were newly diagnosed with cancer, so he began agitating for an update of Philip Cole and Elizabeth Delzell's worker studies from eight years earlier. The company agreed in early 1996, contracting yet again with Cole's industry-oriented group at the University of Alabama at Birmingham. With the completion of each earlier study, the evidence of pollution-induced cancer at the factory had gotten progressively stronger; as the body count increased, the confounding influence of chance diminished. The next study seemed likely to continue the pattern, but its results, company officials told the union, would not be available for another two years.

Ciba's immediate problem was getting through the next few months. The disclosure of the cancer cluster came at an especially sensitive time because the company was finalizing its planned merger with Sandoz, its former partner at Toms River Chemical. The new company was to be known as Novartis, and Ciba had hoped that the rebranding would help improve its image as it shifted from chemical to pharmaceutical manufacturing. Publicity over the Toms River cluster was undermining that effort. There were hundreds of newspaper, magazine, and television stories, most of which featured descriptions of Ciba's waste dumping as well as wrenching interviews with the Gillicks, the Pascarellas, and other cancer-stricken families. The company had another problem, too: Its long-delayed lawsuit against its insurance companies, seeking reimbursement for several hundred million dollars' worth of cleanup costs in Toms River, finally went to trial in the spring of 1996. Ciba would eventually win, but in the meantime the trial was unearthing embarrassing information, including—at long last—the company's hushed-up contamination of the riverside Holly wells in the mid-1960s.

For Toms River, there was only one beneficial aspect to all the attention: It was much easier now to secure the funds needed to complete the state's ongoing tests of the town's water, air, and soil. Those tests were expensive, but the unceasing press coverage attracted politicians like catnip. By summer, Congress had appropriated $900,000, and the state legislature added $600,000. It would not be nearly enough, as things turned out, but it was an impressive beginning.

Mostly, though, the publicity ticked people off. The news coverage, like the public's anxiety about the cluster, seemed to feed on itself. The stories attracted the attention of public officials, who then gave reporters more to write about. So did Linda Gillick. She cooperated with all of the national reporters and kept local journalists well informed of each incremental step in the state's investigation. Gillick became the funnel for all sorts of bizarre information from worried locals, including ex-Ciba workers and even hunters who reported seeing weird tumors in Ocean County deer. While she was usually careful about what she told reporters, Gillick always made it clear that she believed the cluster was real and caused by polluted drinking water. She never missed an opportunity to describe the anguish of the Ocean of Love families and their crusade to prevent more tragedies in Toms River. It was an irresistible storyline, especially because Michael Gillick was a smart, articulate teenager who was unafraid to go on camera, where his appearance made for a compelling visual tableau.

No one in town ever criticized Michael, but his mother was another story. Many locals resented the attention Linda Gillick was getting. News stories often styled her as an ordinary suburban mom thrust into extraordinary circumstances, but there was nothing ordinary about her. The same relentlessness that made her such an effective advocate also infuriated many people who came into contact with her. "I had a lot of respect for Linda, but she was a no-holds-barred kind of person," recalled Gary Casperson, who scrapped with her as chairman of the county board of health. Even Gillick's friends sometimes winced at her badgering of officials she believed were impeding the cluster investigation. "Linda's approach was not the approach that I would have taken, but for the situation we were in, the times that we were in, it was very effective," said Bob Gialanella, who was on the Ocean of Love board as well as the citizens' advisory committee Gillick chaired. "It worked to have somebody stand there and sometimes personally accost these people, verbally, to the point where they would finally do something to assist us."

A key cause of the grumbling was financial. In May, Gillick found an anonymous note in her mailbox: "The water is fine, cancer cluster is probably a freak. Meantime, Ocean County will suffer this summer because you have scared away tourists, home buyers and others." The negative publicity was striking directly at the two fragile industries that constituted the core of the town's post-Ciba economy: real estate and tourism. "Toms River is a summer tourism site, and property values were going down. People couldn't sell their homes," remembered Casperson, the banker and board of health chairman. "The business people at first weren't real vocal about how they felt, but there was an underlying feeling of, 'Why is everyone saying bad things about Toms River when the experts don't know the answers?'" The real estate slowdown had powerful psychological consequences, too, by making some residents feel trapped in homes they could not sell.

Deepening the sense of gloom was another surprise announce-ment, in August of 1996, that Ciba was shutting down what little manufacturing remained at the Toms River plant, sending another 170 jobs south to Louisiana and Alabama. After December, only about thirty employees would be left, mostly technicians monitoring the Superfund cleanup, which was just beginning. The mini-city be-hind the trees would be virtually deserted, a ghost town atop an un-imaginably vast amount of buried toxic waste.

Toms River had little to show from its first six painful months in the national media spotlight except a battered reputation and a thicket of "For Sale" signs. An explanation for the cluster was as distant as ever.

Floyd Genicola intended to change that. He was already making prog-ress. Using an unusual type of spectroscopy called isobutane chemical ionization, he had managed to determine that the mystery com-pound's molecular weight was 210 daltons—not 156, as he had earlier thought.[10] Now Genicola knew enough about the compound that he was ready to give it a name: It was a "nitrogen/phosphorus tentatively identified compound molecular weight 210"—or a TIC for short. But what was it, really? And where did it come from? Answering the sec-ond question, Genicola thought, might help answer the first. Reich Farm was the obvious suspect, since it had already been fingered back in 1988 as the source of the trichloroethylene that had turned up in Parkway wells. If an industrial solvent could make the one-mile trip underground from Nick Fernicola's dumping ground all the way to the Parkway well field, why not the mystery TIC?

Genicola chose an unorthodox way to try to find out. One day in August of 1996, he picked up the phone and called Union Carbide—a highly unusual step for a state agency that usually communicated by memo, often through lawyers. Actually, Genicola called a contracting firm, Radian International, which was monitoring groundwater in Pleasant Plains for Union Carbide. Genicola asked for chromatograms of any unidentified compounds the company had found in groundwater at Reich Farm that had molecular masses of 115, 129, 156, or 210 daltons. It was a cheeky request because there was no regulation requiring a company to supply that kind of information. "I didn't think there was any way they would do it, but they did," Genicola remembered. "Surprisingly, they faxed it over right away."[11] When the chromatograms from Radian International emerged from his office fax machine, Genicola could barely contain his excitement. They were identical to the chromatograms the state had generated from samples of Parkway drinking water. Whatever mystery compound was in the wells was also at Reich Farm. In fact, there was a lot more of it at the dumpsite. In one monitoring well just south of the site, Radian International had measured the mystery compound at a concentration of about one part per million—almost two hundred times higher than what was in the Parkway drinking water.

To Genicola, it was obvious what was happening: The mystery compound was flowing south from Reich Farm straight to the Parkway wells—and was not being removed by the air stripper. Whatever that mystery chemical was, the citizens of Toms River had been drinking it for years. In fact, they were *still* drinking it.

All that was left was to identify the TIC, but even someone as self-confident as Genicola knew by now that the task was beyond his expertise. He needed a world-class mass spectroscopist to crack the puzzle and determine the molecular structure of the mystery compound. Some EPA chemists suggested a candidate: If anyone could do it, G. Wayne Sovocool could. Sovocool's reputation inside the EPA was similar to Genicola's inside the DEP. He was stubborn, quirky, and an absolute whiz at interpreting mass spectra. At Cornell University, Sovocool had taken a class from Fred McLafferty, who had ushered in modern analytical chemistry by pairing liquid-gas chromatography with mass spectrometry and then writing the standard text on interpreting mass spectra. Sovocool went to work for the newly created EPA in 1972 and had been analyzing pollutants for it ever since, working in North Carolina and then Nevada. His triumphs included being the first to find highly toxic dioxins in fly ash from coal-burning power plants and the first to find chlorobenzene in the blood of Love Canal residents. Sovocool liked challenges, and he was about to get one.

It took weeks for Floyd Genicola to get his wary supervisors to agree to let him contact Sovocool and send the data and new water samples to his lab in Las Vegas. The samples had to come from the state health department because the DEP's laboratory had just been shut down in a budget-cutting move. If it had closed just a few months earlier, the mystery TIC would never have been found because the older equipment at the health department could not have detected a compound at levels as low as five parts per billion.

The lucky timing with the DEP lab was one more indication that nothing about the Toms River case—from the detection of the childhood cancer cluster to the decision to investigate it and now the identification of a possible cause—had been preordained by consistent, proactive government oversight. Instead, a fortunate combination of the right equipment, at the right time, in the right hands had found the TIC. Now its identity would be unmasked by the right person.

Wayne Sovocool had spent a quarter-century running pollutants through spectrometers, and he very rarely encountered a chemical he could not quickly identify. The one from Floyd Genicola was tricky, though. So many combinations of atoms could add up to 210 daltons, and shared some of the fragments of the TIC's mass spectrum, that Sovocool decided to try to reduce the number of suspects by making some educated guesses. Like a child with a Tinkertoy set, Sovocool started building models of the molecule, atom by atom, to try to find an arrangement that matched what little he knew about the TIC. He worked in two dimensions, with pencil on paper, but he was thinking in the three-dimensional, submicroscopic world of molecular structure. Sovocool knew that nitrogen was almost certainly an important component of the TIC and that industrial chemicals that contain nitrogen often do so in the form of a nitrile group, which consists of a nitrogen atom and a carbon atom connected by a triple bond. Further, he knew that the mass spectra of compounds containing two nitrile groups frequently yielded spectrometer fragments at 156 daltons, just as the TIC did. So Sovocool guessed that the TIC had two nitrile groups and began searching spectral libraries for industrial compounds that had two nitriles and a mass spectrum similar to the TIC.

By mid-October, Sovocool thought he had fingered the perpetrator: a dye ingredient called 1,2-benzenediacetonitrile. The chemical was a logical suspect for Toms River because it was used in dye manufacture and contained two nitrile groups plus a benzene ring of six carbon atoms, another ubiquitous feature of industrial solvents. The chemical's mass spectrum, including the fragments, was very similar to the TIC's. But it was not a perfect match, which bothered Sovocool. He wanted confirmation, and the EPA's Las Vegas lab had just developed a new technique (it carried the unwieldy name of Mass Peak Profiling from Selected Ion Recording Data) that might be able to provide it. Developed by a colleague of Sovocool's, Andrew Grange, it was a form of spectrometric wizardry that could determine the precise mass and abundance of even the smallest molecular fragments. Sovocool asked Genicola to send him additional superconcentrated water samples from the Parkway wells, and then Grange set to work.

It took only a few days. By the beginning of November, Grange had solved the central mystery of the TIC by deducing its chemical formula: fourteen carbon atoms, fourteen hydrogen atoms, and two nitrogen atoms, or $C_{14}H_{14}N_2$. Even more usefully, he determined the exact molecular masses, to the fourth decimal point, of *ten* molecular fragments broken off during the TIC's trip through the spectrometer, not just the four most common fragments Genicola had tried to match. Several of those fragments were so small they had never even been measured before. When he saw the new data, Sovocool knew that his earlier guess about the dye chemical was wrong. But he was not discouraged; he was excited. At last, he had a formula and a precise spectral signature of the TIC. He was closing in. Floyd Genicola was even more excited. With barely concealed glee, he relayed the discovery to Union Carbide. "I told them that we had the molecular formula, and that we were going to identify it," Genicola recalled.

The final step was not a simple one, however. More than three hundred thousand compounds were listed in digital spectral libraries, including forty-six with a mass of 210 and the formula $C_{14}H_{14}N_2$. None was a perfect match for the TIC, which was too obscure to be listed. Instead, it would have to be found the old-fashioned way, via published abstracts. To conduct the search, Sovocool talked the EPA into hiring a specialist, Joseph Donnelly, who scoured hundreds of old scientific journals in search of an industrial process that would yield a $C_{14}H_{14}N_2$ molecule with the required characteristics. Within a week, he found it. In articles published in the 1960s, Donnelly discovered references to a process in which two molecules of acrylonitrile were bound to one molecule of styrene to form a strong, flexible polymer, suitable for plastics. The process had since fallen out of favor, but during the 1960s and 1970s several Japanese and American companies had produced styrene acrylonitrile for use in products ranging from toys to piping.[12] What caught Donnelly's attention was that the chief waste product of the process was a thick, smelly brown liquid called styrene acrylonitrile trimer that matched all of Sovocool's criteria. "It quickly became apparent that this fit the data perfectly," Sovocool recalled.

On November 13, Sovocool sent Genicola a triumphant fax. "I think this is it!" he scrawled on the cover page. On the very same day, another important fax arrived in Genicola's office. This one was worded much more soberly, which, considering the source, was understandable. It was from Craig Wilger of Union Carbide, and it said that the company had identified a suspect based on the $C_{14}H_{14}N_2$ formula Genicola had given Wilger ten days earlier. A search of old records from the company's Bound Brook factory, the letter said, had discovered that Union Carbide in the early 1970s had made its polystyrene plastic by mixing in acrylonitrile. The process created useless trimer byproducts—formed by accident—that were discarded as waste. "The compound of interest may be trimers," the company's letter advised, adding that the company had no information about its toxicity.

Union Carbide was being truthful about that. No agency had ever ordered it to test the safety of styrene acrylonitrile trimer, and there was no indication that the company had ever tried to find out on its own. SAN trimer was an unwanted byproduct, something that was thrown away instead of resold—why bother testing it? But now that it had been found in Toms River's water supply, even at low levels, there was reason for concern. Its structure was similar to those of toxic aromatic compounds like benzene and naphthalene that were known or possible carcinogens. In fact, both of the trimer's namesake components—styrene and acrylonitrile, the combination of which generated SAN trimer—had long been strongly suspected to be carcinogenic and probably mutagenic, too.[13]

The identification of SAN trimer proved that Reich Farm waste was still tainting the town's drinking water and probably had been for years—long after the DEP, the EPA, United Water, and Union Carbide had all claimed the problem was solved with the 1988 installation of the Parkway air stripper. As those agencies knew, an air stripper could be counted on only to remove volatile chemicals like trichloroethylene that vaporized easily. SAN trimer was a *semi*volatile compound; the way to remove it was carbon filtration. But as far back as the late 1970s, the water company had rejected requests to install expensive filters on the Parkway wells. The agencies did not insist, even after Reich Farm was finally confirmed as a source of TCE contamination in those wells. Instead, the EPA sided with Union Carbide and dropped plans to force the company to intercept the pollution plume before it reached the Parkway wells. The EPA had trusted solely in the air stripper to protect the water supply; the unmasking of SAN trimer now made that decision seem like a foolish gamble.

The identification of SAN trimer in Parkway water was like a perfect fingerprint left at a crime scene. It was definitive evidence implicating both Union Carbide and the agencies that were supposed to be protecting the town's water supply. SAN trimer was so obscure that it could only have come from the five thousand drums of Union Carbide waste Nick Fernicola dumped back in 1971.[14] Its identification also strongly suggested that there were *other* strange compounds from Reich Farm in Parkway water. After all, the chromatographs Genicola examined showed many unidentified spectral signatures, not just trimer. The other mystery compounds were at even lower concentrations, but they were present—and Genicola was already making plans to try to identify them, too.

Now Union Carbide, United Water, and the state faced some very uncomfortable decisions. They would have to consider closing important water wells they had resisted shutting for twenty years. Union Carbide would also have to face the possibility of being sued for its role in polluting the drinking water of a town where a cancer cluster had just been discovered. Even for a company accustomed to dealing with catastrophe—the 1984 gas leak in Bhopal, India, killed thousands—the situation in Toms River was alarming.

Union Carbide's belated admission that the mystery chemical was theirs was unprecedented in the tortuous four-decade history of chemical pollution in Toms River. For years, the company had denied that it was contaminating the public water supply, and state and federal regulators had backed them—and backed Ciba, too. Now those denials had been refuted. The residents of Toms River had drunk low levels of Union Carbide's toxic waste in their tap water. That was no longer in doubt. Now the debate would shift to how much they had consumed and for how long, and to whether that pollution had anything to do with the unusually high numbers of childhood cancer cases in Toms River.

Floyd Genicola, a cantankerous state bureaucrat acting on his own initiative, had made it happen. Like Linda Gillick, the anonymous nurse Lisa Boornazian and a few others, he did not turn away from the evidence in front of him. Because he did not, the Toms River saga was about to enter a new and even more contentious stage.

≈

CHAPTER NINETEEN

Expectations

On the day after Wayne Sovocool's triumphant "I think this is it!" message rolled out of Floyd Genicola's fax machine, Governor Whitman told United Water to shut down the Parkway well field—the whole thing, all eight wells, for the first time ever. Overnight, Toms River lost about 45 percent of its water supply, forcing the water company to make up the difference by pumping its other wells to the very edge of their capacity, as if it were midsummer instead of mid-November. How United Water would cope the next summer—the same question that had plagued its corporate predecessor, Toms River Water, during the water crises of 1965 and 1987—was anyone's guess.

Whitman and her health commissioner, Len Fishman, braced for a fresh outbreak of panic as they announced the discovery of an industrial pollutant in the town's drinking water, but the response was surprisingly muted. Now that state officials were no longer keeping information secret, residents reacted more calmly to each new disclosure. The wild scene in the high school auditorium back in March would never be repeated. Almost no one visited the public liaison office that the state health department had recently opened in Toms River; soon, the office would close for lack of interest. An old pattern was reasserting itself: With each passing month, more residents wanted to move on with their lives. The same thing had happened after brief surges of anti-Ciba activism in the mid-1960s and the late 1980s. Now that the town was getting so much negative attention, the desire to change the subject was stronger than ever. Residents were still worried about the water, and they sympathized—from a safe distance—with the Ocean of Love families, but most seemed to think the authorities now had the situation in hand. As Fishman repeatedly pointed out, the water in Toms River had now been tested more thoroughly than anywhere else in New Jersey. For most people in town, that was good enough.

Bruce Anderson adamantly disagreed. Like some other Ocean of Love parents, he wanted to keep the pressure on. Attendance at the monthly public meetings of Linda Gillick's advisory committee was dwindling, but Anderson made a point of never missing one, even though he rarely spoke up. He wanted Fishman and the other officials to know that the families were expecting action to uncover the cause of the cluster and to prevent it from reoccurring. They had been through too much pain to simply accept their children's affliction as a mystery and move on. The families were a small but formidable political force, with ready access to journalists and politicians, including newly elected U.S. Senator Robert Torricelli, one of their early champions. They had earned their bully pulpit in the most agonizing way imaginable, and they were not going to squander it. Their message, consistent and powerful, could be distilled to a single phrase: We want answers. After everything they had been through, who could deny them that?

Cautious and serious behind his wire-rim glasses, Jerry Fagliano had not endeared himself to Linda Gillick and the other Ocean of Love parents during the first year of the state's environmental testing in Toms River. He was an eleven-year veteran of the New Jersey Department of Health and Senior Services (the health department's new official name, as of 1996), which he joined after earning his master's degree in public health from Yale University. Fagliano was a natural choice to serve as the state's on-the-scene epidemiologist in Toms

River, since he had expertise in assessing the health risks posed by hazardous compounds in drinking water. His doctoral dissertation, which he was still finishing up in 1996 as a part-time student at Johns Hopkins, looked at whether pregnant women who drank water tainted with low levels of industrial chemicals were more likely to have children with brain cancer. (He found a modest increase in risk.)[1]

Fagliano had spent most of the previous year quietly searching for a scientifically legitimate way to conduct the epidemiological study that the Toms River families so badly wanted. His reticence at meetings annoyed the parents, but he did not want to promise what he might not be able to deliver. The decision on whether to go forward with an in-depth study would be made by his boss at the health department, Len Fishman, in consultation with Governor Whitman. After the initial panic in Toms River, they had backed down from their flat-out opposition to an in-depth study but said that they would authorize one only if it were based on a credible, testable theory of how pollution could have caused a cancer cluster that was unique to the town. Until the identification of SAN trimer, that seemed unlikely. The discovery of low-level radiation in the town's drinking water might have qualified as a causal hypothesis, except that similar radioactivity had subsequently been found all over South Jersey. The results from soil tests around town, except at the Superfund sites, were similarly unremarkable. The air quality was normal, too—or, at least, was normal now that Ciba had ended manufacturing.

But there was nothing normal about the discovery of a previously unknown compound in drinking water from the Parkway wells. Fagliano had assumed, like almost everyone in state government except Floyd Genicola, that because the extensive environmental testing the state had undertaken in Toms River was aimed at calming the town, nothing surprising would turn up. Now that his assumption had been proved wrong, Fagliano thought that the case for a full-blown study was much stronger. "At that moment, we knew that there was at least one contaminant that people in Toms River were exposed to on a continuing basis, and it was a chemical that was probably unique to the community," he recalled. That was a "sobering thought" but also an almost ideal hypothesis to explore. "Now we had at least one question we could ask because we had a pathway to a unique set of chemical exposures. That was a really critical piece to justify the study."

Now the question was, what kind of study? The fastest, cheapest option would be a case series, similar to the ones Wilhelm Hueper had conducted at DuPont in the 1930s and urologist Arthur Wendel had undertaken with Mitchell Zavon a generation later at the Cincinnati Chemical Works. Those investigators simply interviewed factory workers with bladder cancer and looked for shared exposures that might have caused their illness. In Toms River, many families wanted to be interviewed; they had their own ideas of what types of pollution might be at fault, and they were eager to share them. But Fagliano thought an interview-only study would yield vague and unreliable results. Unlike the dye workers at DuPont and in Cincinnati, who could describe shoveling BNA and benzidine with their bare hands, the affected Toms River families had no idea which chemicals were in the water they had drunk or the air they had breathed. Plus, there would be no control group of unexposed people for comparison purposes— a crucial shortcoming.

Another option would be a prospective cohort study that would, in effect, transform Toms River into a living laboratory for ongoing epidemiological research. For a study period that could last ten years or longer, investigators would monitor the health and habits of the town's eighty thousand residents and regularly test the water, air, and soil. As the years passed, some children in town would get sick, allowing researchers to then look for correlations between those illnesses and possible environmental causes. Prospective cohort studies were highly credible because they did not depend on the foggy memories of participants, but they were extremely expensive and no one—least of all the Ocean of Love families—wanted to wait another ten years for results. Besides, the emergency closure of the Parkway wells had finally interrupted the last known pathway conveying industrial waste to the residents of Toms River. All of the recent environmental testing had turned up no new risks except radioactivity, now known to be ubiquitous in South Jersey. What the town needed now was a study of past exposures, not current ones.

A case-control study was the obvious choice. Its results would not be as reliable as a prospective cohort because it would depend in part on participants' memories of past exposures: how much water local mothers had drunk while pregnant, for example. But most case-control studies could be completed in a year or two, and even the most ambitious rarely cost more than a few million dollars. (The Toms River study would become an exception as it grew, taking five years and costing more than $10 million.) Like all forms of epidemiology, case-control studies could never determine the cause of any particular case, but they could confirm correlations between a disease and risk factors, as Richard Doll had shown in the 1950s with his smoking studies, which he later bolstered with even stronger results from prospective cohort studies.

A case-control study of more recent vintage played a decisive role in convincing Fagliano, and his bosses, that an epidemiological study was worth undertaking in Toms River. It concerned the *other* heavily publicized childhood cancer cluster in the United States: the leukemia cluster in Woburn, Massachusetts, ten miles north of Boston. The similarities to Toms River were obvious. In 1979, upgraded tests had found trichloroethylene and other industrial chemicals in two public wells in Woburn, near a field where hazardous waste had been dumped years earlier. A local mother named Anne Anderson and her parish priest, Bruce Young, together compiled a list of twelve local children diagnosed with leukemia between 1969 and 1979, marking the locations of their homes on a map. They saw an obvious clump: Within a six-block area near the two contaminated wells, there were six leukemia cases—including Anderson's son Jimmy, who died in 1981. The ensuing investigations in Woburn would continue virtually nonstop for another fifteen years, but until 1996, case-control studies at Woburn had failed to identify a likely cause for the cluster. The investigations had foundered because researchers couldn't tell how much pollution Woburn residents had been exposed to in the past. The problem was that Woburn had eight public wells, and there was no obvious way to look back to the 1960s and 1970s and determine which neighborhoods in the small city were getting water from the two contaminated wells versus the six clean ones.

The big breakthrough came in 1991, when a professor of civil engineering at the University of Massachusetts, Peter J. Murphy, completed an astonishingly intricate "hydraulic mixing" computer model of Woburn's water distribution system as it existed between 1964 and 1979, when the two contaminated wells—known as wells G and H —were in operation.[2] His was not the first attempt to build a computer model of a water distribution system, but it was the most elaborate by far. Murphy started by modeling the current water system, a task that required data on the location, age, diameter, and flow rates of 221,000 feet of water piping. When he was finished with the "current conditions" model, state officials conducted a real-world test by using fluoride tracers to see if the water from each well really went to the neighborhoods where Murphy predicted it would. His model passed the test, but he still needed to push back in time to pre-1979 conditions, when the wells were pumping contaminated water into Woburn homes. Consulting archival city records and weather reports as well as the memories of longtime city employees, Murphy adjusted his model and refined its accuracy so that it could estimate the water mix that had been delivered to fifty-four specific neighborhoods during each of the 110 months that wells G or H operated—a feat akin to reconstructing an entire ancient city based on nothing but old maps and scattered artifacts.

By the time the Massachusetts investigators were ready to attempt a new case-control study in Woburn, they not only had Murphy's water model, they also had nine additional cases and thus could be more confident that their results would not be overly influenced by chance. Twenty-one local children—about four times more than expected for a town Woburn's size—had been diagnosed with leukemia between 1969 and 1986. Using the home addresses of the nineteen cases whose parents had agreed to participate, along with thirty-seven matched controls, investigators used Murphy's model to see if children who drank water from wells G and H were more likely to develop leukemia than those who did not. In May of 1996, at the same time Linda Gillick was pushing for a similar study in Toms River, the Massachusetts Department of Public Health announced the results of the new Woburn study. There was little difference in leukemia incidence between Woburn children who drank the tainted water and those who did not, but there was a huge increase in risk if their *mothers* had consumed the water while pregnant. In fact, mothers who drank any water from wells G and H while pregnant were eight times as likely to give birth to children who later developed leukemia, compared to those who drank none. And mothers who were particularly heavy drinkers of the contaminated water were *fourteen times* more likely.[3] The study, one of its authors declared, shows that "it is likely it was the water, and we believe it was the water during pregnancy" that was responsible for the cluster.[4]

It was a stunning outcome, and not only because it followed a long string of inconclusive studies at Woburn. For the first time, an epidemiological study had identified the likely cause of a residential cancer cluster—a landmark achievement that raised doubts about the cluster-busting stance of the Centers for Disease Control and Prevention. The study confirming the Woburn cluster and identifying its likely cause had its impetus sixteen years earlier with Anne Anderson's observation that there seemed to be too many children with leukemia in her neighborhood. That was precisely the kind of anecdotal report that the cluster-busters, led by Kenneth Rothman of Boston University, declared had "little scientific value" and that Alan Bender of the Minnesota Department of Health called an "absolute, total, and complete waste of taxpayer dollars." Now the Woburn experience suggested that perhaps residential cancer cluster epidemiology was not a complete boondoggle after all.

As soon as he heard about the Woburn study results in May of 1996, Jerry Fagliano saw their significance for Toms River. Woburn had been a touchstone of his career for fifteen years; it was the key reason he had specialized in drinking water epidemiology. He knew all about the difficulties of estimating past exposures, and he considered Murphy's water model to be a brilliant solution. The parallels between Woburn and Toms River became even more obvious six months later, when SAN trimer was found in two of United Water's wells in Toms River but not in eighteen others. If a case-control study could be paired with a water-distribution model in Woburn, Fagliano thought, why not Toms River? With the help of a Murphy-style model of Toms River's water system, he could see which case families had been exposed to water from the tainted Parkway wells and could thus determine whether drinking Parkway water increased the risk of cancer. Woburn could even help with another crucial part of a study: the long questionnaire that would have to be administered to all of the participating families. Devising the questions was difficult because there were so many topics to cover and the wording had to be as neutral as possible. Fortunately for Fagliano, Massachusetts researchers had already developed an elaborate questionnaire—one Fagliano thought he could adapt to Toms River.

All the pieces were falling into place for the study the families had been seeking. In January of 1997, Fagliano and Elin Gursky flew to Atlanta to finalize the arrangements with the Agency for Toxic Substances and Disease Registry, which would fill Peter Murphy's role by constructing a water distribution model for Toms River. A few days later, Gursky announced that the case-control study would go forward. "Thank God," was Linda Gillick's reaction. "It's the only way we will find out what happened here."[5]

Jerry Fagliano was not so sure. Now that his study had been approved, he was more worried than ever that the families' expectations were too high. In Woburn, the key contaminant, trichloroethylene, was a known carcinogen (by the 1990s, at least), but no one knew whether styrene acrylonitrile trimer was dangerous at low levels. Fagliano was also not sure that Murphy's modeling feat could be replicated in Toms River, where the piping system was much larger and had changed drastically with the town's explosive growth. His biggest concern, however, was that there were too few cases for a statistically valid study. He had already decided to limit the study to childhood leukemias and brain and central nervous system cancers, because those were somewhat more common and thus less subject to the uncertainty of random variation. Even so, the state's newly updated case totals showed that in the Toms River "core zone," the bull's-eye of the cluster, there were just five brain cancer cases and four leukemias between 1979 and 1996. That was about eight times more than expected, but it was still just nine cases—too few, Fagliano thought, to generate a statistically meaningful result in an epidemiological study. The other problem was that a case-control environmental study needed to compare exposed and unexposed groups, but it was likely that almost everyone in the core zone had drunk at least some Parkway water and thus would be classified as exposed, since those wells were so close to the middle of town.

A larger study of the entire township, instead of just the core zone, Fagliano thought, would ease these problems. Some parts of the town did not get Parkway water, and there were many more cases: twenty-two childhood leukemias and eighteen brain and central nervous system cancers between 1979 and 1996. Those totals were still low enough to make Fagliano nervous about statistical significance, however. Woburn was less than half the size of Toms River, yet had almost the same number of childhood leukemia cases—twenty-one—over a different seventeen-year period (1969 to 1986). Toms River had 60 percent more childhood leukemia than expected, but Woburn had 400 percent more.

With just forty cases, the Toms River study would need dramatic results to reasonably exclude the effects of chance. Fagliano had calculated that he would need to show at least a three- or fourfold increase in cancer risk among exposed children in order for the results to be statistically significant. "We were pessimistic, and we told the families that," Fagliano recalled. "Whether they heard us, I don't know."

Jerry Fagliano was not the only person who saw parallels between Woburn and Toms River. In 1996, Linda Gillick read *A Civil Action,* a newly published book about the convoluted, excruciating lawsuit that arose from the Woburn investigations. The antihero of author Jonathan Harr's book was a charismatic, tenacious, and ultimately self-destructive young lawyer named Jan Schlichtmann who represented eight families struck by leukemia, including Anne Anderson's. On their behalf, he embarked on a nine-year odyssey through the Massachusetts courts, from 1981 to 1990. Seeking $400 million, Schlichtmann sued three companies whose properties near wells G and H were contaminated with toxic waste. He was a passionate advocate but made a series of tactical errors, including passing up opportunities to settle the case, antagonizing the judge, and underestimating the determination of his deep-pocketed opponents. After a series of adverse rulings, Schlichtmann had to accept a settlement offer of just $8 million from one of the two major defendants, the manufacturer W.R. Grace. The other, Beatrice Foods, paid nothing at all.[6] After deductions for legal expenses and fees, the eight families ended up with less than $500,000 each, and the high-living Schlichtmann ended up homeless and more than $1 million in debt. "I was totally exhausted at the end of that experience," he recalled years later. "I was bankrupt financially, spiritually and emotionally. I had no car, no condo and no clothes. But the most devastating thing for me was that I didn't want to be a lawyer."

Declaring personal bankruptcy in 1990, Schlichtmann moved to Hawaii (a friend bought him the plane ticket), camped out on the beach, and renounced the practice of law—though only temporarily, as things turned out. When *A Civil Action* was published in 1995 and became a bestseller, Schlichtmann was suddenly in demand on the lecture circuit. He also collected a $250,000 consulting fee from Disney, which was producing a movie based on the book, with John Travolta starring as the young knight-errant tilting at corporate dragons. Schlichtmann's tenuously revived law practice got a big boost, too. There were dozens of calls from people who believed they had an environmental case just like the families in Woburn—only winnable.

Linda Gillick called Schlichtmann in January of 1997, two months after the discovery of SAN trimer in the Parkway wells. "Before the trimer, I never heard anyone say, 'Let's sue them!' because there was no one to sue," remembered Kim Pascarella, the local lawyer who became active in Ocean of Love after his infant daughter died of brain cancer. "But once a responsible party was identified, I told Linda you have to start thinking about it. You had a potential defendant now" in Union Carbide and United Water—and maybe even Ciba. So Gillick called Schlichtmann and told him about the cancer cluster and Ocean of Love, and about Ciba, Reich Farm, and the Parkway wells. The families might soon need legal counsel, Gillick said. Was Schlichtmann interested? He was. Schlichtmann was intrigued by the situation in Toms River and impressed by the organizing work Gillick had done at Ocean of Love. Gillick wanted to be an active participant in any future case; that was a quality Schlichtmann loved in a potential client.

Before the discussion went any further, though, Schlichtmann wanted to change Gillick's expectations of him and her potential case. Gillick had called because she admired the Jan Schlichtmann of *A Civil Action,* but that brash gladiator no longer existed, he told her. The searing experience of Woburn, and the emotional breakdown and exile that followed, had triggered a personal transformation. Schlichtmann was now an evangelist for a very different kind of lawyering that valued mediation over combat, reconciliation over victory. The lesson of Woburn, he liked to say, is that the civil justice system is built on a lie—that vanquishing the other guy will solve your problems and make you happy. "As soon as you say, 'I'm going to use the legal system to impose my will on you,' you're buying into the lie," Schlichtmann said. "You'll get chaos, and the system will disappoint you. You'll get decision-making that only makes sense in Bedlam. We think the legal system is there for us, but it's not. It's there only for itself."

Schlichtmann was not sure if he made much of an impression on Gillick. He thought she sounded like a tough, smart woman who had been through a lot and was not ready to talk peace with the companies she blamed for her son's illness. Resolving conflicts outside of the adversarial system required a recognition that neither side had a monopoly on the truth, but the Gillicks—both Michael and Linda—were sure that they already knew the truth: Toxic pollution had caused Michael's illness. What they wanted was proof and vindication—things they could not get from the legal system, Schlichtmann believed. Even so, he and Linda Gillick ended the conversation on a friendly note, promising to keep in touch.

At the Ciba plant, meanwhile, an era was ending in bitterness. The last few production employees left the factory at the end of 1996, including the union president John Talty, his brother Ray, and George Woolley, who had headed the environmental committee. John Talty had started as a laboratory technician in June of 1960 at a salary of $78 per week. At the end, he was making $760 a week as a janitor, pushing a broom around the shuttered buildings. Collectively, the three men had put in 104 years at the Toms River factory. Over and over, they had backed the company in its battles with local activists, despite their worries about their own health. Now, they felt abandoned. A few months earlier, a maintenance worker had found an unsigned memo in a garbage can that laid out the company's confidential strategy for its final negotiations with the handful of union workers left. Ciba's priority now, according to the memo, was to "minimize negative impact on sensitive community relations issues," including "the cancer cluster." For the union employees, there would be no severance pay and no special consideration for rehiring in the pending Superfund cleanup of the factory property, according to the memo. Furious, John Talty called a press conference and accused Ciba of holding back information about cancer cases among the employees. The company denied the charge, citing the ongoing update of the epidemiological study by the Alabama researchers. Talty later met with EPA officials and even with Linda Gillick, telling them what he knew about Ciba's disposal practices.

"At the end, there were very hard feelings," Talty recalled. "I worked there for thirty-seven years, and I didn't get a single penny of separation pay." What was even worse, though, was the chilling expectation, shared by many of his union brethren, that cancer would eventually come for them, too. "People didn't like to talk about it," he said, "but they thought about it. How could you not think about it?"

To no one's surprise, the spring and summer months of 1997 were a struggle for United Water Toms River, thanks to the contamination of the Parkway wells. The spring was drier than usual, and lawn-watering season had begun. The water pressure was so low that United Water could not even flush the hydrants in some sections of town, something it typically did twice a year to remove iron buildup in pipes. As a result, the water in certain neighborhoods looked, smelled, and tasted foul. One angry resident brought a jug of tea-colored water to a meeting of Gillick's advisory committee in June. "I've got a Jacuzzi full of this crap," he complained.[7] Many in town assumed that chemical pollution was to blame for the discoloration, though the actual culprit was iron. Industrial pollutants, measured in the parts per billion, were invisible.

Ignored or mishandled for a generation, the underground plume of pollution from Reich Farm was now an acute dilemma for the water company. United Water would never make it through the summer without the 45 percent of its supply that came from the Parkway well field, but that was only the most obvious problem. The deeper issue was that it needed to restart the two tainted Parkway wells, wells 26 and 28, as soon as possible in order to keep the Reich Farm pollution plume from being sucked up by four other shallow Parkway wells, as had already happened during earlier shutdowns of wells 26 and 28. The plume hidden beneath Pleasant Plains was like a serpent ready to strike at anything that moved; the only solution was to give it a sacrificial target. Wells 26 and 28 would have to start pumping again, but this time only as pollution recovery wells, not water supply wells. The other Parkway wells could then reopen as supply wells, and Toms River would have just enough water to get through the summer.

The irony was that after more than twenty years of delay and denial, carbon filters would finally be installed on wells 26 and 28—even though, for the first time, no one would be drinking the water they pumped. SAN trimer would at last be removed from the Parkway water, but the filtered water would be pumped back into the ground instead of distributed to residents. Union Carbide finished installing the filters in May, and in June the Parkway wells started pumping again—just in time to head off another water crisis.

This jury-rigged plan, however, was not enough to get United Water through July, when demand hit a record twenty-two million gallons per day. The water company responded by pumping the newly reopened Parkway wells beyond their normal limits, a move that risked pulling in pollutants that were now being sucked into wells 26 and 28. But even then, United Water was still running short of water, so it quietly sought and received special permission from the state and town to change the plan and start using water from 26 and 28 too. Without informing its customers, the water company pumped the carbon-filtered water from those two wells into its distribution system for several weekends instead of discharging it into the ground. Gillick

and her fellow committee members were furious when they found
out, but Mayor George Wittman Jr. said that the town council had no
choice but to allow the diversion. Mandatory curbs on water use, he
said, would not have worked because the town had too few employees
to enforce them.

To Gillick and the Ocean of Love families, the secret, if brief, reopening of the two water wells suggested that all of their efforts over
the previous eighteen months were being ignored. The closure of the
Parkway wells had been their most tangible victory so far, but now six
months later all the wells were open again—even wells 26 and 28,
briefly—because the town council would not even try to order homeowners to reduce their lawn watering. "We're getting a cocktail of
contamination in our water again," Gillick complained that summer,
even though there was no longer any evidence that industrial pollutants were actually reaching homes, thanks to the new carbon filtration system.[8]

Even a reconfirmation of the cancer cluster did not shake the town
out of its torpor. The state health department completed its promised
update of the cancer registry in April of 1997, issuing a report confirming Michael Berry's earlier findings and updating his statistics
through 1995, instead of 1991. The new data seemed to show that the
frequency of brain tumors and leukemias had peaked in the late 1980s
and early 1990s and had been easing since then, though the decline
was still too recent to be convincing. For the Ocean of Love families,
the new data provided a vindication of sorts. "I think once and for all
we can put to rest the theory that there isn't a problem," Gillick said.[9]
She was irritated that there had been little progress on the case-control
study. Jerry Fagliano and his team were still designing it with the help
of a group of outside experts, including several Woburn veterans. Fagliano was also consulting with Gillick's advisory committee, which
had many ideas for expanding the study's scope.

The wheels of government were turning, but not fast enough for
the Ocean of Love families. Frustrated by the apathy of the town, Gillick, Pascarella, and others wondered if there was anything else they
could do to spur the investigation while also protecting the families'
rights in case there was a lawsuit. Perhaps, they thought, it was time

to bring in a lawyer. Gillick called Jan Schlichtmann again, this time inviting him to come down to Toms River and meet with eight of the families at her home. The visit did not go well. When Schlichtmann began talking about his post-Woburn transformation, Michael Gillick stopped him. As Schlichtmann remembers it, Gillick said: "We're dealing with wolves, and you need to *be* a wolf to deal with a wolf." Schlichtmann argued back. "I know about wolves," he said. "We have a whole mystique about wolves. But wolves don't kill people. We can be a little more sophisticated about this instead of going back to our old paradigms." Besides, he added, the mortal combat approach did not work. "If it worked," he added, "you wouldn't be talking to me."

It was a deflating message. "You don't want to hear from your lawyer that the other side has more money and you're going to lose. You want to believe that we live in America and justice is going to prevail," said Pascarella. He knew from his own experiences as a lawyer that the system was not always fair, but he was nonetheless irked by Schlichtmann's approach. "He definitely was not very warmly received," Pascarella recalled. The Gillicks also were not immediately sold—"I actually wanted my day in court," Linda Gillick explained later—but she shared Schlichtmann's cynical view of the legal system and agreed that his method could produce a faster result.[10] She knew that many of the families were having money problems and did not have the emotional stamina to wait out a decade-long legal ordeal. Thanks to Woburn, Schlichtmann was experienced in cases involving cancer and groundwater, even if his knowledge had been gained through trial and error (or, perhaps more accurately, error at trial). Thanks to the fame of *A Civil Action,* the movie of which was already in production, he would be a magnet for media attention, which had been essential to the families' success so far. Finally, Gillick saw in Schlichtmann's Woburn odyssey a quality she valued dearly: loyalty. As she later told a reporter, "Here's a man who ended up losing his practice and everything he'd worked for in life to make a commitment to these families."[11]

Jan Schlichtmann was not the only lawyer Linda Gillick spoke to, but most of them wanted a large up-front retainer fee. They did not want to repeat Schlichtmann's mistake of investing millions of dollars in a sprawling case and losing their shirts. (In Schlichtmann's case, it was literally true: After Woburn, he had to sell his clothes.) But Schlichtmann did not demand a retainer. Neither did Mark Cuker, a Philadelphia attorney who had also been in touch. They were willing to be paid on a contingency basis; they would receive 25 percent of any money the families ultimately received.

Mark Cuker was a quietly intense, studious lawyer who took justifiable pride in his mastery of the details of his environmental cases. He was known in Ocean County for his representation of two hundred homeowners in Pine Lake Park, a development just north of the Ciba factory. Dozens of backyard wells there had been contaminated with solvents since the mid-1980s, but Ciba was not the cause because its pollution plumes flowed east toward Oak Ridge Estates, not north to Pine Lake Park. The case had languished for years until Cuker took it over and pursued it vigorously, targeting a nearby asphalt plant that used TCE. He later joked that he knew he had a winner when he visited the asphalt plant's "laboratory," a shed where the solvents were kept, and found a dead animal on the stoop.

The asphalt plant's lawyer, Ellis Medoway, told Cuker that he should be suing Union Carbide, arguing that Reich Farm was the real source of the groundwater contamination at Pine Lake Park. Cuker thought that was very unlikely. Pine Lake Park was almost two miles west of Reich Farm and on the other side of the Toms River. Still, Cuker did not want the asphalt company to be able to blame someone else if there were a trial, so he added Union Carbide to the lawsuit and took sworn depositions from the key characters involved in the 1971 dumping at the egg farm, including Nick Fernicola and his brother, Frank. Cuker was both amused and appalled by the Fernicolas' descriptions of their misadventures in the waste business. In 1996, when Reich Farm was back in the news because of the cluster investigation, Cuker thought that Linda Gillick would be interested in the Fernicola transcripts. He got her phone number from one of his Pine Lake Park clients, called her, and sent her the depositions. He figured that Gillick would want to read them, and he also thought that she might need a lawyer sometime soon.

Almost a year later, in the late summer of 1997, Gillick invited

Cuker to a get-acquainted meeting similar to the one with Schlichtmann. Cuker brought one of his law partners, Esther Berezofsky, a
gregarious trial specialist who would also work on the case if they
were hired. A few months earlier, in May, Cuker had settled the Pine
Lake Park case for an impressive $4 million. (The asphalt company
paid; Union Carbide had already been dismissed from the case, as
Cuker knew it would be.) Their meeting with the families went well,
but Cuker was concerned when Gillick called a few weeks later and
told him that she wanted to hire his firm *and* Jan Schlichtmann, the
attorney from *A Civil Action*. Cuker knew about the Woburn case
and Schlichtmann's subsequent flameout; he wondered whether it
would be possible to work with a person like that.

Cuker's low-grade worry turned to high-grade anxiety after he,
Berezofsky, and their third law partner, Gerald Williams, met with
Schlichtmann to discuss the potential arrangement. Cuker was dubi-
ous when Schlichtmann explained his new passion for non-adversarial
problem solving, and he found it ridiculous that Schlichtmann had
told the families the case could be resolved in eighteen months. In essence, Schlichtmann was arguing that the families and the companies
could arrive at a mutually satisfactory result if all of the lawyers involved, and their scientific experts, made a commitment to listen re-
spectfully to each other's arguments. "The difference between the Jan
we met that day and the Jan of the book was really a shock to me,"
Cuker remembered. "He seemed very idealistic about this nonlitiga-
tion process. It sounded very pie in the sky."

Still, as crazy as some of Schlichtmann's ideas sounded to Cuker
and Berezofsky, they had to admit his new philosophy matched up
well with the Toms River case, which would be difficult to win via a
conventional lawsuit. The children would surely capture the hearts of
a jury, but maybe not their minds. Proving that a company was liable
for toxic dumping was only half the battle in a case like this—the easy
half. The much tougher challenge was to prove that the dumping
caused the plaintiffs' injuries. Proving causation would be particularly difficult in Toms River, for at least four reasons: The childhood
cancer rate in town was high but not extremely so, the pollutant concentrations detected in the water and air were relatively low, the worst

exposures had occurred years earlier, and the health risks of some of the pollutants, including SAN trimer, were unknown.

To make a credible argument for causation in a case like this one, a lawyer would have to hire a small army of scientific experts: chemists to identify the contaminants at issue, industrial hygienists to estimate how those contaminants were produced and used, hydrogeologists and atmospheric chemists to account for their movement through soil and air, toxicologists to assess their potency, epidemiologists to estimate the risk they posed to the exposed population, and physicians to explain the subsequent disease. The exorbitant cost of these experts (some charged $400 an hour) was a key reason why Woburn and similar cases sucked up millions of dollars and took so many years to litigate—and why most plaintiffs' lawyers wanted nothing to do with them.

Schlichtmann's approach offered a cheaper, faster route to a more certain destination. The pot of gold at the end of the road was likely to be smaller than what a sympathetic jury might award. But it was unlikely to be nothing at all, which was always a possibility in the high-risk world of trial litigation. Keeping the case out of court also would eliminate the blizzard of extraneous legal filings that stretched complex cases to absurd lengths. The three likely defendants, Ciba, Union Carbide, and United Water, were big companies with vast resources; if the case ever devolved into a contest to see who could file more motions over a longer period of time, the families would surely lose.

The Toms River case was challenging, but it did have some attractions for the lawyers. Government investigators were already doing some of the work that would otherwise have to be undertaken by experts-for-hire. Both Superfund sites had already been thoroughly assessed, and the state had tested the water, soil, and air all over town. Of course, Jerry Fagliano and the rest of the government researchers were not in thrall to either side; no one knew how their case-control study would turn out, or even when they would finish it. But that uncertainty was just as big a problem for the potential defendants as it was for the plaintiffs. With the famous Jan Schlichtmann involved, the case would generate a stream of bad publicity for United Water, Ciba, and Union Carbide, each of which had strong incentives to avoid negative attention. United Water and Ciba (thanks to its Superfund cleanup) were still active in Toms River, and Union Carbide's reputation had been battered by Bhopal and other accidents. They would not be eager to go to war against grieving families, especially with the assertive Linda Gillick in the forefront.

"It boiled down to a couple of things," Berezofsky remembered. The companies "had a public relations nightmare on their hands, and we had real causation issues. We thought we could prevail, but it was a real risk." Schlichtmann's offbeat ideas held out the hope of at least partially reducing the risks. After some internal debate, Cuker and Berezofsky told Gillick that they would take the case, in collaboration with Schlichtmann. They would start with his conciliatory approach. If it did not work, they would sue.

The lawyers' first task was to figure out whom they were representing. The Ocean of Love of families wanted only children with cancer to be included in the legal case, and the lawyers readily agreed; the sympathy those children evoked was their strongest asset. Besides, the only evidence of cancer clustering in Toms River was for children—the state had not analyzed adult cancer rates in the town. In that sense, Michael Berry of the state health department had already set the parameters of the potential lawsuit. Berry's 1995 incidence study had focused on all forms of childhood cancer, but not adult cases, because that is what nurse Lisa Boornazian had asked for. His study had stretched back to 1979 because that happened to be the first year of reasonably complete records from the cancer registry. Those decidedly unscientific criteria defined the Toms River cluster as a kids-only, all-cancers, post-1978 phenomenon. The lawyers and families together decided to follow the same criteria.[12] There was a geographic restriction, too, again thanks to the health department. Its latest data showed that cancer rates were higher in Toms River than anywhere else in Ocean County, so the lawyers and Gillick decided that only children who had lived at least part-time in the town before they were diagnosed would be in the case—a wrenching decision since Ocean of

Love helped families all over the county. In all, about one hundred children met the case criteria. The lawyers quickly signed up the families of thirty-eight. Sixty-nine would join eventually.

With his clients in place, Cuker picked up the phone and called Vincent Gentile, the Union Carbide lawyer he knew from the Pine Lake Park case. Trying hard to set the nonconfrontational tone Schlichtmann wanted, Cuker called in the late afternoon, when he thought Gentile would be more relaxed. "I've been contacted by people in Toms River, New Jersey, who have kids with cancer," Cuker told him. "I'm investigating, and I don't know if it's a case. We don't want to file suit, but we do want to meet with you." Another lawyer at Gentile's firm, which was one of the largest in Philadelphia, joined them on the line. "You'd better watch out," he told Cuker. "You'd better be sure you don't lose your shirt like that guy from Massachusetts, the one in the book."

Cuker didn't know what to say to that, so he said nothing.

PART IV

CAUSES

≈

CHAPTER TWENTY

Outsiders

In 1854, when John Snow decided to investigate the cause of the cholera epidemic in the Soho section of London, he walked ten blocks from his office on Sackville Street to the heart of the outbreak zone and started knocking on doors. Within four days, he had pinpointed a likely cause and launched modern epidemiology. Jerry Fagliano could only dream of having that much freedom of action. He was the nominal chief investigator of a study that in reality had many masters: the state health department, the federal Agency for Toxic Substances and Disease Registry, and two review panels of outside scientists. Most especially, Fagliano had to be responsive to Linda Gillick's advisory committee. His bosses in the health department were intent on avoiding any more bad headlines from Toms River because, for most of 1997, Governor Christie Whitman was enmeshed in a hard-fought reelection battle. She won in November by just twenty-five thousand votes, 1 percent of all votes cast.

By the time all those groups had signed off on the study protocol, it was January of 1998, twelve months after Fagliano had received his initial approval. By then, the case-control investigation had expanded far beyond his original vision of a Woburn-style study focused on the possible links between the Parkway well water and local cases of childhood leukemia and brain cancer. Everyone seemed to have an idea of what else should be covered, and few of those ideas were turned down (the state did draw the line at including adult cancers, as some town residents had requested). Now that the Toms River cluster had the full attention of the media and state politicians, money was no impediment. More than $2 million in federal, state, and county funds had already been spent, most of it on the emergency testing Governor Whitman had ordered in 1996. She had asked Congress for another $5 million, a request several New Jersey congressmen formalized by introducing the "Michael Gillick Childhood Cancer Research Act." Whitman eventually got every penny she asked for, and the state kicked in millions more—all for a town that was home to just 1 percent of New Jersey's population. All told, the Toms River investigation, including the case-control study, would cost well over $10 million, though no one ever tried to calculate the exact amount; too many people and agencies were involved to ever get an accurate total.

The discovery of SAN trimer in the Parkway wells made Union Carbide an obvious target of any investigation, but the Ocean of Love families were adamant that they wanted Ciba—the town's chief source of environmental anxiety for two generations—included in the study too. That was not an easy call for the state. The health department had information about local cancer cases only since 1979, and the last time Ciba's pollution was known to have reached the public water system was during the 1960s, when dye waste tainted the riverfront Holly Street wells. Because of this disparity in timing, Fagliano knew that very few children in the planned case-control study could have been exposed to the contaminated Holly water. With only a handful of exposed cases, any finding implicating Ciba as a cause of local cancer would be statistically suspect. On the other hand, the ATSDR had already agreed to spend millions of dollars on a computer model of the town's water system to find out which neighborhoods received polluted water from Parkway wells during the 1980s and 1990s. Extending the model back to the 1960s to look at the Holly contamination would not be much additional work. Fagliano thought it was worth trying.

He also supported studying the *other* way that many people in

Toms River had been exposed to pollution from Ciba—from the air—even though an air pollution study would entail even more guess-work than a groundwater study. In groundwater, pollutants moved very slowly, perhaps four hundred feet in a year. In the air, they traveled that far in a few seconds. The groundwater beneath Pleasant Plains was a geological archive holding the evidence of past pollution at Reich Farm, but there was no such record of the airborne emissions from Ciba's smokestacks during the years when the factory was running at full throttle. There was, however, a way to guess at the impact of those air emissions. Researchers could consult old weather records and see which neighborhoods tended to be downwind from the factory and thus bore the brunt of its air emissions. Then they could see if there were more childhood cancer cases in those downwind areas compared to other parts of town. It was a virtually identical strategy to the drinking water part of the study, in which Fagliano and his team would look to see whether cancer rates were highest in the neighborhoods that got the most Parkway water or Holly water during the critical years when the wells were known to be tainted with chemical waste. In all, there would be three windows of exposure to study: the Parkway wells from 1982 to 1996, the Holly wells from 1962 to 1975, and Ciba air emissions from 1962 to 1996.

The next step was to select the participating children. According to the state cancer registry, twenty-two Toms River children had been diagnosed with leukemia and eighteen with brain or other central nervous system cancers during the study years. Each of those forty cases would be matched with four "control" children from Toms River who had never had cancer but otherwise mirrored their designated "case" child in age and gender. Fagliano and his colleagues would use the water-distribution and wind-direction computer models to see if case children were more likely to have lived in highly polluted neighborhoods during the critical years, compared to the matched cancer-free children. Finally, working from a detailed script, researchers would conduct long interviews with all two hundred participating families to see whether other factors such as nutrition, smoking, and occupation were influencing the results.

It was an ambitious plan, but Linda Gillick and her committee were still not satisfied. Some Ocean of Love parents lived in other parts of town where waste chemicals had leaked into groundwater, including near the now-closed town landfill and Ciba's old leaky pipeline to the ocean. Others were worried about radiation from the Oyster Creek nuclear plant, ten miles south of town. There was no direct evidence that children had been significantly exposed to carcinogens from those sites, but the state agreed to investigate them all—again, by looking to see if cancer rates were higher in neighborhoods with the highest exposures. There was almost no chance that there would be enough exposed cases to generate statistically meaningful results for any of these micro-studies, but Fagliano made no apologies for including them. "The mandate was to leave no stone unturned in terms of potential past exposures," he recalled. The state even hired researchers to analyze dust in local homes, to see if its chemical composition was unusual. (Learning that it was not cost several hundred thousand dollars.)

Even after all of those extras were added to the study, some parents still felt left out because their children were victims of other types of cancer, not leukemia and nervous system tumors. Other aggrieved families argued that the study should include children who were born in Toms River but moved away before they were diagnosed. To satisfy them, the state agreed to conduct a *second* case-control study. This one would have the same aim but would rely only on home addresses listed on birth certificates, not interviews. With forty-eight cases and 480 controls, the birth record study would be more than twice the size of the interview study but would be easier to conduct because there would be no questionnaire and no need to secure permission from each family.

The state's politically driven, the-more-the-merrier attitude about expanding the studies pleased the Ocean of Love parents, but it carried scientific risks. Fagliano's planned investigation had grown from testing a single hypothesis—that children exposed to tainted Parkway water had a higher risk of leukemia and brain cancer—to testing more than a dozen hypotheses in two studies. Each addition increased the chance that an association uncovered by the studies might be due to luck instead of a real environmental risk. The typical, if somewhat arbitrary, definition of a statistically significant result was one that had no greater chance than one in twenty of being a random fluke (thus the 95 percent confidence interval). But if researchers looked for many associations between pollution and cancer, the likelihood of at least one "false positive" would grow—in the same way that a dice player would be much more likely to roll snake eyes if he played twelve games of craps instead of just one. "If you search hard enough in the data and analyze it enough ways, you can always find something," said Dan Wartenberg, an epidemiologist at Rutgers University and an outside adviser to the Toms River study. "If you just keep adding hypotheses and doing more analyses, then you have to wonder about any associations you see."

State health officials felt that they had no choice but to run that risk. If they left out of the study anything that the families were worried about, they would be criticized and the study results would be second-guessed. "We didn't want to have a situation where we didn't analyze something and show some results," Fagliano explained, "because then people would say, 'What about this? Why didn't you look at that?' We felt we had to be exhaustive."

Jerry Fagliano and his colleagues were going to look at everything, and in March of 1998 they were finally ready to begin.

Time passed achingly slowly in Toms River now that scientists and lawyers were dictating the pace. In a legal sense, the clock stopped completely. The new phase began on December 12, 1997, when the families' lawyers—Jan Schlichtmann, Mark Cuker, and Esther Berezofsky—appeared at a press conference near the old courthouse in Toms River. Mayor George Wittmann Jr. didn't exactly roll out the welcome mat. "I'm disappointed that the town is going to receive additional negative publicity, when we've worked so hard to create a more positive image," he told the *Asbury Park Press*.[1] The arrival of a celebrity lawyer like Schlichtmann was sure to bring more unwanted attention to Toms River, further fraying the ties between the families and the rest of the town.

But the lawyers and their clients were careful not to play into the money-soaked narrative of shark attorneys and passive victims. They had not come to file a lawsuit, Schlichtmann declared. Instead, they hoped to "open up lines of communication" with Union Carbide, Ciba, and United Water, the companies their clients suspected had harmed their children. "We want to work together as partners to find out what happened here," Schlichtmann said. A lawsuit, he added, would be a last resort. "It's foolish, absolutely foolish, for anyone to choose that road, when we have another alternative," he declared.

As for the families, they were not going to just watch and wait while their attorneys tried to settle the case. With the lawyers' help, they had formed a new organization called Toxic Environment Affects Children's Health. The mission of TEACH, Schlichtmann explained, would be to educate the community about environmental hazards, especially the two Superfund sites, and to lobby for stricter cleanups. It was an advocacy group whose membership consisted solely of the families who were part of the potential legal case. Schlichtmann loved the idea—and not only because he had thought of it. Whether or not TEACH turned out to be an active group (it was, at times), it brought the families closer to each other and to their lawyers, creating a sense of shared mission and mutual support. It also showed the town that the families wanted to work for the betterment of all Toms River, not just for themselves. And perhaps most importantly, it signaled Union Carbide, Ciba, and United Water that the families were determined to stick together and pursue the case. "It sent a message: We're united and we're serious," remembered Bruce Anderson, who became the most active member of TEACH. Anderson designed T-shirts for the group depicting a bald-headed child clutching an intravenous stand. He even created a website, www.trteach.org, which he updated diligently with news and information about Toms River's pollution problems.

Schlichtmann's stay-out-of-court strategy could succeed only if the companies believed that the families were willing to sue if they had to, but the clock was running out on that threat. Under New Jersey law, a plaintiff in a personal-injury case has just two years to file a lawsuit after discovering the alleged cause of the injury. There was an exception for children. For them, the two-year clock did not begin running until they turned eighteen. But some of Schlichtmann's clients were already over twenty (to be included in the case, they merely had to be underage when diagnosed), so the question of exactly when the two-year clock started ticking was a crucial one and not at all easy to answer. A judge could conceivably decide that it was as early as September of 1995, when Linda Gillick first saw a copy of Michael Berry's cancer incidence study. So for the plaintiffs, the most conservative strategy would be to sue as quickly as possible.

A rush to the courthouse, however, was the opposite of what Schlichtmann wanted. His vision called for a period of fact-finding research, followed by expert presentations and discussions among all the lawyers, with the goal of reaching a settlement. It would work only if there was an atmosphere of mutual respect, but building trust among adversaries could take a very long time. Schlichtmann had told the families that there could be a resolution in just eighteen months, but no one believed it. Realistically, the only way his alternative process could possibly succeed would be for the lawyers to negotiate an agreement to "toll," or temporarily halt, the two-year clock on the statute of limitations. If the companies were willing to sign onto a tolling agreement, there was a chance that Schlichtmann's out-of-court strategy might work. If they were not, the case would almost certainly turn into Woburn II, with years of motions, trials, and appeals and millions of dollars in expenses, all for a highly uncertain outcome.

The issue came to a fast boil at an all-day meeting in Princeton on January 15, 1998. Present were more than a dozen mutually suspicious lawyers. Later, many of them would spend so much time together that they would learn the landscapes of each other's lives: the weddings, the illnesses, and the funerals. For now, though, they were strangers playing hardball, and they did not like each other very much. The most aggressive was Robert Butler, Union Carbide's chief litigation counsel, who "basically said our clients were a bunch of greedy opportunists looking to profit from the misfortunes of their own children," remembered Esther Berezofsky. "He was fairly sneering at the notion that our clients had anything close to a legitimate claim. For me, it was hard not to get up and tell them where they could take their condescending arrogance."[2]

Schlichtmann's beautiful dream of a new kind of lawyering was turning ugly fast. The more toxic the atmosphere in the conference room became, the more frantically he spoke about the need for both sides to avoid another Woburn. The chief outside lawyer for Union Carbide, William Warren, had read *A Civil Action* and was surprised at what he was hearing from Schlichtmann. "I think that at the beginning everyone was appropriately skeptical," said Warren, who headed the environmental practice group at one of Philadelphia's largest law firms, Drinker Biddle. Butler, his colleague, was less shocked because he had read the more recent news stories about Schlichtmann's self-exile to Hawaii and fervent conversion to conciliation. That did not make it any easier to negotiate a tolling agreement, however.

"It was a very, very tense meeting, it almost broke up at several points," remembered Mark Cuker. The defense lawyers "were basically saying, 'We know you can't litigate this case because you don't have your proofs, so why should we give you more time?'" In the end, though, the companies decided that the possibility of avoiding an expensive, uncertain, and highly publicized lawsuit was worth the risk of giving their adversaries an extra eighteen months. The two sides agreed on the tolling agreement, and the statute of limitations clock stopped ticking on January 31, 1998. Schlichtmann's unorthodox ideas about how to pursue the case had passed their first test; as a result, there was at least a slight chance it might eventually be resolved without a lawsuit.

Now the work of the case began. Schlichtmann started hiring scientists. In Woburn, he had spent well over a million dollars on experts. In Toms River, there was no need for such extravagance; public agencies were already doing most of the needed scientific work. Still, the lawyers could not afford to wait for the case-control study results. The companies had agreed to a pause of just eighteen months, and no one knew how long Fagliano's team would take to finish the study. Instead, the families' lawyers would need to conduct a parallel investigation, a shadow version of what the government was already doing. Schlichtmann turned to Woburn veterans for assistance. Needing an epidemiologist, he hired Richard Clapp of Boston University, who had worked on the Woburn case and had served as a consultant to

Gillick's advisory committee.[3] For water modeling, Schlichtmann signed up Peter Murphy, whose intricate reconstruction of the pipe system in Woburn had been so crucial. He also hired the aptly named John Snow Institute in Boston to survey the families about their health, habits, and medical history.

Mark Cuker, meanwhile, immersed himself in the scientific intricacies of the Toms River case. He even purchased a stupefying textbook called *Principles of Polymerization* (it weighed three pounds and was 832 pages long) and taught himself the basics of industrial chemistry, especially plastics. He also spent long days in the main branch of the Ocean County Library in downtown Toms River, the official repository of all of the documents associated with the state and federal investigations of Reich Farm and Ciba stretching back to the early 1960s. There were about three hundred thousand pages of documents, filling 168 feet of shelf space. The "wall of shame" is what the librarians called it—and still do.

The first interviews for the state's case-control study began in the spring of 1998, just as a period of relative calm in Toms River was coming to a fractious end. In mid-May, at a meeting of her advisory committee, Linda Gillick announced that five new cases of childhood cancer had recently been reported to Ocean of Love by hospital social workers—the first new cases in sixteen months. Gillick left no doubt that she believed local pollution was to blame. "We can't jump to any conclusions, but common sense would say, what do they all have in common?" she told reporters. Later, she was more specific, saying she suspected that the cases were somehow linked to United Water's decision, during the previous summer, to use water from the two tainted Parkway wells for several weekends at the height of the town's water shortage.[4] She was almost certainly wrong about that because, by mid-1997, water from those wells was being run through carbon filters as well as the air stripper. After filtration and stripping, it contained less than one part per billion of trichloroethylene or SAN trimer and posed no known health threat.

Gillick's announcement provoked a new round of stories identifying Toms River as a cancer town, just as the summer tourism season was about to begin. This time, business leaders did not hide their anger. "A lot of people were very annoyed, as I was, that we were getting beat up nationwide," remembered developer Gary Lotano, who was an officer in the Toms River–Ocean County Chamber of Commerce. "People were saying we were Love Canal all over again. I thought it was an injustice." Soon after Gillick's declaration, Lotano vented to the *Asbury Park Press,* saying it was "irresponsible" for Gillick to "create hysteria" without waiting for the state to confirm the new cases.[5]

Real estate agents were complaining too: The market was slow, and they blamed the publicity. "There were a lot of people who really wanted this whole thing to go away," remembered Robert Gialanella, who served on Gillick's committee. Several of his physician colleagues warned him not to get too involved with Ocean of Love; he lost several patients because of it. Later, when Gialanella tried to sell his house and move to Florida, his first buyer walked away after signing a contract. "We got a call from the realtor who said the buyer was backing out because of all the chemicals in the water in Toms River," he recalled. Lots of people in town had similar stories; emotions were raw.

The strain was also showing within state government. Elin Gursky, Health Commissioner Fishman's top deputy, had been working intensively on the investigation for more than two years. After a disastrous beginning in which she had infuriated the families by saying that a full-blown study of the cluster would be a waste, Gursky had worked hard to gain their trust. She drove across the state to attend the monthly meetings of Gillick's advisory committee and was careful never to repeat her early mistake of sounding insensitive. So when a reporter asked her about Gillick's announcement of the five additional cases, Gursky said she was "very, very concerned."[6] That was the wrong answer, as far as Fishman was concerned. He did not want his staff stirring up any more anxiety and anger in Toms River until the department had officially confirmed the cases. Within days, Gursky had resigned under pressure and left the state.

Linda Gillick had the last word, as usual, because the health department soon confirmed that there were indeed five new cases. No one who knew Gillick was surprised; she was a meticulous record-keeper of childhood cancer cases, and her sources at area hospitals were impeccable. Still, Gillick was wounded by the public criticism. She would remember the incident as the emotional low point of her many years of community activism. From then on, though she remained as blunt-spoken as ever, she would go out of her way to tell visiting reporters that Toms River was a "beautiful town" with "beautiful people"—but also a community with a cancer problem.

That was not a message many people in town were willing to accept. "I think the general attitude was a little worse than just skepticism about the cluster. It was closer to hostility," remembered parent Kim Pascarella. "The main part of the community treated anyone from our group like we were nuts. They thought we were alarmists, and some people thought we were doing it for the money." What sociologists call the "outsider versus insider dichotomy" had taken hold.[7] The town had split into two cultures, one much larger than the other, each with its own language and way of thinking. Across a chasm of mistrust and misunderstanding, the two sides regarded each other warily.

What most people in Toms River craved—what they had *always* craved—was to be perceived as normal. They were trying to get on with their lives. But life could never be completely normal for the Ocean of Love families. For them, moving on was not an option. That was especially true now that the case-control study was under way. They welcomed the investigation because, above all else, they wanted to know why their kids had been stricken, when so many others had not. For almost everyone else in town, though, the study was an unhappy reminder that their town had been branded.

Could Toms River be re-branded, like New Coke or Marlboro Lights? People had tried. Gary Lotano and other business leaders had briefly hired a famous New York public relations firm, Rubenstein Associates, for an image makeover of the town, but soon abandoned the plan as unworkable and too expensive. Earlier, residents of South Toms River had proposed changing the borough's name to Cedar

Pointe, but voters nixed the plan. In August of 1998, however, a group of twelve-year-old boys succeeded where so many others had failed. The Toms River East baseball team won the Little League World Series, defeating a Japanese team and sending the entire town into a state of rapture.

Little League was the ultimate expression of suburban normalcy, and Toms River worked hard to excel at it. Thanks to aggressive fundraising, the fields and facilities were superb, including stadium lights and indoor batting cages that would be the envy of some minor league professional teams. Eighteen hundred boys played. The female equivalent in Toms River was competitive cheerleading, at which local squads also excelled. The Little League team from Toms River East had been to the World Series tournament once before, losing in the 1995 elimination round. The 1998 team stormed through the eight-team bracket without a loss, finishing with a come-from-behind victory over Kashima, Japan, in which the star was a little-used player named Chris Cardone, who came off the bench to slug two home runs. There were more than one hundred thousand Little League teams around the world; Toms River East was the best of them all.

The victory electrified Toms River like nothing before or since. For the first time since Maria Marshall was murdered alongside the Garden State Parkway in 1984, Toms River was in the national news for something other than polluted water and cancer. And for the first time anyone could remember, Toms River was famous for something positive. "Whenever I tell anyone I'm from Toms River, they say, 'Oh, the water and the murder,'" one woman told the *Los Angeles Times*.[8] "Our little guys are undoing all that," another told *The New York Times*. "People are finally going to see Toms River as just another all-American town." Said a third, "It's good to have something to talk about that doesn't make us ashamed anymore."[9] The returning heroes were met by a police escort at the off-ramp of the Garden State Parkway and taken to their home field, where two thousand fans were waiting and hundreds more lined the route. A few days later, more than forty thousand people—almost half the town—turned out for a parade down Hooper Avenue. To mark the occasion, Route 37 was renamed Little League World Champions Boulevard. The same road ran past the old chemical plant, the hospital where many children had first been diagnosed with cancer, and the Ocean of Love office.

There was a special resonance to the victory because Toms River's triumph had come via its children. After the Little Leaguers' victory, residents who were asked by outsiders to explain the town's athletic prowess often responded the same way: "It must be the water." Some of the Ocean of Love families considered this arch reply to be a barb directed at them. They thought that some of their neighbors were using the Little League triumph to mock the notion that Toms River had a genuine pollution problem. Melanie Anderson, Bruce's wife, wrote a letter to the *Asbury Park Press* saying that "the true heroes" were the sick children. "I said that baseball is hardly life and death," she remembered. "If these baseball players were true heroes, they would reach out to these kids" with cancer. The Little Leaguers later attended a party for the Ocean of Love children, but the breach between the families and most of the rest of the town did not heal.

Around the same time as the Little League frenzy, the county legislature authorized a modest monument in memory of local children felled by cancer, after months of lobbying by the Ocean of Love families. Toms River loved memorials. There were more than thirty monuments and memorial plaques along Washington Street, the busy downtown thoroughfare and parade route that is now, by decree of the town council, also known as the "Avenue of Americanism." Veterans, merchant seamen, volunteer firefighters, and first-aid providers all had markers, and there was plenty of room for more memorials on the spacious front lawns of the courthouse, the county administration building, and town hall, where a newly erected sign honored the victorious Little Leaguers.

The county politicians instead decided to place the pink stone slab memorializing the dead children elsewhere: in the back corner of a small and very quiet county park called Riverfront Landing. The dedication ceremony was sparsely attended; most of those present were relatives of the dead. The names carved into the granite slab included Gabrielle Pascarella, Randy Lynnworth, and Carrie-Anne Carter, whose 1995 funeral had drawn nurse Lisa Boornazian to Toms River and spurred her to ask for an investigation of childhood cancer in Toms River.

There was some empty space on the pink granite, enough for a total of fifty names. Not enough space, as it would turn out.

There were more than 2,500 living ex-employees of the Ciba factory, and that summer they received a package telling them they had nothing to worry about. The fourth, and final, worker epidemiological study had been completed, and the results showed that the cancer death rate among ex-employees was slightly less than the statewide rate. "The bottom line is, we're very pleased with the results," a company manager told reporters.[10] Like its predecessors, the study was conducted by the industrial epidemiology group at the University of Alabama at Birmingham, led by Philip Cole and Elizabeth Delzell, whose work was mostly supported by manufacturers and industry associations. Ciba not only paid the $320,000 cost of the study, it was also paying Delzell to serve as its consultant on the state's case-control study; she had already appeared at a public hearing on the company's behalf.

As with the previous studies, there were rough patches beneath the glossy surface. The overall cancer death rate at the factory was about the same as the statewide population due to the healthy worker effect, the well-established observation that industrial workers who are well enough to be hired tend to be healthier than the general population.[11] Yet there were clusters of specific cancers among workers who had the most contact with hazardous chemicals. The lung cancer fatality rate among vat dye and maintenance workers, for example, was about 50 percent higher than expected. Azo dye workers were four times more likely than expected to die from stomach and bladder cancer. Deaths from central nervous system tumors were more than three times higher than expected among vat, azo, and plastics workers. Those associations had been found in earlier studies too but were stronger now because there were more deaths. Even so, Delzell and her coauthor, Nalini Sathiakumar, argued that chemical exposure was not necessarily the cause because death rates for veteran workers with more years of exposure were no higher than for workers hired more recently. However, that argument overlooked two other aspects of the healthy worker effect: Recently hired workers tend to be assigned to high-exposure jobs, and those who are unhealthy tend to leave sooner.

Sathiakumar and Delzell did acknowledge the toll that toxic exposures had taken among one accursed group of longtime Ciba employees: the eighty-nine workers, all now dead, who had toiled at the Cincinnati Chemical Works before coming to Toms River in 1959. They were much more likely than expected to have died from bladder, kidney, or central nervous system cancers. For all cancers combined, their death rate was 33 percent higher than the statewide rate (twenty cases instead of the expected fifteen), without taking into account the healthy worker effect. For these men—and *only* these men—the Alabama researchers were willing to acknowledge that chemical exposures influenced the death rate. Having shoveled benzidine or beta-naphthylamine with their bare hands, and often without dust masks, they were part of an absolutely conclusive chain of epidemiological links between those two dye chemicals and bladder cancer— a chain that stretched all the way back to Wilhelm Hueper's research at DuPont in the 1930s. For everywhere else at the plant where incidence was high, however, Sathiakumar and Delzell gave no ground. For the cancer clusters among dye, resin, and maintenance workers, they would say only that "increases may be attributable to chance, to uncontrolled confounding by smoking or to an unidentified occupational exposure."[12]

Former employees saw the study as a final insult. "I thought it was ridiculous, to be honest," said John Talty, the former union president. He had heard the same rose-colored message back in 1987, when Delzell conducted a similar study that, like the new one, relied solely on the cause of death listed on death certificates. What the workers wanted, then and now, was an *incidence* study that would track cases, not just deaths. There had been just two modest efforts to track cancer incidence at the plant, by Delzell in 1987 and her doctoral student Fabio Barbone in 1989. They covered only cancers of the brain, nervous system, and lung, yet both found unusually high numbers of cases. Barbone even concluded that chemical exposures caused more than half of the nervous system cancers at the plant and about one-third of the lung cancers.[13] Now, ten years later, Sathiakumar and Delzell were again relying on death certificates and were again making a questionable comparison to the general population.

Still bitter about how Ciba had treated its longtime employees, John Talty could only shake his head in resignation at what was being studied in Toms River—and what wasn't. He watched as the government spent millions of dollars scrutinizing the possible role of low-level chemical exposures in triggering cancer in neighborhoods far beyond the factory fence line. Talty did not begrudge Linda Gillick her study, but he wondered why no one would do something similar for his 2,500 fellow retirees. They had faced higher risks than any child or pregnant woman who drank Parkway well water or lived downwind from the Ciba smokestacks. But there would be no factory-based equivalent to Jerry Fagliano's case-control study. Ciba had enough problems in Toms River already, including a hugely expensive cleanup and a possible lawsuit from the families of dozens of sick or dead children. Besides, the company was much smaller now because Novartis (as the merged Ciba and Sandoz was known) had dropped chemical manufacturing for cleaner, and more profitable, pharmaceutical production. In 1997, a year after the merger, the Swiss conglomerate spun off its chemical operations, including the shuttered Toms River plant, into a much leaner company known as Ciba Specialty Chemicals. For the scaled-down Ciba, Toms River represented only costs, not revenue. Revenue was in Alabama, China, and India. Under those circumstances, Ciba was not about to order up another epidemiological study of its former workers in Toms River.

Ever since his stubborn persistence had led to the identification of styrene acrylonitrile trimer in the Parkway wells, Floyd Genicola had been spending his nights and weekends hunting additional TICs. The acronym stood for "tentatively identified compounds," and Genicola was trying to unmask more of them based on the spectral peaks on his chromatograms. His supervisors at the state Department of Environmental Protection did not approve, which is why Genicola was allowed to do the work only during off-hours. He told his bosses that he was doing the extra work because he hoped to make it the subject of a future doctoral thesis. This was true but not the full story. The deeper truth was that Genicola had formed an emotional connection to the Ocean of Love families. His work had paved the way for the case-control study by linking Union Carbide's SAN trimer waste to the Parkway water. Now the former police chemist wanted to find out what else from Reich Farm had seeped into the Parkway wells and reached thousands of Toms River homes.

The analytical work was difficult because the unidentified compounds were obscure and their concentrations were mostly less than one part per billion. Worse still, the original molecules had broken into fragments. Trying to piece together those fragments into coherent molecules was like trying to reassemble the shell of an egg dropped off a second-story balcony. Even so, Genicola was making progress. By the end of 1997, he had identified about 250 fragments in the Parkway water samples. Many were closely related to SAN trimer, but others looked more like flame retardants and other compounds Union Carbide had added to its plastics to improve durability. Still others seemed to be new molecules that formed when compounds mixed during their mile-long underground journey from Reich Farm to the wells. Concentrations of those 250 molecular fragments in Parkway water were very low, but the people of Toms River were not drinking them one at a time. "We had kids at schools who were drinking this water, and we didn't even know what was in it," Genicola recalled.

His supervisors, as usual, thought Genicola's efforts were a waste of time. "Floyd was honestly motivated. However, other people did not see the TICs that he saw, and there was a question as to what level these things gained any significance," remembered Gerald Nicholls, the DEP's director of environmental safety and analysis. By mid-1998, the DEP and the health department were no longer paying any attention to Genicola's TIC work and had quietly stopped all other efforts to identify chemicals in the Parkway water. As far as the state was concerned, the problem was solved. Low levels of SAN trimer and trichloroethylene had been found in two wells, and carbon filters and an air stripper were now removing them. No one was drinking water from those wells anymore, except briefly during the water emergency of the previous summer. Instead, the wells were being used solely to clean up the Reich Farm plume. Why spend time and money looking for more contaminants?

To no one's surprise, Floyd Genicola did not accept his supervisors' decision quietly, and he knew whom to complain to about it. He requested a meeting with Linda Gillick. Nicholls agreed, but only if he could be present to offer a rebuttal. The meeting, held at the Ocean of Love office, was stormy. When Genicola said the Parkway wells had been tainted by other chemicals, not just SAN trimer and trichloroethylene, Nicholls countered that those detections were at very low levels and were irrelevant because carbon filters were removing them. Genicola was not finished, however. He also told Gillick that low levels of trimer had infiltrated two *other* Parkway wells, which were not being filtered and were important supply wells for the town.

A few months later, chagrined state officials had to admit Genicola was right again. The problem, as always, was that United Water could not keep up with peak demand. In July of 1998, during a heat wave, the water company pumped its Parkway wells more heavily than usual and pulled in the Reich Farm plume. The detections of trichloroethylene and SAN trimer were very low, less than one part per billion, but they were there.[14] When the state declined to order the water company to filter the newly tainted wells, Gillick got Senator Frank Lautenberg to come to town and demand the filters. "Something is terribly wrong in Toms River," he declared as he stood beside the new cancer memorial at Riverfront Landing Park. Governor Whitman quickly capitulated, ordering carbon filters for the two additional wells. This time, Union Carbide refused to pay to remove chemicals at such low concentrations; taxpayers picked up the tab: $1.5 million.

With the addition of the new filters, the last known traces of the Reich Farm pollution plume finally vanished from the water consumed by the people of Toms River. Since at least 1982, Union Carbide's waste had infiltrated the Parkway wells, but the chemical composition of the tainted water was still largely a mystery. Only two specific contaminants had been confirmed: the solvent trichloroethylene and the plastic byproduct SAN trimer. Floyd Genicola had some ideas about what else might be in the water, but no one was listening to him, and everyone else had stopped looking.

In its final report, the state "TIC Committee" concluded there were 261 distinct but unknown compounds in the prefiltered Parkway water samples, 122 of which were not related to SAN trimer. "The information contained herein," the report stated, "is meant to provide a starting point for further analysis should such analysis be deemed important to the investigation."[15] But there would be no more analyses of the Parkway water, even though there was much more that could have been done. Genicola had pushed for a cumulative risk assessment of all of the TICs to see whether the mixture of low-level compounds posed a collective hazard, but the Department of Environmental Protection nixed the idea. Understanding the toxicity of SAN trimer would be hard enough without also trying to assess the combined impact of more than one hundred other chemicals at low levels.

The DEP's decision shifted the Toms River studies even farther away from the complicated reality of the town's environment. Later, even a consultant for Union Carbide would regard that as a mistake. "We really should be looking at the total of all that stuff, as well as at the individual contaminants. If you add them all up, the sum total is more than one part per million, which is not a low level," said Jon Sykes, an engineer at the University of Waterloo who mapped the Reich Farm plume for Union Carbide. "You just look at it and you say, 'My God, there's a lot of crap in that water.'" But the State of New Jersey was unwilling to keep looking. As a result, the scientific case that polluted groundwater had contributed to the cluster would rest almost entirely on just one compound: styrene acrylonitrile trimer. There was trichloroethylene too, but TCE concentrations in Toms River water were rarely over the safety limit of one part per billion after 1998. For SAN trimer, though, there was no established limit because no one had ever tried to assess its toxicity. Much would be riding on the question of how hazardous SAN trimer really was, but testing to find out was only just getting under way.

Floyd Genicola, the perennial outsider, soon gave up trying to force his way back into the investigation. In 1999, he moved to a different section of the DEP and then to the state health department, where he focused on food safety and antiterrorism and tried to forget about Toms River. He even put aside his unfinished doctoral thesis about the unmasking of SAN trimer, although years later (after he retired in 2012) he began working on it again, warily. He would always be proud of his crucial role in spurring the Toms River study, but he also believed it badly damaged his career in state government. "There are a lot of painful memories with Toms River," Genicola said. "My stomach still turns."

By the end of 1998, the state health department had completed all but a handful of the planned two hundred interviews for the case-control study. Despite all the delays and compromises, the study was off to as promising a start as Jerry Fagliano could have hoped. All forty case families had agreed to participate, which was a relief because he would need as many as possible to get a statistically significant result. The case families' cooperation was no surprise: They were outsiders, the minority in town who wanted so badly to get to the bottom of the cluster mystery. The surprise was that recruitment of healthy children was going well, too. About 80 percent of the families asked to serve as controls agreed to do so. With such a high recruitment rate, no critic could claim that the control group was unrepresentative. Now Fagliano had one thing fewer to worry about.

A few weeks later, on a rainy Friday night in Toms River, Kim Pascarella and a dozen other TEACH parents went to the movies. *A Civil Action* was playing, with John Travolta starring as Jan Schlichtmann. Linda and Michael Gillick had traveled to Boston for the premiere, appearing at a forum organized by an environmental group. A deluge of publicity accompanied the movie's opening, with many of the stories featuring the real-life Schlichtmann, energized by his return to the spotlight. In interviews, he often brought up the Toms River case, describing it as an opportunity to redo Woburn "the right way." Town officials seethed at this latest eruption of unwanted attention.

On the other side of New Jersey, near Trenton, a group of state officials involved in the Toms River investigation had their own movie night. They got Chinese food beforehand and joked about which actors would portray them if there were ever a movie about Toms River. Jerry Fagliano did not go; he had read the book, and that was enough.

The film took many liberties with the Woburn story. Travolta-as-Schlichtmann ended the movie bankrupt but ennobled, instead of alienated and severely depressed. There was no blizzard of legal motions, no tedious wrangling about the toxicity of trichloroethylene or the risk factors for childhood leukemia. Even so, the emotional core of the story rang true to the Toms River families; they felt that they were living the same story. In Woburn, they saw another town with secretive polluters, tainted water, and unresponsive public officials; they saw another group of suffering families trapped in the twilight between deeply held belief and demonstrable scientific proof. "It was Hollywood entertainment," recalled Pascarella, but it was also "the synthesis of everything we had gone through."

By the end of the movie, some of the Toms River parents were so overcome with emotion that they did not even notice the text that briefly appeared on the screen just before the final credits. It informed viewers that Schlichtmann "is currently representing sixty families in Toms River, New Jersey, in another contaminated water case."

CHAPTER TWENTY-ONE

Surrogacy

He spent the last fifteen years of his life in furious pursuit of scientific immortality, producing 106 manuscripts—only eight of which were published before he died in 1541.[1] So it's a safe bet that Paracelsus would have reacted with his usual churlishness if he had known that he would be remembered best for a phrase he never wrote: "The dose makes the poison." What he actually wrote in 1538, ten years after fleeing Basel, was less pithy: "All things are poison, and nothing is without poison; the dose alone makes a thing not poison."[2] The Paracelsian idea that the *quantity* of a potentially harmful substance matters just as much as its inherent qualities would become the bedrock tenet of toxicology—and a major dilemma for the researchers seeking to untangle the relationship between pollution and cancer in Toms River.

By the twentieth century, all experimental scientists looked for dose-response relationships in their results. It was a simple idea with profound implications. In an environmental health study, the group exposed to the highest dose of a hazardous compound (or mixture of compounds) should have the greatest risk of disease, while those receiving a moderate dose should have moderate risk and those getting the lowest dose should have the lowest risk. The relationship between dose and response was not necessarily linear; sometimes, it formed a curve when plotted as a graph. The precise shape was less important than the underlying relationship: Higher exposure should correlate with greater risk. (Researchers now know that some hormonally active synthetic chemicals can pose bigger health risks at very low doses than larger ones, but in general Paracelsus's rule still applies.)[3] Without a dose-response curve, the evidence of a link between exposure and disease would always be suspect.

By that standard, the Toms River case-control study could never be fully legitimate scientifically because no one would ever know exactly which chemicals, at what concentrations, were in the air and water during the 1962 to 1996 study period. The water may have smelled, and the smoke may have been colored, but the precise causes would always be a mystery because there was so little environmental testing at the time. Lacking a time-traveling machine, the researchers trying to understand what had happened in Toms River would have to rely on extrapolation and inference, and they would need surrogates.

Still, there was no stopping now. The politicians had promised to leave no stone unturned in searching for the cluster's causes, and so had the lawyers and scientists. This was the sociological equivalent of the dose-response curve: As the publicity, expense, and time accrued (by early 1999, the investigation was three years old and had cost taxpayers roughly $10 million), the investigators felt that they *had* to respond with credible answers. If there were data gaps about past hazards, they would just have to be circumvented.

On a typical summer day in Toms River, sixteen million gallons of water coursed through 488 miles of pipes. Water entered the system from twenty-three wells in eight well fields and exited via the faucets of more than forty-five thousand homes and businesses. Morris Maslia's formidable task was to determine the provenance of each gallon—where it came from and where it went—not only currently, but in every single month all the way back to January of 1962. The success of the Toms River studies depended on it. An environmental engineer at the federal Agency for Toxic Substances and Disease Registry, Maslia specialized in estimating contamination levels at Superfund sites, but he had never tried to tackle anything as complicated as the Toms River investigation. He had never tried anything as costly, either: The water model's price tag would ultimately top $5 million, with the funds coming from special congressional appropriations. "Being naive, I thought we could do this pretty easily and quickly," he remembered. "That was wrong."

The computer model Maslia started building in 1997 was an attempt to cope with the data gaps in Toms River. Lacking reliable records of what was actually in the town's drinking water in those earlier years, Maslia would have to calculate risk indirectly, based on which wells supplied which neighborhoods. The more than seven hundred Toms River children in Jerry Fagliano's two studies were born between 1963 and 1996. If Maslia could figure out which wells had supplied the homes of each child and pregnant mother, then Fagliano could test his hypotheses about the role of water pollution in triggering childhood cancer. A team from Rutgers, similarly, was trying to develop estimates of which neighborhoods had borne the brunt of air pollution from Ciba and downwind radiation from the Oyster Creek nuclear plant. They had even less information to work with than Maslia did, and they had to do even more guessing. For historical records on wind direction and air pressure, the Rutgers group had to rely on a weather station fifty miles away. And since there were no reliable records on the volume and composition of Ciba's air emissions over those thirty-four years, the group had to guess based on the factory's production records.

Maslia soon recognized his folly in thinking that his reconstruction of the water distribution system could be done quickly. Before he could even consider building a model of the piping network as it existed as far back as 1962, he had to create an accurate model of the present system. In Woburn, Peter Murphy had gone through the same process, but Maslia faced a much tougher challenge. Toms River's twenty-three wells pumped four times as much groundwater as Woburn's eight wells and served more than twice as many homes. There were six pumping stations and nine storage tanks in Toms River, which allowed the water company to react to huge swings in demand from season to season by drastically changing the way water circulated through its piping network. During the winter, the water company would shut down as many as half of its wells and then bring them back online during the spring, one at a time. By midsummer, every well would be pumping at close to capacity, followed by a reduction through the fall. The yearly oscillations meant that a family's water in February probably came from a different combination of wells than in August. For his model to be nimble enough to account for the seasonal shifts, Maslia would have to include all sixteen thousand pipe segments, incorporating variables as obscure as the size of each storage tank and the composition of each pipe. (Water moved more slowly through rough concrete than smooth plastic.)[4]

By the time Maslia finally finished his model of the current piping system, two years had passed. It was early 1999, and he now knew which water well, or combination of wells, was supplying each household in Toms River. But he still faced the even more difficult task of adjusting his present-day model to mimic the pipe network as it had functioned as long ago as 1962. In fact, he would have to create 420 models, one for each month from January 1962 to December 1996. Maslia would have to look back in time, to the days when Toms River was just a very small town with a very big factory. In the meantime, the present-day town would have to keep waiting.

Bruce Molholt first heard Linda Gillick on the radio in early 1999; it was love at first sound bite. "You could just *feel* her mind galvanizing, getting everyone together to work on this thing," he remembered. "I knew I wanted to try to help." Molholt's academic training was in microbiology and genetics, but his passion was studying the health effects of industrial chemicals, having spent four years at the EPA as a toxicologist in the Superfund program. Earlier, as a young professor at the University of Kansas, Molholt had led teach-ins against the Vietnam War. His politics had not changed since then, even though he was almost sixty now and worked as the director of toxicology and risk assessment for a large environmental consulting firm with many industrial clients.

So when a friend told him that Gillick was going to be interviewed on a Philadelphia radio show, Molholt stepped out of his suburban office and into his car to listen. "I was curious because I had already heard about the Toms River case, and had told my staff that this was something that really should be pursued," he recalled. By the time the interview was over, Molholt was ready to follow his own advice. He contacted Gillick, who put him in touch with lawyer Mark Cuker, who made Molholt an unpaid consultant to the legal team. That did not sit well with Molholt's bosses at Environmental Resources Management, which listed both Ciba and Union Carbide as occasional clients. When he refused to stop assisting with the Toms River case, Molholt was told to empty out his desk and leave. He didn't mind; he wanted to work on something he believed in.

The day after he resigned, Molholt showed up at Cuker's law office. It was the summer of 1999, and Cuker was glad to see him because it was time to figure out whether the families had a case, and against whom. Over the previous eighteen months, as Morris Maslia worked on his water model and Jerry Fagliano collected his questionnaire results, Cuker had amassed enough documents to fill three hundred file boxes. There were technical assessments as thick as phone books, yellowed lab reports, correspondence, plume maps, and all sorts of other records relating to the forty-five-year history of chemical pollution in Toms River—not just at Ciba and Reich Farm but all over town. Some documents, surprisingly, came directly from Ciba, Union Carbide, and United Water. Company attorneys had sent those records to Cuker, who had responded in kind by sending them his clients' medical records—all without being ordered to do so by a judge, whose intervention would have been required if a lawsuit had already been filed.

The highly unusual voluntary exchange of documents was the most tangible sign of the changing relationship between the two sides. The lawyers had toned down the bombast of their initial gathering back in early 1998. There had been more than a dozen meetings since then, with no more insults or threatened walkouts. To the shock of everyone but Jan Schlichtmann, they seemed to be making progress—though exactly what they were progressing toward was never defined. One of Schlichtmann's rules was that the "s-word"—*settlement*—could not be discussed. This was not a negotiation or even a prelude to a negotiation, Schlichtmann insisted. It was, instead, a "process," a mutual search for answers to the families' questions about whether pollution was responsible for their children's cancer. Perhaps because neither side yet had much to lose, the opposing attorneys discovered that they were actually learning from each other. "It takes time, it takes interaction, to build up that trust, but over a period of time we began to trust each other," remembered William Warren, the Union Carbide lawyer. Things were going well enough that the attorneys extended their clock-stopping agreement. Later, they would agree on five additional extensions.

The families, however, were less enthusiastic than their lawyers. Schlichtmann had touted his "process" as a way for the parents to participate in the search for answers in Toms River, yet many of them still felt excluded—even Bruce Anderson, who saw the lawyers more than anyone except Linda Gillick. Anderson had volunteered to scan thousands of pages of medical records and technical reports onto computer disks for Cuker. Even so, he did not feel like a full participant in the case. "We had no real involvement," Anderson remembered. "It was disappointing." Schlichtmann had told the families back in 1997 that the case could be resolved in eighteen months. That interval had passed, yet the lawyers had not even begun to talk about a settlement. Everyone else in town seemed to have long since moved on (the big local topic during the summer of 1999 was Toms River East's second consecutive trip to the Little League World Series—they lost in the U.S. championship game), but the TEACH families were still trapped in an endless cycle of information gathering. Every few months, Cuker and his law partner Esther Berezofsky would drive to Toms River to update them. As time passed, the meetings grew tenser. "I would say pretty much all of them were impatient," Berezofsky recalled. "After we would explain what we were doing they would come around, but we would always say that we weren't going to hold off forever on moving to the next step."

So Bruce Molholt's arrival in Cuker's office in mid-1999 came at a propitious moment, just as the lawyers were trying to focus their case where the evidence for causation was strongest, in preparation for settlement negotiations. Cuker had spent the previous eighteen months investigating all sorts of theories about what might have caused the childhood cancer cluster, including radiation from the Oyster Creek nuclear plant and leaks from Ciba's ocean pipeline and the old town landfill. Now he and Molholt studied a map showing the home addresses of his sixty clients. There seemed to be a grouping near the Parkway well field, but it was not an obvious cluster; collectively, the sixty cases had just one exposure in common: They all got their drinking water from the Toms River Water Company and its successor, United Water of Toms River. Back in 1996, when he was designing the case-control study, Jerry Fagliano had concluded that if pollution were responsible for the cluster, the water network was the likeliest exposure route. Now, Cuker and Molholt were coming to the same conclusion, spurred by the knowledge that Ciba's waste had tainted the Holly Street wells in the 1960s and that Union Carbide's waste had done the same to the Parkway wells in the 1980s and 1990s.

Just a few weeks after Molholt's arrival in Cuker's office, the government gave the families' case a powerful boost. The state health department and the ATSDR finished a new health study for Reich Farm, concluding it was "a public health hazard due to past exposures." A few months later, they issued an identical finding in a report about Ciba. In both cases, the agencies concluded, dumped chemicals had traveled down a "completed exposure pathway" from the sites to backyard water wells and public wells—and then to the people of Toms River, who were also directly exposed to toxic air emissions from Ciba.[5] There were still more steps to go in the government's investigation, including the completion of Morris Maslia's water model and its merging with all of Jerry Fagliano's data on the study children. But the ATSDR's declarations that toxic pollutants from Ciba and Union Carbide had definitely reached the people of Toms River suggested that the families and their lawyers might be able to wring a financial settlement out of the companies even if the government investigation ultimately did not uncover a strong correlation between cancer risk and exposure to contaminated air or water. Fagliano's study results would be scientifically meaningful only if he could show that they were statistically significant and therefore very unlikely to be due to chance. If there were a lawsuit, however, the families' attorneys would not have to meet such a high standard. A circumstantial case tying the dumped chemicals to the sick children might be enough to convince a jury, which is why the ATSDR's findings were so helpful to the families.

Mark Cuker set about trying to build a case that he could press no matter how the government investigation turned out. As a first step, he put Molholt to work going through stacks of chemical manifests and government reports to try to identify the specific ingredients of the chemical goulash that had been dumped at Reich Farm and Ciba. One old EPA document—Cuker called it the "purple sludge" report—described a six-foot-thick layer of waste beneath one of the sludge dumps at Ciba. The EPA tested the purple muck and did not detect any of the 130 organic chemicals on its agency's "target compound" list for Superfund sites.[6] The agency's contractors could see the sludge and they could smell it, but under Superfund rules it did not exist. In fact, only a few of the 156 chemicals and six dyes Ciba was producing in the mid-1980s were on the Superfund target list.[7] What that meant to Molholt was that the government's by-the-book tests of the river water probably missed many Ciba contaminants—just as routine testing of the Parkway wells had missed styrene acrylonitrile trimer, another pollutant too arcane to appear on an EPA checklist.

It took a few months for Molholt to get his bearings amid the avalanche of paper, but he was encouraged by what he read. From a toxicological standpoint, he thought, the families had a pretty good case. There was solid, if indirect, evidence that many of the compounds dumped at Ciba and Reich Farm were capable of inducing cancer. Not one compound out of the hundreds Molholt identified in the wastes dumped at Ciba or Reich Farm had been conclusively shown to cause leukemia or nervous system cancer, the two categories of cancer that were most highly elevated in Toms River.[8] But there was plenty of evidence that more than a dozen of those compounds *might* cause tumors in either category.[9]

The wild card was SAN trimer, which was now being scrutinized for the first time. Toxicity testing of this obscure waste product was supposed to have started in early 1997 but was delayed for more than

a year because Union Carbide had none on hand—indeed, it had not
had any since the 1970s, when it stopped combining styrene and acrylonitrile to make plastics. Instead, the company had to synthesize a
small amount of SAN trimer from material purchased from one of its
competitors, Bayer. The families wanted the EPA to test this newly
synthesized trimer, but Union Carbide wanted its contractors to conduct the tests, and the EPA did not insist on maintaining control. Despite being responsible for the safety of more than sixty thousand
chemicals in commerce, the EPA almost never conducted its own toxicity tests—and still does not. Instead, it typically restricts its oversight to evaluating tests conducted by manufacturers or manufacturers'
contractors.

A contractor for Union Carbide finished the first toxicity tests on
SAN trimer in mid-1999 and concluded that trimer was toxic to rats
only at high doses and that it was probably not mutagenic. Actually,
the very first round of tests suggested that trimer might damage bacterial DNA and induce chromosomal aberrations in hamsters. But the
company had a different lab redo the tests, using a purer sample and
rats instead of hamsters, and got the results it was hoping for: no evidence of mutagenicity.[10] These were not the tests that mattered, however. Toms River was a cancer cluster; the key question was whether
styrene acrylonitrile trimer was a carcinogen. Linda Gillick was determined not to allow Union Carbide to provide that answer, too. She
convinced New Jersey's congressional delegation to petition the federal National Toxicology Program to take over cancer testing of the
trimer, and in 1999 the agency agreed.

The twenty-year-old NTP, based in North Carolina, was the most
prominent exception to the leave-it-to-industry attitude toward chemical testing that prevailed elsewhere in government. It was a research
program that tended to operate like a reform school, taking on the
most troubled cases—the most toxic compounds in commerce. The
agency specialized in a costly test known as the two-year bioassay, a
distant descendant of the cancer induction tests Katsusaburo Yamagiwa had devised for his rabbits in 1913. The NTP's standard bioassay
called for rats or mice to be fed, injected, or gassed with a suspect

compound at one of four dose levels: high, medium, low, and none. After twenty-four months, the survivors were asphyxiated and autopsied to see if animals that received the highest doses had the most health problems, especially malignant tumors. The process was so complex and required so many steps before and after the two-year bioassay that it often took eight years or longer for the agency to render a final verdict. By 1999, the NTP had finished screening more than five hundred compounds, most of them chosen because they were already suspected to cause cancer.[11] Slightly more than half of the time, the NTP concluded, a screened chemical was carcinogenic in rats or mice.[12]

The four hundred mice or rats in a typical two-year bioassay (fifty males and fifty females per dose level) were supposed to serve as surrogates for a much larger population of humans. But it was a highly imperfect surrogacy. The rodents had to be genetically identical to ensure that chemical exposure, not individual variation, was responsible for any detected tumors. Yet they were supposed to be stand-ins for a genetically diverse human population in which some individuals were much more susceptible than others. Just fifty animals were typically tested for each sex and dose, yet they were supposed to be an indicator of risk for a condition so rare that fewer than one in six thousand children were diagnosed with any type of cancer each year. The animals were kept in a carefully controlled environment in which they were exposed to just one hazardous chemical at a time, yet they were supposed to be a surrogate for a human population exposed to hundreds of potentially hazardous compounds every day, though almost always at very low levels.

The NTP resolved these contradictions by giving rodents much higher doses than humans would confront in a real-world environment, even in Toms River. The agency had no other choice; the process was already so complex that a typical two-year bioassay cost at least $2 million, with about half of the cost associated with handling the animals. An improved system, one based on real-world exposure levels, would require tens of thousands of rats or mice instead of a few hundred, at astronomical cost. The only practical choice was to give very large doses to small groups of rodents, even if that would make the studies less realistic and more vulnerable to criticism from industry.

To Linda Gillick and the rest of the TEACH parents, it was obvious what the NTP should do: It should feed those four hundred rodents a more potent version of the same cocktail of hundreds of contaminants that the families of Toms River had unwittingly drunk via the Parkway wells. True, the contaminant concentrations in that mixture were low—their collective concentration was probably a few hundred parts per billion, though no one knew for sure—but tens of thousands of people, including pregnant women and children, had drunk that low-level mixture for many years. If four hundred genetically identical rodents over just two years were supposed to be surrogates for a much larger and more diverse human population exposed over a much longer period of time, then those animals should get higher doses of the same mixture the families had drunk, the parents thought.

But the NTP was not in the business of testing chemical mixtures. Its two-year bioassay was designed to look for the all-important dose-response relationship, but how could anyone establish three consistent, controlled doses of a mélange of more than 250 compounds, most of them unidentified? How could anyone tell which compounds were benign and which were dangerous? When Union Carbide objected to the families' proposal to test the entire Parkway mixture, NTP managers quickly sided with the company. The agency had always conducted its toxicology tests one chemical at a time and it was not going to change now, not even for a very high-profile case like Toms River. The New Jersey Department of Environmental Protection had already given up trying to find out exactly what was in the Parkway water mixture, and now the National Toxicology Program would not even try to find out whether the mixture was carcinogenic.

Instead, everything would come down to styrene acrylonitrile trimer, a chemical that under normal circumstances would be too obscure even to be considered for testing by the NTP. But Toms River, of course, was anything but a typical case. An entire state congressional delegation had taken the very unusual step of asking for the tests, and the NTP was not about to say no to the same lawmakers who set its budget every year. Besides, SAN trimer did have a disturbing lineage: Both of its key ingredients, styrene and acrylonitrile, had already been linked to cancer in factory studies or animal tests.[13]

The NTP did agree to one change that made the four hundred caged rats slightly closer surrogates for the families of Toms River. Normally, the agency's two-year tests involved only adult rodents, which made sense when the program was looking at compounds that were mainly risks to factory workers. But in Toms River, the fear was that SAN trimer might be triggering cancer in children before they were even born. So for this study, the NTP decided to conduct a multigenerational test. Pregnant rats would be fed daily doses of trimer, and after birth the dosing would continue for three weeks while the mothers nursed. The pups would then be fed trimer-laced food every day for two years until they were euthanized and autopsied. The Toms River parents liked this idea: They had been dosed while pregnant, so the rats should be, too.

Still, there were many reasons to doubt whether the rat study would end in a meaningful result. A key concern was that so little SAN trimer was available for testing: only twenty-one kilograms of the smelly, brown gel. This was a major handicap because the most alarming kinds of cancers in Toms River—leukemias and brain tumors—were extremely difficult to find in rodents. Fischer 344 rats, the kind the NTP chose for the trimer tests, were especially poor models for human leukemia because they were naturally prone to a different type of leukemia almost never found in people.[14] The NTP picked Fischer rats anyway because they were the best hope of finding brain tumors.[15] With more money and more trimer, the NTP could have tested two kinds of rats, or rats and mice, and thus had a chance of finding leukemias, too. It could even have doubled the number of rodents tested and thus increased the tests' ability to detect even rare brain tumors. But funds were scarce, and Union Carbide was not providing any more trimer.

What the families had envisioned as a broad inquiry into the carcinogenicity of the Parkway well water had narrowed drastically to just one group of cancers—brain and central nervous system tumors—and just one compound: styrene acrylonitrile trimer. With each narrowing of the investigative lens, the chances of discovering a carcinogenic effect diminished. And now the families would have to wait eight years or longer for any results at all—almost certainly far too long to influence the outcome of the legal case. Floyd Genicola's discovery of SAN trimer in Parkway water back in 1996 had been electrifying, but now the slow-grinding realities of contested science had set in. Linda Gillick and the TEACH families faced a very long wait for a very uncertain result. They would have to pin their hopes elsewhere.

The first, very preliminary results of the Toms River case-control studies were released in December of 1999, and they were tantalizing. Under pressure to show progress after three years, Jerry Fagliano and his colleagues had finished analyzing the questionnaire, looking for differences in the responses of the forty case families compared to the 159 healthy controls.[16] Interview-only studies were a weak form of epidemiology because they depended on fallible memories. If everyone's memory were equally unreliable, this might not be a major problem. But families struck by traumatic illness tended to have much more detailed memories about past exposures than control-group families who had been spared. This well-known phenomenon was known as recall bias. The air and water distribution models, which were still under construction in 1999, would minimize the distorting effects of recall bias because their exposure estimates were based not on memories but on reconstructions of historical emissions patterns. Since the models were not ready, Fagliano warned the families not to make too much of the interview results.

   Still, the results were intriguing. The case and control families had been asked more than two hundred questions, on topics ranging from miscarriage history to hot dog consumption. They were asked about so many possible risk factors that it was likely that for a few of those questions the differences between the cases and controls were going to look significant even if they were really due to nothing but chance. As it turned out, though, there were only a few examples of these flukes.[17] Instead, almost all of the potential nonenvironmental causes explored in the interviews—including prominent theories such as maternal smoking, alcohol consumption, and family history of cancer—were no more common in case families than in control families. Case and control families tended to have similar diets and were similarly likely to have used nail polish, pesticides, and paint thinner, among many other possible risks. By process of elimination, water and air pollution now loomed larger than ever as possible explanations for the cluster. This was especially true because when the interviewers asked the study families whether they drank tap or bottled water, children who later developed cancer were 8 percent more likely to have drunk tap water and 60 percent more likely to have been born to mothers who drank tap water during their pregnancies. In fact, for both mothers and kids, there was a dose-response relationship: Cancer risk was highest for those who drank more than five glasses of tap water per day, lowest for those who drank none, and in between for those who had just a few glasses.[18] As usual, Fagliano kept his feelings to himself, but he could not help being excited about this preliminary finding, even if it was influenced by recall bias. "It suggested we were on the right track," he later explained.

The lawyers for the families were excited too; the drinking water correlation was another bit of leverage they could use in possible settlement negotiations. By early 2000, the tone of their meetings in Princeton had shifted again. No longer just exchanging information, the opposing attorneys were now focused on convincing each other of the correctness of their arguments. The stridency of their initial meetings was gone for good—as Schlichtmann had hoped, the lawyers knew each other too well now for overt hostility—but both sides also knew that the time for casual research was ending, especially because the case-control study was at last approaching completion. Fagliano had told Gillick's committee that the study results would be available by the end of 2001—almost six years after the rowdy public meeting that had launched the investigation.

From the start, the case-control study (now two studies: the interview study and the birth record study) had loomed over the lawyers' meetings as an omnipotent, if seldom discussed, presence. If, in the end, the study results linked the cluster to air pollution from Ciba or to water pollution from Reich Farm and Ciba via United Water's pipes, then the families would obviously have a very strong case for a huge payout, either from a settlement or—years later—a jury verdict. On the other hand, if the study did not find an association between pollution and cancer, the companies would surely offer nothing more than a token amount, and the families would have to decide whether to take it or risk a court battle premised on a greatly weakened case. A crucial question for both sides, then, was whether to take an all-or-nothing gamble by waiting for the case-control study results or instead try to negotiate a settlement before the study was completed.

"It got to the point where it was clear that the case-control study really was going to come to an end, and that we on the plaintiff's side were not going to go away," said lawyer Mark Cuker. "So the discussion became, how do we resolve this?" William Warren, a lawyer for Union Carbide, made it clear that his client would not settle the case merely out of sympathy for the families or to avoid bad publicity. He and his counterparts at Ciba and United Water hinted that a settlement was possible, but first the families' lawyers would have to show that they could present a legitimate case for causation.

Essentially, what the companies wanted was a mini-trial that would be a surrogate for the real thing, but with no judge or jury—not even a mediator. Union Carbide, in particular, had resisted mediation because it implied that a settlement was in the offing. However, the company's combative general counsel, Robert Butler, did have a soft spot for Eric Green, a law professor at Boston University and one of the best-known mediators in the United States. (He would soon achieve even more notoriety for brokering the massive 2001 antitrust settlement between Microsoft and federal prosecutors.) So when Schlichtmann heard that Green was going to be in Princeton for a different case on the same day the Toms River lawyers were meeting there, he convinced Butler and a few others to interrupt Green in the middle of a solitary lunch at the Forrestal Hotel (he had a napkin under his chin and was busy tucking into a steak) to see if he would be willing to consider getting involved in the Toms River case.

They were not looking for a mediator to engineer a settlement, the lawyers told Green. Instead, they wanted a "facilitator" to organize and oversee a series of meetings at which experts from each side would present information and opinions about the key issues, just as they would do if they were witnesses at a trial. Instead of trying to persuade a judge, jury, or mediator, however, the lawyers and their experts would be trying to convince their counterparts on the other side that they could present a winning case if there ever were a trial. Later on, the lawyers told him, Green might play a more formal role in brokering a settlement—but that would depend on the persuasiveness of the expert presentations. Each side would first have to be convinced that a deal was preferable to the risk of waiting for the case-control study results or going to trial. Though irritated at having his lunch interrupted by the ever-pushy Schlichtmann, Green was intrigued; he had never been involved with anything quite like this. The lawyers agreed to hire him, splitting his fee, which would ultimately top $100,000.

Almost immediately, there were complications. In May of 2000, four other lawyers sued Ciba on behalf of six hundred residents, mostly of Oak Ridge, who lived downwind from the factory's smokestacks or atop its groundwater plumes. About two hundred had cancer or other health issues they blamed on the factory.[19] (A second class-action suit, seeking compensation from Ciba for diminished property values in Oak Ridge, was filed eight months later on behalf of about seven hundred homeowners.)[20] Cuker, Berezofsky, and Schlichtmann had been dreading the prospect of other lawyers entering the fray because they feared that it would shut down the ongoing exchange of information and end any chance of a settlement. Union Carbide was already considering pulling out of the talks because it was being acquired by Dow Chemical, which had a history of taking a hard line in lawsuits. After a tense few weeks of uncertainty, the company lawyers said that they would keep participating while reserving the right to drop out at any time.

The expert presentations in front of Eric Green were shaping up as critical; they were perhaps the last time the attorneys would control the fate of their own case, since the case-control studies would be completed soon after. Both sides prepared intensively. Cuker pushed his experts to prepare a visual demonstration showing how contamination and cancer washed across Toms River in successive waves. He envisioned a series of maps that would serve as a sort of time-lapse movie documenting the cluster over time as well as in space. They would show how Union Carbide's plume spread south from Reich Farm, reached the Parkway wells, and was then distributed all over town, year by year. For each cancer diagnosis, meanwhile, a dot representing the affected child's home address would be added; by the 1996 map, there would be dozens of dots.

The maps neatly summarized the families' powerful but circumstantial case. Toms River had an extraordinary amount of toxic pollution and a discernible cluster of childhood cancer, and the two seemed to line up, roughly, in what looked like a cause-and-effect relationship. But the families' case was based on epidemiology, and epidemiology was a science of probability, not certainty. Its practitioners looked for correlations between exposure and disease and then tried to assess the likelihood that they were causal. Even with all the pollution and cancer in Toms River, the apparent association could never be confirmed definitively because of the unanswerable questions about long ago exposures and also because of the enigmatic nature of cancer, which struck so unpredictably and had so many possible causes. If the case ever went to trial, a jury might be persuaded by Cuker's maps, but the lawyer would have to make an indirect argument on the families' behalf, one that relied on surrogacy and correlation—and plenty of emotion, too.

Bruce Molholt, lawyer Mark Cuker's volunteer toxicologist, thought that there might be a more direct way to link the children's cancers to local pollution. His training was in molecular genetics, and he was accustomed to assessing the impact of pollutants on individual cells and genes, not on large groups of people, the traditional domain of epidemiologists. Molholt no longer did research, but he kept up with the field. He knew that by the mid-1990s, scientists all over the world were looking for—and finding—distinctive changes in genes that seemed to be associated with exposure to specific environmental pollutants. In moments of enthusiasm, the new field's practitioners sometimes claimed that these genetic biomarkers, as they were called, were like fingerprints—unique identifiers of specific chemical exposures. They were not. Very few biomarkers were caused by only one chemical, and not everyone who was exposed to a particular compound carried the associated biomarker in his or her DNA. Lacking this one-to-one correspondence, researchers instead looked for patterns, probabilistic associations between biomarkers and exposures, which is why the new field was called molecular epidemiology.

The new discoveries were exciting, even if their meaning was unclear. New analytical tools were allowing scientists to peer into the architecture of the double helix, where they could see how certain industrial chemicals wreaked genetic havoc. Some pollutants slipped in between the tightly coiled base pairs of DNA like playing cards in a door spring; others fused with DNA and formed adducts that disrupted cell replication. No one was sure exactly what this meant for cancer causation, though it certainly seemed significant, in light of what Alfred Knudson and others had already proved about the importance of multiple "hits" on DNA in triggering cancer. If biomarkers proved to be sufficiently reliable—a big if—they might even save lives by serving as early indicators of cancer, long before tumors could be detected by conventional means, and by identifying populations that were particularly susceptible to cancer.[21]

What particularly excited Bruce Molholt about molecular epidemiology were its obvious implications for investigations of cancer hotspots like Toms River. So far, the research projects undertaken in the town were squarely within the twin realms of classical toxicology and epidemiology, fields that could trace their lineage back to Paracelsus. At the end of the National Toxicology Program's two-year rat bioassay, pathologists would look for physical evidence of tumors, just as Katsusaburo Yamagiwa had done with his rabbits almost a century earlier. Similarly, Jerry Fagliano and his team were conducting a textbook epidemiological study, using exposure mapping and case-counting techniques that were much more sophisticated yet essentially similar to what John Snow had utilized during the London cholera epidemic of 1854. Fagliano identified his forty cases solely on the basis of a diagnosable tumor, not a genetic mutation or other biomarker. But tumors were only one way to document past exposure, and not a very accurate one, since most cancers had many possible causes, and most exposed people did not get cancer.

A biomarker study of the Toms River children, Molholt thought, held out the hope of uncovering clearer evidence of exposure—not quite a fingerprint, but something more convincing than the correlative evidence of Mark Cuker's maps. It would also be a way of expanding the focus of the investigation beyond SAN trimer, which Molholt worried was getting too much attention compared to other compounds in the town's air and water. He thought that many of the chemicals dumped in Toms River—especially benzidine and anthraquinone dyes—were good candidates for a biomarker study. The dyes, he thought, were probably intercalators, which meant that they could damage genes by slipping between two base pairs of DNA. Several carcinogenic pollutants had already been shown to intercalate, including benzo(a)pyrene, the coal tar ingredient that Ernest Kennaway in 1932 had confirmed as the first known synthetic carcinogen. Benzidine and anthraquinone were excellent clothing dyes because they were highly reactive, binding tightly to fibers, and because they did not dissolve in water. Those same qualities, Molholt thought, probably made them DNA intercalators. They would be attracted to the hydrophobic environment of the inner surface of the double helix, binding there and, perhaps, triggering cancer-promoting mutations.

If he could convince a molecular epidemiologist to look for biomarkers in Toms River children, Molholt thought, new evidence strengthening the families' case might surface, maybe even in time to give them extra leverage in the looming settlement negotiations. He already had someone in mind for the job.

CHAPTER TWENTY-TWO

≈

Blood Work

The call from Bruce Molholt surprised Barry Finette. Although he studied how synthetic chemicals damage DNA—a potentially litigious topic if there ever was one—Finette was not called often about legal cases. Many scientists involved in the Toms River investigations were veteran expert witnesses in environmental lawsuits, specializing in one side or the other. Molholt and Richard Clapp almost always testified on behalf of alleged victims of pollution, while Philip Cole and Elizabeth Delzell of the University of Alabama usually worked for manufacturers. But Finette was not a professional witness. He did not particularly like lawyers, and he worked at the University of Vermont, well off the beaten path for big-time biomedical research.

Back in 1998, two years before Bruce Molholt called him, Finette had caused a stir in the world of environmental cancer research by discovering that children of pregnant women exposed to secondhand cigarette smoke carried rare genetic mutations associated with leukemia and lymphoma.[1] It was a novel finding, but it did not make Finette a well-known figure in the field. He did not mind in the least. Finette was a pediatrician as well as a molecular geneticist, and he made a point of spending almost as much time seeing children as he did analyzing DNA in his lab. Even if he had given up seeing patients to focus on research, the monomaniacal lifestyle of a high-profile investigator would have been incompatible with his family choices. Finette and his pathologist wife shared an old farmhouse south of Burlington with their blended household of six children, three adopted from difficult homes. He was the legal guardian of a seventh child. In Vermont, a state suffused with altruists and unconventional families, the Finettes fit right in.

Bruce Molholt's unexpected proposition resonated with Finette in every way he cared about. In his lab and at the Vermont Cancer Center, where he saw young patients, Finette puzzled over the enduring riddles of childhood cancer. If some common chemicals are mutagenic and carcinogenic, why are malignancies so rare in children? If cancer is closely linked to aging (the average age at diagnosis is sixty-seven), why is there any childhood cancer at all? Scientifically, what Molholt was proposing sounded awfully interesting—and feasible, too. Molholt had access to a population of children with two attributes that were almost never this well documented: They had been exposed to toxic chemicals via the local water supply, and they were much more likely than expected to have developed cancer at a young age. Finette's work on secondhand smoke led him to believe that genetic changes could be important clues for understanding the causes of childhood cancer. He envisioned setting up a comparative study of exposed and unexposed children, another case-control design. But while Jerry Fagliano's case-control was aimed at finding out whether cancer risk was affected by the proximity of a child's home to air or water pollution, Finette's study would look for evidence of exposure directly in the bodies of the children, in the DNA of their white blood cells. Based on his earlier work, Finette thought there was a decent chance he might find something interesting.

After several conference calls with Jan Schlichtmann and the families' other lawyers, Finette said yes. Like many professionals who had gotten involved, he was already developing an attachment to the families, even though he would never meet most of them. Molholt, Schlichtmann, Linda Gillick, and Bruce Anderson were just voices on the telephone, but Toms River was turning into more than just another research project. "I felt if I could help answer any of these questions then it would be worth it," Finette recalled. "This was potentially a unique opportunity, but potentially a minefield as well."

One of the biggest landmines was the difficulty of getting enough human subjects to participate to yield a statistically significant result. If Barry Finette wanted to find out whether there really was something different about the DNA of the children who drank Toms River water, he would need to test as many children as possible. But getting blood samples from the kids would not be easy. The Toms River children were already all too familiar with the agony of needles, and many of the families whose children were now in remission were trying to forget that they had ever been touched by cancer. They would not be eager to participate in a research project that would bring back such painful memories.

The chemotherapy and radiation therapies the children had already endured posed another difficult complication. Those excruciating treatments had severely damaged the children's DNA, making it impossible to determine which mutations might have been caused by environmental exposures that predated treatment. Finette's solution was to draw blood also from their healthy brothers and sisters, who presumably had drunk the same tainted water as their sick siblings and were genetically similar to them. Since he was using a case-control model, Finette would also need samples from a control group: children who were not from Toms River and had not drunk the water but who matched the ages and sexes of the exposed children. Later, he would also include another matched control group: out-of-town children with cancer. Finette made this last addition so that if he found a pattern of genetic anomalies in the exposed Toms River children, he could be reasonably confident that it was caused by local pollution and not by the disease itself.

Finding enough willing children in all those categories would have been difficult even if Finette were blessed with a huge budget for recruitment, which he was not. The lawyers had budgeted just $30,000 for the entire project. Finette could afford to hire a local phlebotomist to draw the blood samples, but everything else in Toms River— including recruiting participants—would have to be handled by volunteers.

In the fall of 2000, Linda Gillick began contacting the TEACH families and urging them to give blood for Finette's research. Within weeks, the phlebotomist had specimens from thirty-two Toms River children who had been diagnosed with cancer, forty-nine of their healthy siblings, and forty-three healthy children from out of town (matched to the siblings by age and sex). The totals were lower than Finette had hoped, but they would have to do. Bruce Anderson took charge of shipping the vials to Vermont. For the first shipment, he made the mistake of writing "Fragile: Blood Samples" in large letters on the box. Late that night, he got a call from a supervisor at United Parcel Service informing him that UPS would not ship such dangerous material. "After that," Anderson recalled, "I learned to take all the labels off the outside of the box."

By November, the cryogenic freezer in Finette's lab was full of small plastic tubes, each containing a New Jersey child's blood, processed into dried pellets and frozen at exactly 80 degrees below zero Fahrenheit. Two young researchers at the lab were already busy thawing out some of the samples and beginning the painstaking search for telltale biomarkers, one chromosome at a time.

At the same time that Barry Finette's team was analyzing the first blood samples from Toms River, the lawyers were gathering in downtown Newark, not far from Nick Fernicola's old stomping grounds, for the first of the expert presentations that would determine whether Jan Schlichtmann's dream of a negotiated settlement was a hopeless fantasy or a realistic possibility. Loosely presiding over this informal "science court" was Eric Green, one of the country's most famous mediators and a founder, in the early 1970s, of what became known as the alternative dispute resolution movement. In Toms River, Green was not technically serving as a mediator; he had no legal power and no authorization to try to broker a settlement. For now, Green saw his role as pushing both sides to make their most persuasive case, without holding anything back.

It was Green's idea to schedule thirteen all-day sessions, scattered over four months. Each side would bring in experts to address every disputed fact in a case that had hundreds of them. What were the causes of childhood leukemia? How fast can SAN trimer move through sandy soil? Exactly what did the Toms River Chemical Company dump into the river in the 1960s and 1970s? How can you tell whether a cancer cluster is real or a random fluke? Green wanted the lawyers to hear each other's best arguments and most persuasive expert witnesses on all of those questions and many others.

None of the attorneys involved had ever been through anything quite like it; nor had most of the experts they flew in from as far away as California and England. "That was the scariest room I've ever been in. It seemed like there were five hundred lawyers in there," remembered epidemiologist Richard Clapp. Actually, there were only about twenty lawyers and consultants present most of the time, but the room was so cramped that it felt like more. The host, a water company lawyer named Steve Picco, quickly learned that no matter how many bagels he ordered, they would be devoured in less than five minutes. By mutual agreement, the proceedings were confidential and completely off the record, which meant that the lawyers had to take their own notes. There would be no official record.

What they witnessed was an extraordinary series of clashes over scientific issues that seemed hopelessly arcane when considered individually but that collectively would determine whether the Gillicks, the Andersons, the Pascarellas, and all of the other Toms River families would get the satisfaction—and compensation—they had sought for so many years. Schlichtmann kept reminding everyone that the process was supposed to be a cooperative search for answers, not an adversarial slugfest, but he often seemed to be trying to convince himself as much as everyone else. By now, the lawyers around the table knew that if there was going to be a negotiated settlement, its terms would almost certainly be determined by how impressed each side was with the other's arguments—and by how impressed Eric Green was. Green could not force a settlement, but he would be listening to every word and giving his assessment, privately, to each side.

At one of the presentations, Mark Cuker finally got to show off the color-coded maps that encapsulated the families' case. The maps purported to show how pollution and then cancer spread (or "metastasized," in Schlichtmann's words) across Toms River. Emma Ansara, a twenty-four-year-old assistant to Clapp, had created the maps. She

was terrified to be showing them to a roomful of hostile attorneys. "I
felt very much out of my league, and very underdressed," Ansara remembered. The maps, covering a period of about fifteen years, were
simple: Colored lines superimposed on a street map of Toms River
represented the pipes carrying tainted Parkway well water; colored
dots marked the addresses of newly diagnosed children. Clapp
thought the pattern was obvious: A colored line would extend into a
section of town and then, on the next few maps of the sequence (representing the next few years), dots would begin to accumulate in the
same area. "It was just sort of jaw dropping how well they lined up,"
he recalled. Schlichtmann was even more excited: "When I saw Dick
Clapp's maps, I knew this was going to be another Woburn. I knew
we had it; this was compelling," he remembered. But the pattern, if
there really was one, was far from perfect. There were cases in uncon-
taminated neighborhoods, and no cases in some areas where the water
had been tainted. And even if there really was a discernible pattern,
the maps were merely a visual representation of the same indirect,
correlative evidence that had frustrated environmental epidemiologists for almost a century. The apparent, if partial, alignment between
pollution and cancer could be merely coincidental.

Were the maps convincing enough to sway a jury? More to the
point, was the other side now worried enough to consider a settlement that would keep the case out of court? Certainly, the lawyers for
the families thought they were making progress. So did Clapp, after
what he heard his counterpart on the other side of the case say in an
unguarded moment. Jack Mandel was an epidemiologist at Expo-
nent, a huge firm that often supplied expert witnesses to the chemical
industry. Inside the conference room, Mandel had argued that the
cluster was coincidental. But out in the hallway during a break, as the
two men chatted, Clapp was surprised to hear Mandel agree with his
observation that the case was looking a lot like Woburn. As Clapp
remembers it (but Mandel does not), Mandel then said something
even more surprising: "I think they should settle, don't you?"[2]

The meetings with Eric Green would continue, intermittently, for
three more months. After each long session, Green and his technical
adviser, Doug Allen, would meet privately with each side and offer

their confidential opinions about the strengths and weaknesses of the day's arguments. Exactly what Green said to each side would be forever secret, under the terms of his consulting contract. But by the end of the thirteen expert presentations, in early 2001, he was apparently telling both sides that a financial settlement of the case was both possible and desirable—and the companies were listening. More than three years after Schlichtmann began his unlikely quest for a lawsuit-free resolution of the case, attorneys for Union Carbide, Ciba, and United Water signaled that they were finally ready to start talking about a settlement. They might not be convinced that the families were in the right, but the prospect of a high-profile lawsuit and maybe even a huge jury verdict was more frightening than ever. There was also the wild card of the long-awaited case-control studies. If their results were clear, one way or the other, there would be little point in further settlement talks. At that point, the companies either could be facing hundreds of millions of dollars in liabilities, or none at all.

Jerry Fagliano tried not to think about how much was riding on the outcome of the studies that had been the center of his working life for more than four years. He cut himself off from any news about what the lawyers were doing but could not help being aware that the outcome of his investigation could add or subtract many millions of dollars from any out-of-court settlement. The studies could even help determine which families would benefit, if they ended up linking certain types of cancer to local pollution while excluding others. "We were certainly aware of that, and it put a tremendous amount of pressure on us to get the study right," Fagliano remembered. Linda Gillick made it impossible to forget what was at stake. She almost never spoke in public about the legal case—that was a private matter for the families and their lawyers, as far as Gillick was concerned—but she never missed a chance to remind Fagliano that much more than money was riding on the outcome of his studies. The families were counting on them to explain what had so far been unexplainable: the cause of the cluster. Fagliano still attended monthly meetings of Gillick's advisory committee to explain, over and over, why the studies were taking so long. Gillick usually greeted him with a sharply worded reminder that

the families were out of patience. By now, she and Fagliano had attended nearly one hundred public meetings together and dozens more
in private. They had developed a mutual respect. But in public, Gillick
was as relentless as ever, even when the audience was sparse, as was
usually the case now that most residents and reporters had lost interest. Easing up was not her style.

By 2000, Fagliano was waiting for just one thing: the completion
of the computer model of the town's water system, which would tell
him which wells had supplied which homes. The lawyers had used a
much simpler model in the maps they devised for the expert presentations in front of Eric Green, but their version was a mere stick figure
compared to the masterpiece Morris Maslia was devising for Fagliano's studies, at a cost of more than $5 million. Maslia made no apolo-
gies for the slow pace. Creating 420 water distribution simulations,
one for each month of the thirty-five-year study period, and then testing their reliability, was extremely complicated and difficult.[3]

Finally, in the fall of 2000, Maslia was ready. In November, at the
same time that Barry Finette was analyzing the first blood samples
from Toms River and the lawyers were holding their first session with
Eric Green, Maslia showed his model to a panel of expert advisers,
who suggested only minor changes. In January, he sent his simulation
results to Fagliano, who already had the results of the air pollution
dispersion model created by the team at Rutgers. As with Maslia's
water model, the air model did not try to determine which particular
pollutants were in specific parts of town at various times; the historical records were far too vague for that. Instead, the models were comparative tools that showed how pollutants *in general* were dispersed
around Toms River and thus could be used to estimate the *relative*
burden borne by each neighborhood, not the *actual* burden.

Jerry Fagliano did not need to know what was actually in the air
and water all those years ago because his case-control study, too, was
based solely on comparisons. What he needed to know was which
neighborhoods, based on the models, got more pollution and which
got less during the three critical periods: 1962 to 1996 for Ciba air
emissions, 1962 to 1975 for Holly Street well water, and 1982 to 1996
for Parkway water.[4] Ever since the shocking release of Michael Berry's

cluster study in 1996, there had been no compelling new information about cancer and pollution in Toms River, despite the millions of dollars spent on investigations. With the completion of the computer models, Fagliano and his collaborators were finally in a position to say something new.

There was just one more preliminary step: The researchers needed to calculate a numeric score, an "exposure index," for every address at which a case or control child lived. They would need index scores not only for historical exposure to air emissions from the Ciba factory and the nuclear plant, and to water from each public well field, but also to water drawn from the private backyard wells that had been so common in town until the 1970s.[5] And they would need to calculate these scores in two different ways because there were two case-control studies: a birth record study and an interview study.

Fagliano knew almost nothing about the 528 children in the birth record study. Forty-eight had been diagnosed with cancer, and the rest were healthy controls matched to case children by age and sex. He did not even know whether they still lived in Toms River.[6] With so little information about residential history, it would be impossible even to guess at the relative pollution burden each child had faced before being stricken. But because he knew the dates and home addresses on their birth certificates, it *would* be possible to use the air and water models to estimate the relative exposure each child's mother faced during pregnancy.[7] That seemed especially useful now that the interview results were suggesting that prenatal exposures might be very important.

For these birth record study children, calculating an exposure index would be straightforward. For drinking water exposure, Morris Maslia would identify the location in the pipeline network closest to the birth address of each child and then run the model to see which combination of wells had supplied the home's water for each month of the thirty-five-year study period.[8] Fagliano's team would then take the model results for the nine months that preceded each child's birth and calculate a prenatal exposure score. For example, a child whose prenatal water supply was 60 percent Parkway water, 5 percent Holly water, and 35 percent other wells would be classified as having high exposure to Parkway water but low Holly exposure.[9] The Rutgers team devised similar scales for air emissions from Ciba and the nuclear plant, and the health department used them to generate prenatal air exposure scores for each child and each type of exposure.

The scores were a simple and sound way to compare exposure but were vulnerable to criticism because they did not take into account where the children lived after they were born and which type of water (tap or bottled) their mothers consumed while pregnant. Depending on the study's outcome, critics on one side or the other would have legitimate reasons to complain. But there was a smaller group of Toms River children that Fagliano knew much more about. These were the 199 children whose families had been interviewed in depth by state health investigators. They were the subjects of his second study, the interview study. Forty of the interview study children had been diagnosed with leukemia or brain or nervous system cancers. One hundred and fifty-nine were healthy controls matched to cases based on sex and age.[10]

Thanks to the interviews, Fagliano knew where each of these children lived all the way from one year before birth until the month of diagnosis. He knew how many glasses of water per day each mother and child had drunk and whether there was a filter on their tap. True, his information depended on memories and thus was subject to recall bias, but it was far better than having no information at all. So for the interview study, Fagliano's team was able to devise a more elaborate exposure formula that took into account changing addresses and varying personal behaviors. A habit of drinking a lot of tap water, for example, would raise a child's water exposure score.[11] These adjustments made the interview study, despite its smaller size, less vulnerable to criticism than the birth record study.

By the beginning of 2001, Jerry Fagliano and his collaborators at the state health department had a full set of exposure scores for all 702 children in the two studies.[12] After five years, the preparation was over. There were no more decisions to be made about which hypotheses to investigate, which children to include, or which formulae to apply. There were no more interviews to conduct or computer models to test. There was nothing left to do but run the numbers and interpret the results.

The biomarker Barry Finette had decided to look for in the Toms River blood samples was a gene with a name only a biochemist could love: hypoxanthine-guanine phosphoribosyl transferase, or HPRT. He wanted to find out whether the white blood cells of children who drank Toms River water had more HPRT mutations than the cells of unexposed children who lived out of town. If genetic mutations were the key to triggering cancer—as Theodor Boveri's century-old speculations, Hermann Muller's fruit fly experiments, and Alfred Knudson's retinoblastoma studies all strongly suggested—then mutation frequency was a logical indicator of cancer risk.

Biomarkers like HPRT were at the heart of the young field of molecular epidemiology. The term had been coined in 1982 by a junior researcher at Columbia University, Frederica Perera, in a paper she wrote with her mentor, I. Bernard Weinstein.[13] An admirer of Knudson's ideas about multi-hit carcinogenesis, Weinstein studied interactions between pollutants and DNA at the most intimate level: the chemistry of individual molecules. Perera and Weinstein thought it might be possible to use the pattern-recognition tools of traditional epidemiology to identify which of those molecular couplings were associated with disease. Again, the evidence would be indirect—correlation, not causation—but it would be powerful because these biomarkers could be measured directly in human blood and tissue without requiring extrapolation from animal tests.

If the molecular epidemiologists were successful, they would be pioneering a new way to assess the health risks of chemicals—and also to win legal cases like the one in Toms River. Even more importantly, their work might lead to the development of early warning medical tests to save the lives of undiagnosed cancer victims. As the people of Toms River knew all too well, conventional cancer epidemiology was like a fire engine that arrived long after your house had burned down. By the time answers were available, many people would already be sick or dead, and it would be too late to do anything but guess which decades-old exposures might be responsible. Molecular epidemiology held out the hope of working much faster—but only if people like Perera could find biomarkers that could be reliably measured and were true surrogates for disease risk.

Perera's first biomarker candidate was benzo(a)pyrene, the same compound Ernest Kennaway had identified back in 1932 as the crucial carcinogen in coal tar, the fountainhead of the chemical industry. A potent carcinogen, B(a)P had a distinctive way of making mischief in human cells: Its molecules would bind tightly to DNA strands by sharing pairs of electrons in tight, covalent bonds. When a cell replicated, these adducts could trigger genetic mutations. Hunting for these "B(a)P-DNA adducts" in human cells, Perera thought, could be a very useful way of assessing an exposed person's risk of getting cancer. Starting in 1980, Perera and her collaborators at the National Cancer Institute looked for adducts in B(a)P-exposed mice, rabbits, and dogs and also in the white blood cells and lung tissue of humans, some of whom had lung cancer. What they found would establish the tantalizing pattern of molecular epidemiology: hints of significance cloaked in ambiguity. Animals injected with B(a)P did indeed have more DNA adducts, but the correlation was murkier in humans with cancer who smoked.[14] Adducts were not a surefire "dosimeter" of exposure and risk, as Perera had hoped. Carcinogenesis was too complex, involving too many steps and too many possible pathways, for any one biomarker to be an accurate predictor of risk. Still, the results were interesting enough to keep going, she thought.

Soon, Perera and dozens of other researchers were setting up larger case-control studies and counting DNA adducts in highly exposed people all over the world. The field soon expanded far beyond adducts to embrace other biomarkers, too. Investigators tested populations for mutations in oncogenes (including the "Philadelphia chromosome" that had inspired Knudson) and also tumor-suppressor genes—especially the TP53 gene on the short arm of Chromosome 17, the most frequently mutated gene in human cancer.[15] There were other genetic biomarkers, too, including indicator genes that did not play a direct role in causing or preventing cancer but seemed to be useful surrogates for assessing overall risk. Today, researchers can quickly and inexpensively scan the entire human genome for mutations, so indicator genes have become less important. But in 2000, when Barry Finette was launching his Toms River studies, they were vital.

HPRT, the biomarker Finette was using in his initial Toms River study, was one of the most popular indicator genes—though not necessarily for the right reasons. No one was sure if HPRT was an important gene for carcinogenesis; the evidence from factory studies was mixed, though there was good evidence that radiation and chemotherapy treatment triggered HPRT mutations in cancer patients.[16] But the gene had other qualities that made it a splendid biomarker, including the fact that it was easy to propagate, or clone, inexpensively in the laboratory.[17] That was an essential trait for human studies, in which every blood sample was precious and had to be carefully conserved. Even so, focusing a cancer epidemiology study on HPRT was a little like looking for lost keys under the streetlamp because the light was better there. It was a very convenient biomarker gene, but hardly the most promising of the thousands of possible candidates.

Finette was handicapped not only by a questionable biomarker but also by an experimental design that relied on some very speculative assumptions. He planned to compare HPRT mutation rates in the siblings of TEACH children to rates in children of the same sex and age who lived out of town. The plan assumed that *all* of the siblings, no matter where they actually lived in Toms River, had been highly exposed to contaminated water and that none of the out-of-town children had been similarly exposed. But water contamination was a fact of life throughout South Jersey, and within Toms River varied significantly in different parts of town. Indeed, Jerry Fagliano's case-control study was premised on this variability, since it compared various neighborhoods within the township. Just as questionably, Finette's study had to assume that siblings were apt surrogates, since the DNA of the TEACH children was already too damaged by chemotherapy to analyze directly. Finally, the blood samples in Finette's study were not collected until 2000, many years after local water and air pollution were at their worst. Finette thus had to assume that the children's DNA still showed damage from many years earlier and had not been masked by subsequent mutations that accumulate naturally as the cells in a child's body keep dividing with age. For all those reasons, Finette worried that he would not find the answers the Toms River families sought.

He was right to be pessimistic. It took Finette's research team five months to count HPRT mutations in white blood cells from the forty-nine Toms River case siblings and the forty-three control children from out of town. In early 2001, he reported his results to the lawyers: There was no appreciable difference in mutation frequency between the two groups.[18] The experiment, Finette thought, had been crippled by the cascade of assumptions built into it, especially the long gap between the years of peak pollution and the collection of the blood samples in 2000. If there had ever been a spike in mutation frequency in the Toms River children—and Finette still believed there had been—the study was too tardy to detect it. As usual, science had arrived too late to make a difference in Toms River.

Finette had other ideas for Toms River research that he would pursue for years to come, returning over and over to the plastic tubules in his lab's freezer, like a pilgrim in search of enlightenment. He would go beyond merely counting mutations and instead look to see if specific genetic changes were present in cells of local children but not out-of-town controls. His ideas would take many years to test—far too long to influence the outcome of the legal case—and there was no good reason to think they would ultimately bear fruit. Molecular epidemiology, in its simplest form, was premised on the idea that diseases could be predictably associated with single, specific genetic variations, but most cancers did not play by those rules. Still, the families and their lawyers wanted Finette to keep going. They were interested in more than just leverage in the upcoming settlement negotiations. If there was any chance to someday learn something new about the cause of the cluster, they wanted to pursue it. So Finette and a dwindling team of assistants would keep working, quietly, as the Toms River drama built to a climax. And then they would keep working for years afterward, searching for faint clues in a dark sea of genetic data.

The families had put their faith in three kinds of case-control studies, and now it was clear that two of them—Finette's blood work and the National Toxicology Program's rat study of SAN trimer—would take many more years to complete and were so weighed down by scientific complexity that they were unlikely to end in a meaningful result. Realistically, there was just one hope left.

CHAPTER TWENTY-THREE

Associations

At first, Jerry Fagliano could not tell whether there was a discernible message buried within the stacks of computer printouts crowding his office. It had taken five years to collect all of the information he had sought about children and chemical contamination in Toms River. By the beginning of 2001, he had everything he needed to see whether kids diagnosed with cancer were truly more likely to have been exposed to pollution than healthy children. But there were so many ways to cut the data that it was hard to avoid getting lost in the blizzard of numbers. Prenatal exposure or postnatal? Parkway well water or Holly Street? Chemical plant or nuclear plant? Interview study or birth record study? Infants or schoolchildren? Boys or girls? Which years? Which cancers?

With so many potential associations to analyze and so few cancer cases (just sixty-three, between the two studies), the data was unlikely to sort itself neatly. Instead, it was likely that some apparent links between exposure and illness would turn up for no reason other than chance, while others would stay forever hidden. This was a perpetual danger of small-number epidemiology: Even an association that passed a one-in-twenty test of statistical significance might still be a fluke. Fagliano knew that finding one or two isolated links between exposure and disease would not be scientifically convincing; there would need to be a *pattern* of associations, a pattern consistent with a prior hypothesis about what might have caused the cluster. On the other hand, associations between exposure and disease that did *not* pass a statistical significance test could not necessarily be excluded as suspects. The numbers were too small to be definitive about anything.

Still, there *was* something that caught Fagliano's attention almost immediately. A surprisingly high number of women who had been heavy consumers of Parkway well water while pregnant had children who developed cancer. No matter how he analyzed the data, children who were exposed to Parkway water after their birth did not face a large extra risk, but those who were exposed prenatally did. Similarly, mothers of case children were more likely to have drunk Parkway water than mothers of healthy control children. This observation was consistent with what Fagliano already knew from the interview results: There was a dose-response relationship between how much tap water women remembered drinking during their pregnancies and their risk of having a child with cancer. In other words, more glasses correlated with higher risk. That interview data was probably influenced by recall bias; mothers who had undergone the trauma of having an afflicted child were more likely to remember drinking a lot of water during pregnancy. But now the much more objective water dispersion computer model, which did not rely on anyone's fuzzy or wishful memories, was backing those subjective interview results.

At a staff meeting in Trenton early 2001, Fagliano and his collaborators at the health department looked at all of the water data collectively for the first time and were startled by the apparent clarity of the results. Instead of the inconsistent findings they expected from a study of such a small population, the Parkway results seemed to tell a coherent story. "That was a 'wow' moment," Fagliano remembered. "When we looked at the wells that were not contaminated, we didn't see any differences between cases and controls, but for the Parkway wells a strong association was there. It was pretty dramatic."

What was especially impressive about the Parkway well data was that there was an internal logic to the results. Fagliano had hypothesized that young children of women who had drunk a lot of Parkway

water while pregnant would be at greatest risk, and now he found that
the more tightly he focused his analysis on those children, the greater
the calculated risk, as expressed by a standard statistical measure
called an adjusted odds ratio.[1] For instance, the cancer odds ratio for
children in the interview study whose mothers, while pregnant, drank
water that was mostly from the Parkway wells was 1.68, which meant
that the odds that a child with cancer had been highly exposed prenatally to Parkway water were 68 percent greater than the odds for a
healthy child of the same age and sex.[2] That was not a huge amount
of extra risk, but what caught Fagliano's attention was that the odds
ratios kept rising as he zeroed in on those subgroups that logically
would be at greater risk if prenatal exposure to Parkway water really
were triggering cancer.

For example, when Fagliano looked only at prenatally exposed
children who were diagnosed with cancer before age five, the odds
ratio jumped to 2.51. That made sense, since cancers in older children
were less likely to have been caused by exposures during pregnancy.
The odds ratio jumped again to 3.01 when he added two other filters
by taking into account how much water each woman remembered
drinking during pregnancy and by counting only Parkway water consumed after 1981 in the "exposed" category (1982 was the year he
assumed Union Carbide waste first reached the Parkway wells). This
meant that the odds that young Toms River children with cancer had
been prenatally exposed to post-1981 Parkway water were three times
higher than the odds that healthy local children of the same age and
sex had been. Finally, when he cut the data even finer by zeroing in on
types of cancer known to have environmental triggers and also by
separating the sexes, the associations grew still stronger: For girls diagnosed with leukemia or nervous system cancers before age five, the
odds ratio was 4.60. For boys, it was 1.64. (There was no apparent
explanation for the difference in genders.)

Other permutations of the data yielded even higher odds ratios.
For example, when Fagliano changed his assumption of when the
Parkway pollution began from 1982 to 1984 and looked at prenatally
exposed girls who were diagnosed with leukemia before age twenty,

the odds ratio soared to 14.70. In other words, for the thirteen girls with leukemia born in Toms River between 1984 and 1996, the odds were almost *fifteen times* higher that they had been highly exposed prenatally to Parkway water, compared to the odds for healthy Toms River girls of the same ages.[3]

Those were scary numbers, and they posed a dilemma very similar to the one Fagliano and Michael Berry had faced back in 1995 when Berry first identified the cluster in Toms River. The dilemma was this: Risk numbers were high, but so was the uncertainty. As in Berry's 1995 analysis, each adjusted odds ratio in Fagliano's studies came with a 95 percent confidence interval; the wider the interval, the more unreliable the result. Almost all of Fagliano's confidence intervals were exasperatingly wide, just as Berry's were. The problem was well illustrated by one of Fagliano's most striking findings, concerning children in the interview study who were exposed prenatally to Parkway water between 1984 and 1996 and diagnosed with leukemia or nervous system cancer before age five. The results table looked like this:

Exposure	Cases	Controls	Adjusted Odds Ratio	95% Confidence Interval
Low	10	42	1.00	-
High	6	11	3.01	0.78–11.60

According to the table, the odds that Toms River children with leukemia and nervous system cancers had been highly exposed prenatally to Parkway water were three times greater than for similar but cancer-free children. But because there were so few highly exposed children in the analysis—just six case children and eleven controls—random variability could be heavily influencing the results. The only thing Fagliano could confidently conclude is that if he re-conducted the study twenty times, nineteen of those times the odds ratio would be somewhere between 0.78 and 11.60. That confidence interval not only was very wide, it also dipped below 1.00, which meant that there was a small but noteworthy chance that mothers who drank a lot of Parkway water while pregnant might actually be *reducing* their odds of having a child with cancer. It was another un-helpful snowy-or-sunny forecast, just as in Berry's 1995 cluster study. In fact, because the interval dipped below 1.00, it did not meet the traditional definition of statistical significance, despite the high odds ratio.

Almost all of the Parkway analyses that had so intrigued Fagliano and his colleagues shared this same problem: They dipped below 1.00 on their lower boundary. There were only a few exceptions, most no-tably when they looked specifically at leukemia risk in prenatally ex-posed girls in the interview study who were diagnosed before age twenty, getting these results:

Exposure	Cases	Controls	Adjusted Odds Ratio	95% Confidence Interval
Low	8	39	1.00	-
High	5	3	5.96	1.12–31.70

Even here, with a very high odds ratio and a confidence interval that was entirely above 1.00 (with a sky-high upper bound of 31.70) there was a great deal of uncertainty, as indicated by the wide confi-dence interval. There were only eight highly exposed girls in this anal-ysis, which meant that adding or subtracting just one case would drastically alter the results. Furthermore, the results of Fagliano's other study, the birth record study, did not yield odds ratios for Park-way exposure as high as the interview study did. Nor did the birth record study results form a distinct pattern of rising odds ratios as Fagliano honed the data for the hypothesized greatest risks.

More frustration awaited when Fagliano and his colleagues tried to extend their analysis to the long-ago contamination of the riverside Holly Street wells in the 1960s, when Toms River Chemical had dumped billions of gallons of diluted dye waste into the Toms—enough to tint the river brown, give it a repulsive odor, and contaminate doz-ens of private and public wells. More than thirty years had passed, but until Fagliano's effort, no one had ever tried to assess the health consequences of the Holly Street well contamination, which had been hushed up at the time. Fagliano soon learned that his attempt had come too late to supply meaningful answers. The wells had been polluted so long ago that the health department could find only a few exposed survivors to include in the studies. Fatal cases were even harder to find, since there was no cancer registry at the time. As a result, there were just three case children in the interview study known to have been highly exposed to Holly Street water between 1962 and 1975—too few to calculate a credible odds ratio. The numbers were slightly better for the birth record study: seven highly exposed cases, with a worrisome odds ratio of 2.73 and an alarming ratio of 10.00 for girls only. But the girls' odds ratio was based on just two cases, generating an absurd confidence interval spanning all the way from 0.03 to 372.00. With so few cases, it was impossible to draw any conclusions.[4] The toll from Toms River Chemical's secret fouling of the riverside wells all those years ago would forever remain a mystery.

That left just one major hypothesis to test as a potential cause of the cluster: air pollution, which had always taken a backseat to drinking water as a source of public anxiety in Toms River. The air dispersion model built at Rutgers University was intricate but relied heavily on guesswork, including its dependence on historic wind direction records from almost fifty miles away in Atlantic City. Fagliano doubted that the air model would turn up anything interesting, and his prediction was borne out when he analyzed windborne radiation dispersion patterns from the Oyster Creek nuclear plant ten miles south of town: Case families were not, on average, exposed more heavily to radiation than cancer-free control families.

When Fagliano ran the numbers for airborne emissions from the Ciba plant, however, he saw something surprising: The data lined up in another dose-response pattern.[5] For example, when he looked at downwind prenatal exposure to Ciba air emissions as a risk factor for leukemia and nervous system cancers, in interview study children under age five, the odds ratio for high-exposed children was 1.59, with a confidence interval that ranged from 0.80 to 14.90. When he homed in further, the odds ratios grew, just as they did with the Parkway wells. When he looked at air exposure as a risk factor for just leukemia, and only in girls under age five, he got these results:

Exposure	Cases	Controls	Adjusted Odds Ratio	95% Confidence Interval
Low	2	18	1.00	-
Medium	3	8	5.21	0.48–56.50
High	2	1	18.90	0.90–397

Those odds ratios were very high, and higher exposures to Ciba air generated higher ratios. Furthermore, the lower bound of the confidence interval for high exposure was very close to 1.00, which meant that living directly downwind from the chemical plant very likely increased leukemia risk in young girls. On the other hand, there were just two highly exposed cases, and the confidence interval was extremely wide, suggesting extreme uncertainty. Still, it was a pattern, and it showed up not only in the interview study but also in the birth record study, unlike the Parkway results. For prenatally exposed girls in the birth record study diagnosed with leukemia before age five, for example, the results for Ciba air emissions displayed this dose-response pattern:

Exposure	Cases	Controls	Adjusted Odds Ratio	95% Confidence Interval
Low	1	28	1.00	-
Medium	1	13	2.04	0.12–35.40
High	3	9	7.78	0.78–77.40

All in all, the computer printouts in Fagliano's office carried a historic if bewildering message. Neighborhood cancer clusters had been an object of scientific fascination and public dread for a century, spawning hundreds of investigations around the world. Yet only once before, in Woburn, had a credible epidemiological study identified a likely environmental cause of a cancer cluster outside of the workplace. Now Fagliano thought that he had found *two* causes: air pollution from the chemical plant, and contaminated water from the Parkway wells. But had he really? The pattern of elevated odds ratios was strong, but the confidence intervals were wide and mostly dipped below 1.00, suggesting that there was still a nontrivial chance—unlikely, but certainly possible—that the associations he had uncovered were caused by nothing but random circumstance.

"In a way, it was the worst possible result because we thought we had found something, but we couldn't be sure," Fagliano remembered. "We knew right away that this was going to be controversial. We knew it was something we would want to check forward and backwards." Back in 1995, the health department had reacted to similarly disturbing results in Michael Berry's incidence analysis by keeping quiet—with disastrous consequences. This time, staying quiet was not an option, not after six years of high anxiety and intense anticipation. This time, the State of New Jersey would eventually have to disclose everything it knew and everything it did not know.

For the lawyers on the Toms River case, negotiating a potential settlement in the first months of 2001 was like playing high-stakes poker without being able to look at your cards. Thanks to the expert presentations they had heard at the meetings with mediator Eric Green, both sides had a sense of the strengths and weaknesses of their arguments. But the most important, make-or-break information would not be available until the results of the case-control studies were released—and Jerry Fagliano was not divulging anything yet.

There were obvious advantages to settling now. Jan Schlichtmann had learned the perils of all-or-nothing brinkmanship in the Woburn case the 1980s: His insistence on pursuing it to the bitter end—spurning several settlement offers along the way—left him homeless and bankrupt. This time, against their initial instincts, Linda Gillick and the TEACH families had bought into Schlichtmann's new strategy of staying out of court and seeking a negotiated solution. If it failed to bear fruit now, they would essentially have to start over or just give up. The companies, too, had good reasons to settle. Dow Chemical, the new owner of Union Carbide, was eager to move on. So was Ciba, which was focused on cleaning up the factory site and then leaving Toms River for good. William Warren, the Union Carbide lawyer who was leaving the case now that Dow was in charge, still believed that the companies would win if there were a trial, but the case-control study was a looming unresolved risk. "I often tell clients that they may look at what's being put on the table [as a settlement offer] and find it horribly offensive and unacceptable, but you can't compare that with what may happen" at a trial, Warren would later explain. "You can stand on principle, but there are potential costs to standing on principle."

The families' three principal lawyers—Mark Cuker, Esther Berezofsky, and Schlichtmann—were trying to present a united front to Warren and the other company lawyers. That was not easy. Schlichtmann felt disrespected and underappreciated, while Cuker and Berezofsky thought Schlichtmann was too eager to settle. They admired his creativity and persuasive skills, but he had done much less work on the case than they had, in part because he was distracted by the seemingly endless repercussions of his Woburn-related financial meltdown ten years earlier and also by an unfriendly split from his most recent law firm, San Francisco–based Lieff, Cabraser, Heimann, and Bernstein, to which he had handed off some of his Toms River work. Cuker and Berezofsky had visited Toms River dozens of times to meet with Linda Gillick and other leaders of the families; Schlichtmann had made only a few trips and was not close to the families. Yet Schlichtmann was the lawyer who received most of the press coverage associated with the case, thanks to his celebrity status as the flawed hero of *A Civil Action*. As the negotiations began, the three lawyers' uneasy partnership was fraying.

Settling the case would require assigning dollar figures to human lives, a fraught process that would be especially challenging in Toms River because of the heterogeneity of the sixty-nine children. The heart of their legal claim was that pollution from Ciba or Union Carbide, conveyed by United Water's pipes, had triggered their cancers. But that argument was much stronger for certain children than for others. Some were exposed to tainted water for only a few months; others, for their whole lives. Some were diagnosed as infants; others, as teenagers. Some suffered through years of agonizing treatments; others recovered faster. For fifteen children, the suffering was over; cancer had killed them.

The most difficult distinctions were by cancer type. There were more than a dozen kinds of cancer among the Toms River children, from acute lymphoblastic leukemias (the most common) to a single case of rhabdomyosarcoma, a rare soft-tissue tumor. Since cancer is not one disease but more than 150 unique conditions, it made little sense to treat them all the same. Leukemia victims, for example, had relatively strong claims because there was credible evidence of a leukemia cluster, thanks to the state's 1995 incidence analysis. There was also a large, if controversial, body of scientific research associating leukemia with industrial chemicals—especially trichloroethylene, which was known to have been present in Parkway water. A young child with leukemia would have an even stronger claim because prenatal chemical exposure would be one of the few credible explanations for what might have caused the disease. For other kinds of cancers, however, there was nothing at all in the medical literature to suggest pollutants might be to blame and no evidence of an unusually high number of cases in Toms River.

As they devised their opening proposal in the settlement negotiations, the lawyers for the families had to consider all of these subtleties. There was little to guide them; the Toms River case was unique because so many children with so many types of cancer were involved, and also because the state had already identified a cluster. Settlements in past lawsuits suggested that a nonfatal cancer in a young child might be worth about $1 million if there was strong evidence of causation. In Toms River, $1 million per child would mean almost $70 million—or maybe more, if siblings and parents got awards, too. But how strong was the evidence, really? Eric Green had heard all of it at the presentations and had given both sides his confidential assessment. Now that negotiations were starting, he was a very aggressive go-between. Under the terms of their mediation agreement, the two sides could negotiate only through Green, who made it clear that he would only pass along proposals he regarded as reasonable.

Green relayed the families' first settlement proposal to the companies in February of 2001; a few weeks later, a counterproposal came back. "We were braced for something insulting, but it was much higher than I expected," Cuker remembered. The companies were offering a little more than one-third of what the families' lawyers had proposed. They were still tens of millions of dollars apart (the specific amounts have never been divulged), but they were closer than had ever seemed possible. Eric Green was one of the most esteemed mediators in America; he knew an opportunity when he saw one. He stepped up the pressure, wringing concessions from each side. But after a few more rounds, negotiations stalled. By then it was May, and the prospects for a deal were fading; the study results could be announced any time, and then all bets would be off. The lawyers returned to the work of building a court case, looking for evidence wherever they could. Cuker even asked Bruce Anderson to tear up some old, disconnected pipes in Anderson's basement floor, hoping to find standing water from the 1980s that he could test for pollutants. Anderson, who loved handyman projects, plunged into this one with gusto but found that the pipes were dry. As he was relaying this disappointing news to Cuker, who was in a conference room in Newark, Eric Green walked into the room smiling. The companies had raised their offer significantly; the two sides were less than $1 million apart. A few weeks later, they had a deal. To the shock of almost everyone involved, Jan Schlichtmann's idealistic dream of a resolution without a lawsuit was coming true.

The deal was still just a proposal, however. The companies were not going to pay unless all sixty-nine families renounced their right to sue, and it was not at all clear that every family would agree. While the overall settlement probably topped $35 million (the total has never been publicly disclosed), the proposed payments to individual families were fairly modest, considering what they had been through.[6] The average, after deducting attorney fees, was probably about $300,000 per family. Even more problematically, the payouts ranged widely: more than $500,000 to some families, less than $100,000 to others.

For a decade, the families had stayed united because Linda Gillick made no distinctions among them: Any Toms River child with cancer was welcome. Theirs was an association of equals, with Gillick careful always to emphasize the primacy of the collective. Now that money was on the table, however, distinctions were inescapable. It was as if all sixty-nine families had crossed the ocean together in a lifeboat, only to find that they would not be rewarded equally for their ordeal when they finally reached land.

The lawyers called the May 31 meeting for five o'clock in the afternoon; they wanted to leave plenty of time for debate. By 4:55, nearly two hundred people were inside the hotel ballroom in Lakewood, just up Route 9 from Reich Farm and Ciba. Almost all sixty-nine families were represented. That had never happened before, but this was unlike any previous meeting. The parents and children in the room had been waiting for this for a very long time. At the front of the room stood Jan Schlichtmann, Mark Cuker, and Esther Berezofsky, along with their law partner Gerald Williams and Steven Fineman, a Lieff, Cabraser attorney who had worked on the case. This was their show, and they were nervous. Since the negotiations had started in January, they had had little contact with their clients. Now they were going to have to explain the proposed deal to families who knew almost nothing about it except that some of them would be getting a lot more money than others.

From the lawyers' point of view, there was nothing radical about an allocation formula that treated each family differently. The Toms River case was about environmental causation, and the evidence of causation was stronger for some children than others. Therefore, the companies were not equally liable for the injuries to each child. Even before the negotiations started in January, Eric Green had insisted that Cuker, Berezofsky, and Schlichtmann devise a "matrix" that assigned proposed awards to each family based on many factors, including the type of cancer, length of exposure, and severity of injury. This matrix served as the basis for the subsequent rounds of proposals and counterproposals, with each back-and-forth whittling down the size of the overall total but preserving the disparities in allocations to individual families. After the deal was struck, the families' lawyers were free to do any kind of reallocation they wanted, subject to the unanimous approval of their clients. The companies cared about the overall size of the pie; they did not care how the families divided it.[7] So the three lawyers conferred and devised a plan that was similar to Green's matrix but reduced the disparities between the families somewhat. Even after the adjustment, however, the highest compensated family would get nearly six times more money than the lowest. "The cancers were equally tragic, but we couldn't ignore the fact that some cancers had much stronger legal cases than others," Cuker would later explain. "That was a difficult thing to tell a client."

It was, for about four hours, a very angry meeting. Almost no one criticized the overall size of the settlement or even complained about the lack of an explicit apology by the companies. The families understood that chemical manufacturers would never admit liability voluntarily. But almost everyone in the room was infuriated by the allocation formula their own lawyers had devised. Most thought that each family should get an equal share; a few thought the families of the fifteen dead children should get the biggest shares. No one spoke in favor of the lawyers' formula, which took many factors into account but gave the greatest weight to the causal evidence for each child.

"A lot of people were resistant. We had always said that this was not about the money, but at that meeting, when it came down to it, it *was* about the money for some families," remembered Kim Pascarella, whose infant daughter had died ten years earlier. Pascarella was sympathetic to what the lawyers were trying to do, especially since he was an attorney himself. On the other hand, Bruce Anderson, who had spent more time assisting the lawyers than anyone else, was strongly opposed. "At the meeting, I said no way," he remembered. "I'm sure some people thought it was great. It's all relative. I thought it was a joke. It didn't come anywhere near compensating us for what we had been through. . . . Any amount of money wouldn't suffice."

Jan Schlichtmann listened to the recriminations with dismay. This was not the ending he had envisioned for his noble experiment in alternative dispute resolution. "That night was more painful and difficult than it had to be," he remembered years later. Ever the dreamer, Schlichtmann had initially hoped to reserve a large share of the settlement to create a foundation, administered by the families, to support environmental education and advocacy in Toms River and elsewhere. "It would have been a chance to connect them to each other and to the future of the community, to say that we've learned from what happened in Toms River. But in the end, the money poisoned it, and we fell back on the traditional model," he said. Ever skeptical of Schlichtmann's offbeat ideas, Cuker and Berezofsky rejected his foundation idea as unserious and unfair to the families, who had been waiting a very long time to be compensated for their ordeal. It also would have reduced the payout to the lawyers, who had devoted more than three

years to the case. Instead, the lawyers had agreed on a settlement proposal that would pass through almost all of the money to the families,
minus their fees and some costs. It did include one innovative element:
a "cancer recurrence trust fund" of almost $4 million (after fees) that
would be tapped to compensate any of the surviving children who
relapsed. After fifteen years, whatever was left in the fund would be
distributed to the survivors.

The deal was carefully balanced, and now it was coming undone—
at least, the lawyers feared that it was. "People were yelling at each
other, fighting. It was an unbelievably tense situation," recalled Esther
Berezofsky. "I'd say we were seconds away from dispersing." Not everyone was quite so worried. Kim Pascarella believed that the dis-
senters would come around, after they realized that there were only
two other choices: wait another five years or longer to take a chance
on a jury trial, or give up and get nothing. He was right. As the evening wore on, the mood shifted from bitter denunciation to grudging
acceptance, spurred by the speech of a young cancer survivor who
urged the group to rise above self-interested bickering and stay united.
The room fell silent as she sat down, and after a few more minutes the
lawyers asked for a show of hands. Not everyone voted for the deal,
but no one raised a hand to vote against it—not even Bruce Anderson,
who was still seething but did not want to be the one to sink the agreement.

Many of the parents were crying as they filed out of the hotel and
into their cars to go home. What they were feeling was something far
short of elation, something much less satisfying than the sweet revenge Michael Gillick had craved since he was a boy. At last, the families had gotten an indirect acknowledgment from Ciba, Dow, and
United Water that their children had been harmed, but it did not come
with an admission, an apology, or even an explanation—
just a dollar sign, and an unequal one at that.

The next six months were a strange and secret time. Jerry Fagliano
and the rest of the cluster investigators had found something important in Toms River but could not divulge the results yet to anyone,
including the families. Linda Gillick and the families had reached a

large legal settlement but could not tell anyone about it, including the investigators. Each group had no choice but to keep its secrets. Before they could announce the results of their case-control studies, Fagliano and his team needed to recheck all of their calculations, write up a final report, and present it for review by a panel of outside experts, all of whom were sworn to secrecy. The families and their lawyers, meanwhile, could not divulge the existence of the out-of-court settlement until a judge approved all of the payment arrangements that included children under age eighteen.

The mind-numbing process that preceded those judicial hearings was, for the families, a bleak reminder of how their private anguish had become a business transaction in which they were mostly passive observers. The attorneys told them how much money they would be getting, and a financial adviser (secured by the lawyers) told them how they should manage it. The cancer survivors and their siblings (the latter were modestly compensated for emotional distress) would start receiving payments when they turned eighteen. Their parents, who received less, were told to calculate their expenses in the form of missed workdays, uninsured medical bills, and even mileage and tolls for long-ago trips to out-of-town hospitals. The lawyers also made sure that the families signed the liability release, which was phrased in the form of a letter to Ciba, Dow, and United Water in which the signers agreed to "give up any and all claims and rights" and to keep the terms of the settlement secret. There was no reciprocal letter from the companies expressing regret about what had happened in Toms River over the previous fifty years.

Money was the way corporations apologized—at least, that is what the families had been told by their lawyers. But the nondisclosure agreements the families were required to sign blunted at least some of the satisfaction. Still, $35 million was enough to make the Toms River settlement one of the largest payouts ever in a toxic exposure case. It is unquestionably the largest in a residential cancer cluster case, dwarfing the $8 million Woburn settlement of 1986.[8] The Toms River families had made history, even if they were too wrung out to savor their accomplishment.

The looming court hearings offered the potential for at least a partial catharsis for some of the families, the ones with underage children at home, because they would get their long-sought day in court. But those hearings kept being delayed—first by the need to prepare so much paperwork and then by a shocking intrusion from the outside world: The hijacked American Airlines jet that plowed into the North Tower of the World Trade Center on September 11, 2001, struck just four floors above the New York offices of William Warren's law firm, which had represented Union Carbide. No one from the firm was killed, but the disruption necessitated more delays. The postponements could not continue indefinitely, however, because in October, Jerry Fagliano told Linda Gillick's committee that the state had finally set a date to announce the results of the case-control studies: December 18. If the settlement was not final by then, it might never be. And if there were any last-minute hitches, super-mediator Eric Green would not be available to resolve them. Green was tied up in Washington, D.C., where he was hammering out one of the most famous mediated settlements in United States history, in the federal government's antitrust case against Microsoft.

A week after Thanksgiving, the hearings finally began in Superior Court Judge Marina Corodemus's courtroom in New Brunswick, an hour north of Toms River. Linda Gillick did not speak. She had always said that she wanted a chance to confront Union Carbide, Ciba, and United Water in a courtroom, but Michael Gillick was now twenty-two and too old to be summoned for the "friendly hearing," as the proceeding was called in New Jersey. The judge's only job was to make sure the settlement was in the best interests of each child; the adults were on their own. Some parents and kids made short speeches to Judge Corodemus, telling wrenching stories about the anguish of chemotherapy and the terror of late-night visits to emergency rooms. After a few minutes, each family's day in court was over, and the next set of parents and children entered to tell their story.

The judge approved the settlement after the last hearing, on November 30. That left just one more issue for the lawyers to negotiate: the wording of a joint statement that would announce the agreement without disclosing its financial terms. Released December 13, the statement was a marvel of lawyerly art: Its seven paragraphs managed to communicate almost nothing of substance. The last four sentences read: "With the assistance of numerous technical, scientific, and medical experts, the parties exchanged detailed factual, scientific, and technical information and studied the scientific basis for the childhood health concerns. Ultimately, the detailed scientific inquiry failed to result in any agreement that any of the companies were responsible for the conditions that gave rise to the families' claims. All involved agreed that a settlement would best advance the community's interest and involvement in the public health issues being studied, and would foster the productive and cooperative approach the families and companies have experienced through their dialogue since 1998. Based primarily on these factors, the families and companies reached a settlement to bring closure to the families' claims."

The final sentence was not quite accurate: The Toms River case was over, but it was not closed.

The disclosure of the settlement came as a huge relief to Jerry Fagliano and his colleagues at the state health department, who were about to make their own blockbuster announcement and had worried about how it might affect each family's legal claim. "The announcement of the settlement seemed to take considerable tension off the table," remembered James Blumenstock, then the deputy state health commissioner. "We felt a little less pressure, but it was still there. We still had to do a gut check on how confident we were, from a scientific point of view, that our findings were correct."

The message the state health officials were about to deliver was nuanced; it would please some families but frustrate others. Fagliano had finished writing a first draft of his conclusions in August and had stuck pretty closely to what he had concluded six months earlier, when he first analyzed the data. The two key findings applied to girls with leukemia. For all girls exposed prenatally to Parkway water, there was a "statistically significant association and consistency in multiple measures of association," Fagliano wrote. There was also a "consistent elevation in the odds ratios and an apparent dose response effect" for girls under age five exposed to Ciba air emissions.[9]

In September, when he presented his draft report to a panel of outside experts for a confidential review, its members asked the obvious question: Why only girls? There was no good answer; not even a good guess. Several epidemiologists on the review panel had been deeply involved in the investigation at Woburn, which had also found an association between childhood leukemia and prenatal exposure to contaminated water. But in Woburn, the association was stronger with boys. The most likely explanation was that gender was irrelevant, and that the preponderance of male victims in Woburn and females in Toms River was coincidental. For that reason, some of the reviewers urged Fagliano to lump boys and girls together when he presented his data. But he resisted; if the sexes were combined, the picture would get more confusing. For both sexes together, the odds ratio for children who were highly exposed prenatally to Parkway water after 1981 was 2.57 but the result was very uncertain, with a confidence interval ranging from 0.72 to 9.10. For children under age five, the odds ratio was 1.80 and the confidence interval was 0.39 to 8.37. In both instances, the odds ratio suggested that the association was not due to chance, but since the lower bounds of both confidence intervals dipped below 1.00, neither met the generally accepted definition of statistical significance.

Statistical problems had already doomed Fagliano's efforts to try to find out whether drinking water from the riverside Holly Street wells in the 1960s raised cancer risk. Now Fagliano could make the Parkway data go away, too, if he took the panel's advice and lumped the sexes together. If he did, the State of New Jersey and the Agency for Toxic Substances and Disease Registry could close out their joint investigation by declaring they had found nothing significant; after almost six years and well over $10 million—nothing. Yet there *was* a pattern of elevated odds ratios in Parkway data, even if most of the confidence intervals dipped below 1.00. Back in 1995, when they saw a similar pattern in Michael Berry's incidence study, Fagliano and Berry had concluded that the results were too uncertain to warrant a follow-up investigation. Now, six tumultuous years later, Fagliano did not want to make the same mistake—especially because he had now identified a similar pattern in *two* environmental exposures: Ciba air emissions and Parkway water. He had found a series of associations between pollution and leukemia, and he was going to report them even if he could not explain why they were so much stronger in girls than boys.

Fagliano would not do the same for other kinds of childhood cancer, however, even though the results for brain tumors and other central nervous systems cancers were also tantalizing. For children diagnosed with those cancers before age twenty, the odds ratio for prenatal exposure to high levels of Parkway water was 4.51. But that odds ratio was based on just three high-exposed cases, and its wide confidence interval dipped as low as 0.39. The story was similar for exposure to Ciba air emissions, except that this time the odds ratios were highest for exposures in the womb and the first four years of life: 9.00 for high-exposed young children with brain or nervous system cancer. Again, though, there were few children in the highest-exposed group: only four cases—almost enough to achieve statistical significance, but not quite.[10] The birth record study results, which could assess only prenatal exposures to Ciba air emissions, were even closer to statistical significance for the seven high-exposed children diagnosed with any form of cancer before age five. One more case, one more child, would have pushed it over.[11]

His two studies included sixty-three families in which there was a child with cancer, but Jerry Fagliano, for all his exertions since 1996, had found an answer likely to satisfy just thirteen of them—the ones that included girls with leukemia. The families were already getting unequal payouts from the settlement; now they would get unequal explanations, too. Fagliano had reached the outer limits of small-number epidemiology. He could go no further.

The families got the news in a private briefing on a Monday evening, December 17, in an austere meeting room at the county health department offices on Sunset Avenue, on the north side of town, about halfway between the chemical plant and Reich Farm. It fell to Jerry Fagliano to explain the mixed results of the studies that had taken four years to complete and cost millions of dollars. This time, not all of the families were present, but there was still a large crowd. They were silent as Fagliano explained what he had found—and not found—in the two case-control studies.

It was hard for the families to know how to react. In a sense, they had been vindicated. The naysayers were wrong; there *was* a likely environmental cause for at least some of the cancer cases. A century-long streak of negative findings from residential cluster investigations (all but Woburn) had been broken because the Gillicks and others had refused to defer to nay-saying experts and instead insisted on a full-blown investigation. But their vindication was less than fully satisfying. The parents interpreted Fagliano's words through the filter of their own experiences, as if his findings were Rorschach inkblots instead of statistical charts and tables. In a strictly scientific sense, the families were wrong to personalize the study conclusions because the associations Fagliano found could never explain the cause of any specific child's cancer. He could only assess the extent to which an environmental exposure was associated with a pattern of multiple cases. Like any epidemiological study, Fagliano's was about correlation, not causation.

Even so, it was impossible for parents not to view the study results through the prism of their own child's ordeal. How could they not? Kim Pascarella thought of his daughter, Gabrielle, dead for ten years, when he heard Fagliano say that the association between brain and other central nervous system cancers and prenatal exposure to polluted air and water had fallen just short of statistical significance. As an infant, Gabrielle had been diagnosed with neurocutaneous melanosis, an extremely rare cancer that struck the brain and spinal cord but was classified medically as a melanoma because of the cells it affected. As a melanoma victim, she was excluded from the interview study, which included only leukemias and nervous system cancers. If Gabrielle had been included, Pascarella thought, Fagliano would probably have found a significant association with nervous system cancers, just as he had with leukemias. "My daughter's case should have been included," he remembered years later. "Of course, it should have been."

Melanie and Bruce Anderson heard Fagliano talk about girls and thought about their son, Mike. He was twenty now, having survived three brutal years of treatment for leukemia in the early 1990s. "I was annoyed that they only found an association with girls and leukemia. I had a son with leukemia," remembered Melanie Anderson. Bruce Anderson, who had devoted hundreds of hours to the families' cause, was deeply disappointed at what he regarded as an inadequate investigation. "I thought they should have been able to bring out a lot more information with all of the millions of dollars they spent on it," he said years later.

Linda Gillick said little during Fagliano's briefing, but afterward she angrily unloaded on him. During the twelve years she had been an all-out activist, she and her son, Michael, had always expressed absolute certitude that contaminated water—and maybe air, too—had caused Michael's neuroblastoma. Whenever cynics would point out that residential cluster studies almost always failed to identify a cause, she would respond that the evidence in Toms River—evidence she had helped to assemble on her pushpin map—was so overwhelming that Fagliano could not possibly miss it. But now his work was over, and Fagliano was telling her that he could not confidently identify an environmental cause for the type of cancer she cared the most about, Michael's kind, which was classified as a central nervous system cancer. "I remember Linda Gillick being quite upset with me personally that night, saying we didn't look hard enough and asking where we went wrong. She was very upset," Fagliano recalled. "That was the hardest thing for me. So much of what was motivating the families was to find answers. We did everything we could, but some families came away from that meeting terribly disappointed and there was nothing we could do about it."

The next day, in the same building, Fagliano publicly announced the study results and answered questions at a press conference. About twenty reporters were present—a good crowd but nothing like the hordes of journalists who had descended on Toms River back in 1996 when the existence of the cluster was first disclosed by the *Star-Ledger*. Frank Lautenberg, who had championed the families' cause until he retired from the U.S. Senate in 2000 (he would be elected again in 2002), sat alone in a folding chair in the hallway, waiting for someone to recognize him. Linda Gillick was there, too. For the reporters, she had toned down her furious reaction of the night before. Now, she was saying she was "disappointed but not dissatisfied" with the results.

As soon as Fagliano finished his briefing, many of the reporters moved out into the corridor, where a lanky, gesticulating man with salt-and-pepper hair was holding court. It was Jan Schlichtmann, and he was telling anyone who would listen that the Toms River study was "an earthquake" that would "impact public health and environmental policy-making for a very, very long time." For Schlichtmann, it had been a disheartening weekend. The settlement of the legal case, announced just five days earlier, had not gotten the national media attention he thought it deserved. He had wanted a much splashier version of the vague statement the lawyers had negotiated. Now Schlichtmann was trying to make up for it by buttonholing as many reporters as he could and giving them his ebullient message. Together, the legal settlement and the study were going to set a new standard for resolving environmental controversies in communities across the country, he insisted. The reporters dutifully recorded his enthusiastic sound bites, but the stories they wrote and broadcast did not get much attention outside of New Jersey and Philadelphia. In the wake of the September 11 attacks, the country was now consumed with terrorism and war; *A Civil Action* felt like a long time ago. In fact, some of the key players in the Toms River drama, including Floyd Genicola and Elin Gursky (who was now in Washington, D.C.), were now working full-time on coping with anthrax and other bioterrorism threats.

There was a community forum later that day, and the health department had made elaborate plans for it—even arranging for a simultaneous broadcast on a local cable channel. The meeting site, at Toms River High School East, was just a few thousand feet from the corner of Bay and Vaughn avenues, where cracked asphalt and a chemical stench had heralded the first stirrings of environmental consciousness in Toms River way back in 1984. This time, though, only about seventy seats were filled in the huge auditorium, and the audience was subdued—a faint echo of the more than one thousand screaming locals who had mobbed similar meetings in 1988 and 1996.

Linda and Michael Gillick had played critical roles at those earlier meetings, contributing black-ribboned roses, impassioned speeches, and a hushed recitation of the names of sick or dead children. Now they were again in the audience, and this time Linda Gillick's speech had a contemplative tone that matched the much sparser crowd.

She addressed the parents who, like her, were disappointed that the investigation had failed to identify a likely cause for their child's cancer. "Stand proud," she told them. "We know what has happened here. Don't let the ball drop."[12] The findings were less definitive than she had hoped, but they were still groundbreaking. Moreover, Toms River's water and air were unquestionably safer now. The chemical plant had shut down, the Parkway wells had been filtered at last and the water in Toms River had been tested more thoroughly than anywhere else in New Jersey—maybe anywhere else in the world. The pathways of pollution that for decades had conveyed industrial chemicals into the bodies of town residents were blocked at last. As a result, Gillick predicted, the number of cancer cases would surely decline. In fact, she thought the decline was already under way, though Jerry Fagliano—ever the scientific killjoy—told her it was too soon to know.

Gillick finished with a vow: The Citizen Action Committee on the Childhood Cancer Cluster, the committee she had created and chaired for almost five years, would not be going away, and neither would she. The families had forged an unbreakable association, and there was still work to be done. Someone would need to monitor the cleanup at the chemical plant and keep a close eye on the water company. Someone would need to pressure the state health department to keep its cancer registry up to date and continue monitoring local childhood cancer rates. Besides, the investigation of the Toms River cluster was not quite over, even if its chief component, the epidemiological study, was now finished. Someday, there would be results from Barry Finette's genetic tests, and from the federal government's SAN trimer rat study, too.

Almost everyone in Toms River was finished with cancer clusters now. The town that had tried to move on so many times before did not have to try anymore. It was really over. "The feeling was that the state did everything it could have done to study this," remembered Gary Lotano, the local real estate developer who headed the chamber of commerce and the hospital board. "At some point, it had to end, and it did. The stigma had to end." But for the Gillicks, Andersons, Pascarellas, and other families, it had not ended. It could never end.

CHAPTER TWENTY-FOUR

≈

Legacies

"A good working definition of a public health catastrophe," a well-known Boston University environmental epidemiologist named David Ozonoff likes to say, "is a health effect so large even an epidemiological study can detect it." By that arch description, the Toms River childhood cancer cluster was a manmade catastrophe, and a preventable one.

An ordinary town in so many ways, Toms River owed its unwanted status to an extraordinary combination of human folly and perseverance, along with a great deal of luck—both bad and good. The confirmation of the cluster and the identification of its likely causes were the highly improbable results of a decades-long process in which things went terribly wrong and then spectacularly right. First came the unprotected wells, inattentive officials, supercharged growth, and obscure poisons discharged by the ton into a shallow river and sandy soil. Then came people like Lisa Boornazian, Linda Gillick, Floyd Genicola, and Jerry Fagliano, all of whom went far beyond what anyone could have expected in forcing an investigation and driving it forward. In order for Toms River to be identified as a "true" nonrandom cluster, there had to be an extreme number of cases—enough to be noticeable to the people who lived there and also enough to clear the very high bar of statistical significance. For the cluster to then be linked credibly to local pollution, the environmental damage had to be severe and the epidemiology sophisticated and extremely expensive—and therefore backed by extraordinary political support.

It was hard to imagine a similar set of circumstances occurring elsewhere, and even harder to imagine that a government agency would ever again willingly embark on an investigative process that in Toms River took almost six years, cost well over $10 million, and embarrassed a boatload of public officials on the way to its deeply unsettling conclusion. In that sense, lawyer Jan Schlichtmann's bold prediction that the Toms River epidemiological studies and legal settlement would "impact public health and environmental policy-making for a very, very long time" has turned out to be correct, but not in the way he expected or wanted. The chief legacy of Toms River instead has been to solidify governmental opposition to conducting any more Toms River–style investigations. And because only governments have unimpeded access to cancer registry information (due to privacy concerns), if public agencies do not investigate clusters, then no one will.

A few health agencies—including even the biggest cluster-buster of all, the Centers for Disease Control and Prevention—have reluctantly undertaken residential cancer cluster studies in recent years, though there have been only a handful of comprehensive investigations. Almost invariably, their studies have followed the Toms River/Woburn template: They were launched only after intense pressure from Linda Gillick–like civic activists and from politicians who rushed in amid widespread media coverage. At least three of those investigations, all in the United States, ended with a disturbing conclusion: The cluster was almost certainly *not* a chance occurrence. The most famous was in the Nevada desert town of Fallon and surrounding Churchill County, where sixteen children were diagnosed with leukemia between 1997 and 2002—more than eight times the expected number.[1] (By one credible estimate, the odds that the Fallon cluster could have occurred randomly were just one in 232 million. Stated another way, a cluster that extreme would occur by chance in the United States only once every twenty-two thousand years.)[2] A second was in Sierra Vista, Arizona, where at least eleven children were diagnosed with leukemia between 1997 and 2003 instead of the expected five.[3] The third was a brain cancer cluster among children in a South Florida area called The Acreage, where four cases were diagnosed between 2005 and 2007 when just one case was expected.[4]

In all three communities, researchers failed to identify a likely cause, even after conducting case-control studies that included testing local air, water, and soil. In fact, *all* of the residential cancer cluster investigations undertaken since the completion of the Toms River studies in 2001 have failed to identify a likely cause. There are still just two lonely names on the roster of non-occupational cancer clusters that have been scientifically associated with specific chemical exposures: Woburn and Toms River.

Why haven't there been more? There are at least three possible reasons. The Toms River cluster may have been a statistical aberration, a random one-in-a-million event that actually had nothing to do with local pollution, despite the findings of Jerry Fagliano's two case-control studies. Or perhaps Toms River really was a nonrandom cluster triggered by toxic exposures, but a vanishingly rare one because industrial chemicals may be a trivial cause of cancer compared to more powerful risk factors like tobacco and family history, as the late Richard Doll and some other prominent epidemiologists have long argued. But there is also a third possibility, the one Boston University's David Ozonoff was hinting at with his wry description of the feebleness of epidemiological studies: Clusters of rare cancers like the one in Toms River may actually be much more common than we can discern with the crude statistical tools of small-number epidemiology. In other words, many more pollution-induced cancer clusters may be out there, but we don't see them and we rarely even bother to look.

Public health agencies are shunning cluster studies at a time when researchers need all the help they can get in understanding why and how cancer begins. The most important cancer breakthroughs in recent years have been about treatment, not causation—reflecting the priorities of research funders in government and the pharmaceutical industry. (One of the most successful new cancer drugs, Gleevec, introduced in 2001 by Ciba's successor company, Novartis, has helped to raise the five-year survival rate for the most common type of pediatric leukemia to 90 percent.)[5] For scientists who are still trying to understand carcinogenesis, meanwhile, there has been a major shift since the 2001 completion of the Toms River studies. Moving beyond DNA mutations, they are now studying all of the steps by which genetic instructions are executed within cells, including protein synthesis regulated by DNA's single-stranded cousin, RNA. Researchers are also looking at a galaxy of epigenetic changes—modifications that alter the way genes function without changing the underlying genetic sequence—in hopes of finding clues about why cells turn malignant.

This new complexity has made it much more difficult for molecular epidemiologists like Frederica Perera and Barry Finette to identify specific changes—biomarkers—within human cells that are reliable indicators of cancer risk across populations. The list of candidate biomarkers keeps growing. Some appear to alter susceptibility to particular cancers, conferring extra vulnerability or extra protection. Others are clinical indicators that hold out the promise of aiding the early diagnosis and treatment of tumors. A few can even serve as "dosimeters"—rough indicators of the extent to which someone has been exposed to a particular toxic compound.[6] What almost all of these biomarkers have in common, however, is that they are helpful only some of the time. In the Fallon, Nevada, cluster investigation, for example, researchers found that all eleven local children with leukemia carried a distinctive variation in a gene called SUOX—but 40 percent of cancer-free children did too.[7] An ideal biomarker for childhood leukemia would be so sensitive that every child with the biomarker would have leukemia, yet so specific that every child with leukemia would have the biomarker. After decades of intensive searching, however, it now seems likely that ideal biomarkers may never be found for leukemia and many other cancers because there are just too many possible routes to carcinogenesis. Even now, more than 160 years after Rudolf Virchow first saw malignant white blood cells through his microscope, the biochemical pathways to cancer are still mostly unmapped.

The research that Barry Finette undertook for the Toms River families is a telling example. Working intermittently for more than ten years, whenever he had enough staff and money available to pursue it, Finette and a rotating cast of junior colleagues kept reanalyzing the Toms River blood samples that still filled two freezer shelves in his Vermont laboratory. Searching, as ever, for biomarkers that correlated to chemical exposure and cancer risk, they looked again at the HPRT gene—this time tracking the types and locations of mutations within the gene's region of the X chromosome. They also tracked alterations in eighteen other genes involved in repairing damaged DNA or expelling toxic chemicals that invade cells. Finally, they scrutinized the children's blood cells for biomarkers that were much larger than individual snips of DNA—so large, in fact, that they could be easily seen through an optical microscope. They were chunks of malformed chromosomes, floating free in the nucleus or reattached in the wrong place or even upside down.

In the end, none of it worked. Finette's biomarker studies revealed no significant differences between the DNA of Toms River children and that of children who lived elsewhere. He struggled to come up with reasons for the multiple failures. Perhaps, Finette thought, he had searched for the wrong biomarkers. Or maybe he found nothing because the chemicals in the air and water of Toms River were too insignificant to have a measurable impact on the DNA of the town's children, even though the state's case-control studies suggested otherwise. Or possibly those genetic abnormalities had once been present in the children's blood cells, but were gone by the time he collected his samples in 2000, many years after local pollution was at its worst. It was even possible that the DNA defects were still there but were subtle and thus too difficult to differentiate from random variation within such a small study population. Finette had samples from just forty-three Toms River children with cancer, plus their healthy siblings, which meant that his study was only slightly larger than Jerry Fagliano's interview study and smaller than Fagliano's birth record study—both of which had been plagued by small case numbers.

Whatever the reason, Finette's ten-year search for evidence that chemical exposures in Toms River had altered the genes of the town's children and raised their cancer risk had come up empty.[8] Even the chromosome malformation study, the one Finette had been the most optimistic about, failed to turn up any associations.[9] In a staggering feat of ocular stamina that ended only when the grant money for her salary ran out, a lab technician named Heather Galick had squinted into her microscope and inspected exactly 201,844 white blood cells from eighty-one children. Because each cell was on the verge of dividing and had been stained, Galick could see the chromosomes inside. She counted 466 cells—roughly one out of every four hundred she examined—that had aberrant chromosomes: translocations, loose fragments, or other signs of damage. That seemed like a lot of malformations to Finette, but the ratio of aberrant-to-healthy chromosomes in the Toms River children turned out to be almost identical to the ratio in nonresidents. In fact, the out-of-towners were slightly *more* likely to have malformed chromosomes, a perplexing result Finette could not explain. After a decade of work, there was very little about any of his results that he could explain.

Linda Gillick and a few other Toms River parents got the disappointing news from Finette on a stifling July afternoon in 2010, in the same Ocean of Love office where their needle-wary children had lined up to give blood samples ten years earlier. Back in 2000, dozens of families were involved; by 2010, just three were represented around the conference table. Besides Gillick, there were Bruce and Melanie Anderson (they lived in Pennsylvania now but had made the long drive) and their former neighbor Joseph Kotran, whose daughter Lauren had survived neuroblastoma as a toddler. Most of the other families had not been in touch since the legal settlement nine years earlier. "They have other things in their lives now," Bruce Anderson would later explain with a shrug.

There was one other big study the Toms River families were waiting for, and many of them cared a great deal about it. Styrene acrylonitrile trimer, the obscure plastics byproduct that Floyd Genicola had discovered in the Parkway wells back in 1996, had acquired outsized importance in the Toms River drama because it was so exotic and apparently unique to the town (it had not been identified as a pollutant anywhere else). To some of the families, SAN trimer felt like an answer—maybe even *the* answer—to their most burning questions about the cancer cluster: Why here? Why us? Even so, there were reasons to doubt SAN trimer's true importance: The concentrations in Parkway water were low, the compound's health risks were unknown, and there were many other toxic exposures in town, especially back in the 1960s and 1970s. Still, "the trimer," as they called it, retained its almost mystical hold on the families as time passed. "Really, in our hearts, we all feel it's the cause," explained Kim Pascarella.

The answers on SAN trimer were supposed to come from the federal government's National Toxicology Program, but they were painfully slow to emerge. The NTP's first attempt at its two-year, multigenerational rat study flopped because too few rats were willing to breed, but a second try, in 2005, succeeded. After two years eating trimer-laced chow, the surviving pups were asphyxiated. Pathologists pulled thirty-two organs out of each rat, dunked the organs in formaldehyde, embedded them in wax, and cut them into more than ten thousand slides, each just five microns wide—thin enough to reveal a single layer of cells under a microscope. The slides were then scrutinized for tumors by panels of experts convened by the NTP, in a laborious process that took an additional three years and was quite contentious at times. The debates over the most ambiguous slides—the ones in which it was very hard to tell whether a dot was a tumor or a noncancerous lesion—at times sounded like a roomful of highly opinionated art critics arguing about the meaning of an abstract painting.

The chemical's defenders got to do their own slide review, too. Dow Chemical and the Saudi Arabian company SABIC, a petrochemical behemoth that still made plastics from acrylonitrile and styrene, had formed an organization called the SAN Trimer Association ("SANTA" was its incongruously jolly acronym) and hired their own pathologists to second-guess the judgments of the government's reviewers. For Dow and SABIC, the stakes were high: Regulatory agencies all over the world, starting with the U.S. Environmental Protection Agency, routinely rely on NTP studies in setting limits on chemical exposures.[10]

The seemingly endless review process for SAN trimer finally ground to a conclusion in early 2011, nearly fifteen years after Floyd

Genicola's triumphant unmasking of the mystery compound in the Parkway well field. Two groups—the NTP senior staff, and then its outside review board—would convene at the agency's North Carolina headquarters to pass judgment on the trimer's carcinogenicity. It was a tough call. The slides showed that, except for one type of tumor, cancer incidence among the three hundred trimer-fed rats was similar to the one hundred unexposed controls. The exception was brain and spinal tumors: Eight of the trimer-fed rats had them, compared to just one unexposed rat. That was startling because brain tumors normally were very rare in rats. They had been found in just four of the almost thirteen hundred Fischer rats used as unexposed controls in recent NTP studies. That meant that the tumor count in trimer-fed rats was almost nine times higher than expected. Even more interestingly, there seemed to be a weak yet discernible dose-response pattern in which rats fed higher doses were more likely to get tumors.[11]

The Gillicks and the other families had always thought of themselves as involuntary guinea pigs in an uncontrolled experiment, which is why they had pushed so hard for the multigenerational rat study. Now its results were showing that in at least one important sense, the trimer-fed rats really did resemble the thousands of Toms River children who had been exposed prenatally to much lower levels of SAN trimer in their drinking water. In both populations, certain tumors were much less rare than expected. Further, there were suggestions in the rat study, just as there had been in Jerry Fagliano's case-control studies of the Toms River children, that those with the greatest exposure faced the highest risk.

Was SAN trimer truly a cancer-causer? As usual, the study population was too small to know for sure. If the pathologists had found a brain tumor in just one more high-dosed male rat—three instead of two—then the dose-response pattern would have been clear. But with just fifty male rats in that dose category, the statistical ambiguity could not be resolved.[12] Nor was there any way to know whether SAN trimer caused leukemia—always the key concern in Toms River—because the type of rat the National Toxicology Program had selected for testing was prone to leukemia and thus a poor test model.[13] Despite these shortcomings, there was no chance the agency was going to further delay its decision by conducting additional tests to try to clear up the ambiguities—not after a decade of work and several million dollars already spent. After arguing about the data for hours at a meeting on January 24, 2011, the NTP's senior staff reached its verdict: equivocal evidence of carcinogenicity, the middle of five possible ranks. It was a compromise, a nuanced response to provocative but highly uncertain evidence.

It held sway for exactly two days. When the NTP's Board of Scientific Counselors convened on January 26 to consider the case, a raft of industry consultants were waiting to testify. Dow offered four speakers, including James Swenberg of the University of North Carolina, a former chairman of the advisory board he was now trying to influence. Batting cleanup was Joseph Haseman, a consultant who had spent thirty-three years at the NTP, most of them as its chief biostatistician. When it was his turn to speak, Haseman launched a stinging critique of the staff's decision. Based on past precedent at NTP—precedent Haseman had established—the correct call for SAN trimer was "no evidence of carcinogenicity," he insisted.

The families of Toms River had no hired guns to match them. When the panel's chairman called on the next speaker, a familiar voice—both pleading and reproachful—issued from a speakerphone. It was Linda Gillick, calling from Toms River. "I really believe our children were the real rat study," she said, her voice crackling across the silent room. "Trying to take what really happened in Toms River and duplicate it is really impossible. . . . I'm just concerned that at the end of all this time, effort and money, and all the lives that have been affected and lost, we won't come out of here with true answers." Bruce Anderson had put his comments in a letter to the NTP. "This cannot be allowed to happen to any children again," he wrote. "We hope that the results of this testing will lead to more proactive protection." By that standard, the staff's proposed finding of "equivocal evidence" could only be seen as a disappointment, the same ambiguous mush the families had been hearing from government experts for almost twenty years.

The Board of Scientific Counselors was not interested in equivocation either, but its solution was not what Gillick or Anderson had in mind. The coup de grâce came from board member Jerry M. Rice, who had been a senior scientist at the International Agency for Research on Cancer before becoming an industry consultant and academic. The Toms River cluster, he asserted, was probably a chance occurrence—no matter what Fagliano's case-control studies had concluded back in 2001. "It's well known that rare events cluster in space and time," he said, adding that "an occasional brain tumor" in rats was trivial. He proposed changing the official classification to "no evidence of carcinogenicity." The vote came a few minutes later: six to one for Rice's proposal, with several board members abstaining because they had business ties to Dow.

In the back row of the room, a man named Bob Fensterheim shook hands with his expert witnesses and then left for his flight back to Washington, D.C. A former top lobbyist for the American Petroleum Institute, Fensterheim now was a consultant who specialized in helping chemical manufacturers in their battles with regulatory agencies. He ran no fewer than ten industry groups, with names like the Chlorinated Paraffins Industry Association and the Alkylphenols and Ethoxylates Research Council. His latest creation, SANTA, the SAN Trimer Association, could now fade into obscurity. Its work was done.

In the morning, styrene acrylonitrile trimer had been a possible carcinogen, at least worthy of future study to clear up the uncertainty. By the end of the day, it was not. For the Toms River families, another door had slammed shut. Science, and scientists, had let them down so many times before that it hardly even hurt anymore.

There is one way—one very important way—in which the passage of time has strengthened the Toms River families' conviction that industrial chemicals played a role in their children's illnesses: The leukemia cluster is gone.

As he promised the families he would back in 2001, Jerry Fagliano has carefully tracked new local cases of childhood cancer, just as Linda Gillick still does at Ocean of Love. As always with a rare disease in a small population, the case counts vary so much from year to year due to chance that the underlying trend is not always clear. To smooth out this instability, Fagliano uses five-year running averages.

What he has found for childhood leukemia is that incidence in Toms River, after peaking in the late 1980s, has fallen sharply since then and is now below the statewide average. When charted through 2009 (the last year that fully verified data is available), the graph looks like this:[14]

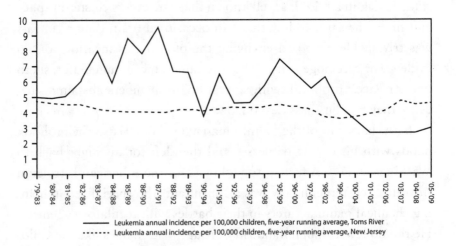

——— Leukemia annual incidence per 100,000 children, five-year running average, Toms River
········· Leukemia annual incidence per 100,000 children, five-year running average, New Jersey

It is an encouraging graph. Leukemia was the type of cancer that Fagliano's case-control studies had most strongly associated with pollution, so if the study were valid, the number of leukemia cases presumably would fall as environmental conditions in Toms River improved. That is what seems to have happened: Rates declined at the same time the factory wound down its operations and the Parkway wells were filtered or shut down.

It is just another correlation, not a proven cause-and-effect relationship, but in the eyes of the families, it is no coincidence. The leukemia decline, many of them believe, demonstrates that pollution was responsible for the cluster and that their activism has thus saved lives. "There are so many fewer cases than we had before all the changes that took place in this town, and that's the way we intend to keep it," Gillick declared at a public meeting in 2011. Jerry Fagliano is less sure about the reasons but is encouraged by the trend. "It's somewhat comforting," Fagliano said, "in that leukemia was the one kind of cancer that really stuck out as being elevated when we did our case-control studies. The fact that it's come back down to normal can mean one of two things: Either the association we saw at that time was causal and we've removed the cause, or else we're simply seeing a return to background rates because of chance. We hope the actions we took have made a difference for public health, but I can't say we've proven it."

One reason for Fagliano's uncertainty is that the trend for all childhood cancers, not just leukemia, is murkier. After falling steadily for a decade, there was an abrupt spike: Eight cases were diagnosed in Toms River in 2004 and nine in 2005, up from the usual four or five. Was the spike just another random consequence of the natural variability of small-number statistics? The fact that the yearly counts have fallen back somewhat since then suggests that it was. So does the fact that no one type of cancer appears to be unusually elevated—unlike in the 1980s, when leukemia incidence jumped. But 2009 was another unusually bad year, with nine cases reported in Toms River, so Fagliano's latest five-year running averages for all types of cancer counted together look especially alarming when charted:

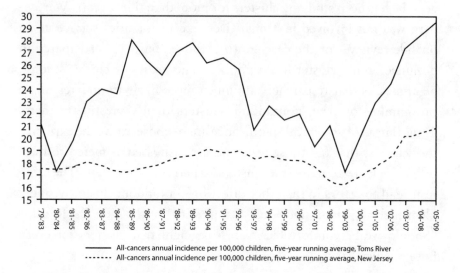

——— All-cancers annual incidence per 100,000 children, five-year running average, Toms River
· · · · · · · All-cancers annual incidence per 100,000 children, five-year running average, New Jersey

The curve is likely to bend back down again soon because the last few years have been good ones: just two childhood cancer cases in 2010 and three in 2011. But those recent numbers are still unofficial and uncounted in Fagliano's five-year running averages. Where rates will go next is anyone's guess. "It is frustrating because the rates are so unstable that there's only so far you can take any interpretation," Fagliano said. "It really points to the difficulty of making sense of cancer rates in relatively small populations over relatively short periods of time. The rates are going to bounce around, even over five-year intervals. With more time, it may become clearer. Maybe."

So what was it, really? Was the Toms River childhood cancer cluster a mirage, an aberration, or a warning? Was it a consequence of nothing but a stunningly bad run of luck, like rolling snake eyes six times in a row? Or was it the product of pollution so horrendous and governmental neglect so extreme that the combination has never been replicated anywhere but Woburn? And what about the third possibility, the one raised by David Ozonoff's dictum that a public health catastrophe is an effect so strong that even an epidemiological study can detect it? Could it be that the only unique thing about the Toms River and Woburn clusters is that anyone managed to recognize them?

Daniel Wartenberg thinks he knows. An epidemiologist at Rutgers, he has been studying clusters for more than thirty years. Wartenberg was not involved in Toms River, except for brief service as an outside reviewer of the case-control studies, but has participated in dozens of other cluster investigations. The reality of cluster studies, he argues, is that researchers are much more likely to underestimate the number of true, nonrandom clusters than overestimate them. Typical tests of statistical significance are so conservative, he explains, that clusters that fail those tests and are dismissed as mere products of bad luck are actually true clusters 20 percent of the time. The same tests lead to errors in the other direction—a random cluster wrongly identified as a true one—just 2.5 percent of the time.[15] He asks, "Should we be comfortable in missing eight true clusters for every false one we detect?"

Even more importantly, the vast majority of clusters are never even tested for statistical significance, Wartenberg notes. More than a century after the creation of the first cancer registry, no government anywhere in the world has a comprehensive program to proactively sift through registry data and identify possible clusters worthy of further analysis. Various proposals in recent years to set up surveillance systems to track the distribution of cancer and to create "response teams" to investigate potential clusters have made little headway in the U.S. Congress or in Europe.[16] Instead, cancer registries are used reactively, in response to press coverage or calls from anxious citizens who happen to notice what looks like a cluster and are motivated enough to call their health department and push for an investigation. In a country as large as the United States, thousands upon thousands of neighborhood-sized clusters would be expected to occur for solely random reasons, in addition to an unknown number of true clusters. Yet almost all of them, random or true, are never even noticed. Health departments apply tests of statistical significance to only a few hundred reported clusters every year. Just two or three of those, in a typical year, are then followed up with a full-blown environmental study aimed at identifying possible causes. Agencies base their decisions on whether to authorize those rare multimillion-dollar studies not on how high the case numbers are but on how successfully the affected population campaigns for attention from the press and politicians.

With so few residential cancer clusters fully investigated, no one should be surprised that just two scientifically credible studies—in Woburn and Toms River—have found a likely environmental cause, says Wartenberg. He adds that it makes no sense to conclude from such a small, biased sample that pollution-induced neighborhood clusters either do not exist or are so rare that they do not matter. The more likely conclusion, he asserts, is that many true clusters go undetected or are found by citizens but wrongly dismissed as random by local health officials who are overworked and don't want to rock the boat. "There are probably quite a number of clusters out there that people haven't noticed but are real clusters," he said. "Even when they are noticed, many clusters aren't studied because no one has any idea what the exposures are and also because it's hard for people to get traction with the health department."

This disconnection between citizens and their government is at its worst in poor communities where pollution and illness are endemic but take a backseat to more urgent social ills like hunger, homelessness, unemployment, and crime. And that is why today's Toms Rivers, Wartenberg suggests, are likely to be in places where people are worried about issues more immediate than cancer clusters.

Places like the industrial boomtowns of inland China.

At thirty-nine, Liu Yu-Shu looks twenty years older. She has the heavily lined face of someone who has not slept well in a very long time. Every night, she spreads a blanket on the floor of the overcrowded hematology ward at Chongqing Children's Hospital, in a narrow space beside the bed of her nine-year-old son.[17] Many parents and even patients sleep in the hallways of this sprawling hospital in central China, so she counts herself fortunate. Liu and her husband are construction laborers; he lays industrial tile, she mixes the adhesive. The previous year, they had left their son with an aunt near Chongqing while they looked for work one thousand miles away in Guangdong Province, in the booming south, where foreign-owned factories sprout like mushrooms after a monsoon rain. Liu found a job there but could not stay long. Back home, her son had developed a cold he could not shake, and tests at a neighborhood clinic showed that his blood platelet counts were dangerously low. A bone marrow test at Chongqing Children's confirmed the diagnosis: acute lymphocytic leukemia, the most common childhood cancer.

A shadow hangs over the wards at Chongqing Children's and across China, and not only because cancer treatment is still substandard by American and European standards.[18] In the last few years, the same kind of amateur epidemiology that captivated Toms River has broken out in provinces like Guangdong, Hunan, and Chongqing. Independent Chinese journalists and activists are collecting their own unofficial, unscientific data about pollution and cancer and are making maps and identifying "cancer villages," posting their findings on the Internet.[19] The hotspots they find tend to be outside cities because many of the worst-polluting factories in China are in exurban areas, and also because semirural communities still have enough social stability for residents to recognize a potential cluster. (Patterns of noncontagious illness are almost impossible to detect in the churn of the teeming cities.) Even the Ministry of Health acknowledges that pollution—industrial waste dumped into rivers and groundwater, plus air emissions from factories and power plants—has made cancer the leading cause of death in China.[20] But because there are no credible cancer registries and the government rarely investigates reported clusters, the unofficial "cancer village" investigations are usually the best information available about local cancer patterns and are given wide credence, including by doctors. A physician on the hematology ward at Chongqing Children's, Liu Xiao-Mei, said that she treats many children of factory workers. Is pollution the reason? "I think so," she said, "but it is very difficult to determine."

What is beyond dispute is that China has zoomed past the United States and is now the largest producer and consumer of many of the world's most heavily used toxic chemicals. The chemical industry is booming in other developing nations, too, but China's position is dominant. In 1996, the year the Toms River factory shut down, the United States and Europe *each* produced eight times as much plastic as China; by 2010, China was making almost as much plastic as the United States and Europe *combined*. During the same period, Chinese production of benzene, ethylene, and sulfuric acid quadrupled. There were similar leaps for dozens of other chemical industry staples—including aniline, the dye molecule that started it all in 1856.[21] Familiar names have fueled the boom. BASF, the old German aniline maker that is now the world's largest chemical company, has seven thousand employees and forty factories in China. BASF is building a fifty-acre plant in Chongqing to make aniline, nitrobenzene, and methylene diphenyl diisocyanate—the latter produced from phosgene, the extremely poisonous gas once used in Toms River to make azo dyes.[22] Dow is on a similar path, with four thousand Chinese employees and twenty factories.

Liu Yu-Shu and her husband have not been lucky enough to land jobs with a big multinational company like BASF or Dow. Like millions of Chinese, they are itinerant laborers who go wherever there is work. Family is important to them. Their older son has severe epilepsy, so they sought and received government permission to have a second child, in contravention of China's one-child policy. But Liu could not afford to rest while pregnant; instead, she worked long hours mixing adhesive for her husband, who was laying tile at an expanding factory that made decorative plates for export. Could prenatal exposure to epichlorohydrin in adhesive, or arsenic in glaze, have triggered her younger son's leukemia? Liu does not attempt to answer the unanswerable question. Unlike the physicians on the ward, she has not thought much about possible links to pollution. Instead, she is focused on her son, and is grateful for the treatment he is receiving. "Since we came here," Liu said, "his spirit is up and he is more cheerful." How long he will be able to stay is an open question. Chongqing Children's Hospital treats more than one hundred kids with leukemia every year, but many are sent home with medicine after only one month of chemotherapy because their families cannot afford a full treatment regimen.

A tall, handsome woman, a visitor from New York, stands out in the crowded hallway outside the hematology ward. Frederica Perera has not lost any of the boldness that led her, thirty years earlier, to coin the term *molecular epidemiology* and help spur a worldwide rush by researchers to look for genetic biomarkers in the bodies of people who live or work in highly polluted places. She has worked all over the world but says the toxic exposures in China are the highest she has ever seen. Collaborating with doctors at Chongqing Children's, Perera and a Columbia University colleague, Deliang Tang, have measured adducts—molecules of benzo(a)pyrene and other invading pollutants that fuse with DNA and disrupt it—in the blood cells of several hundred children in a nearby city. They have found strong correlations between the presence of adducts and an array of developmental problems, including delayed speech and poor motor skills.[23] There are also weak associations between adducts and childhood cancer, and Perera thinks that if she had the millions of dollars it would take to study thousands of highly exposed children, the statistical links to leukemia and other cancers would be clearer.

"We already know enough," Perera said, "to conclude that these exposures are causing developmental delays and probably raising cancer risk." Since there is no sign of a minimum threshold exposure below which there is no risk, she added, whatever is happening in the factory towns of China is happening everywhere else, too—just much more subtly. The question, as always, is whether the imperfect tools of epidemiology are capable of finding it.

There is a time-warp quality to the meetings of the Citizen Action Committee on the Childhood Cancer Cluster, which still convenes once or twice a year at the Toms River Municipal Complex on the Avenue of Americanism, one block from the spot where apocryphal town founder Tom Luker supposedly shared his wigwam with Princess Ann. At the front of the room, Linda Gillick still occupies the center chair. Sometimes, when his fragile health allows it, her now-adult son, Michael, sits to her left, still waiting for answers, as he has ever since the committee began meeting during the frenzied weeks of early 1996.

Nowadays, the audience in the overlarge meeting room is always sparse and consists mostly of people who are paid to be there: lawyers, company representatives, and government officials. Jerry Fagliano still drives over from Trenton for most of the meetings, and when he does he can count on another ritualistic scolding from Gillick about the slowness of the state cancer registry in compiling and releasing case counts in Ocean County and around the state. Thanks to her network at Ocean of Love, which still assists families all over the county, Linda Gillick still knows about every local case of childhood cancer at least a year before the state registry does.[24]

There is a familiar rhythm to the committee's agenda items, too. If it is summertime, United Water is sure to be struggling to meet the water demands of its customers, and Gillick is certain to be furious that town residents are not turning off their lawn sprinklers and otherwise conserving water. "People have forgotten what this community went through," she will say, shaking her head. Many of the issues that made the township so vulnerable to pollution are still unresolved. Having shed its cancer town label, Toms River is growing again, though not at its previous breakneck pace. Years of fiscal austerity, however, have cut environmental enforcement to the bone in the New Jersey departments of health and environmental protection, and there are still no mandatory water conservation rules in Toms River—not even after the two senior water company managers in town were caught faking safety tests in order to try to keep up with ever-increasing local usage.[25] (The parent company caught them, not state regulators.)

Much of the time at committee meetings is given over to relentlessly cheerful updates on the local Superfund cleanups, delivered by smiling "community relations" managers from Dow and BASF. (Ciba's ancient German rival in the aniline industry, BASF, acquired what little remained of Ciba's chemical business, including its cleanup liabilities, in 2009.) At the Parkway well field, SAN trimer readings in the two still-closed wells have fallen steadily and are now barely measurable at less than thirty parts per trillion. Soon, Dow officials suggest, it will be time to reconnect those wells to the drinking water system—a move Linda Gillick vows to oppose. "No one should be exposed, no matter what your studies show," she told a Dow consultant at a 2012 committee meeting. With the EPA's permission, Dow is already making plans to shut down the air stripper that was supposed to have solved the Parkway well contamination problem back in 1988—but didn't.

Over at the shuttered chemical plant, meanwhile, the pine forest is slowly reclaiming the property, six decades after the Swiss cleared the land for their mini-city devoted to dye production. For now, the water tower that Greenpeace seized for those three crazy days in 1984 still stands; town volunteer firefighters use it to practice rescues. And while there are still about thirty-eight thousand drums buried in the portion of the thirty-year-old landfill known as Cell One, most of the even older dumps on the factory site have finally been excavated.[26] The Smudge Pots, the Moon, the Acid Pits, and all of the other open-pit dumps whose evocative nicknames were once part of the everyday lexicon of workers like George Woolley and the Talty brothers have at last been cleared out, as part of a $92 million cleanup operation overseen by the EPA and funded by Ciba's insurers.[27] By the time the project ended in 2010, more than 343,000 cubic yards of soil had been scooped and treated—enough to cover twenty-seven football fields with six feet of tainted dirt.

The completion of the soil cleanup seemingly clears the way for a momentous step that Toms River politicians have been talking about for years: the redevelopment of the last large open tract left in the township. While the worst areas of the factory site would be fenced off, there have been many ideas for what to build in the less contaminated areas: ball fields, a large office park, maybe even housing. So far, no plan has moved forward, mostly because billions of gallons of contaminated groundwater still lie beneath the property. Thirty-seven recovery wells on or near the factory grounds are still pumping, treating, and re-injecting almost two million gallons of groundwater every day, as they have since 1996. The pumps will run until at least 2025, the earliest the site could be taken off the Superfund list. The legacy of contamination will linger even as the Ciba name, which dates from 1884, fades from memory—a casualty of corporate consolidation.[28] Over at Reich Farm, meanwhile, similarly lingering contamination means that Samuel Reich, who turned ninety in 2011, almost certainly will not live to see the redevelopment of his old egg farm, now a ramshackle site for industrial storage. He and his wife, Bertha, have been trying to sell it since 1975, four years after their calamitous decision to let Nick Fernicola bring waste drums onto their land for $40 a month—a pittance Fernicola never bothered to pay them.

A few blocks from town hall along the river, in the far corner of Riverfront Landing Park, a mass of unkempt shrubbery is encroaching on the granite memorial to the lost children of Toms River. But the names etched into the stone are still starkly clear, and the families—those who have not moved on or moved away—are as determined as ever to protect their legacy. There has been no talk of dissolving the citizens advisory committee, even though its meetings are canceled more frequently now. "It is hard to keep it going after all this time, but we've learned that you can't assume somebody else is going to take care of things for you and make all the right decisions," said Kim Pascarella, who still co-chairs the committee with Linda Gillick.

"I feel that's the best way I can honor the memory of my daughter, by staying involved, even if we never get definite answers from the scientists," Pascarella said. "In the end, what I've learned is that when it comes to this type of science, you really can't be arrogant and assume you know the answers. Arrogance by these companies and by the government is what got us into all this in the first place. We know that now."

Acknowledgments

~~

Nonfiction writing is a community endeavor camouflaged as a solitary one. If you have made it this far, you know that *Toms River* is the story of a very large number of extraordinarily passionate people. I will always be grateful that so many of them were willing to extend their passion to the creation of this book. The events recounted here were often painful and always intensely emotional for everyone involved; the passage of time has done little to dull the pain. That 140 people were nonetheless willing to talk to a stranger—several times, in many cases—demonstrates a level of trust and commitment that I have tried very hard to honor in these pages.

I owe a special thanks to Linda Gillick for initially helping to orient me in Toms River. Her decision not to sit for a formal interview does not diminish my deep admiration for what she has accomplished in her community. Even now, after twenty-five years of service to local families in crisis, she and her staff at Ocean of Love are continuing to do outstanding work providing emotional and financial support to Ocean County children with cancer and their siblings and parents. You can learn more about their work at www.ocean-of-love.org and can assist them directly by sending contributions to Ocean of Love for Children with Cancer, 1709 Route 37 East, Toms River, New Jersey, 08753.

The list that follows includes all those I interviewed who were willing to be named. Others asked me to withhold their names and I have done so. I am profoundly grateful to them all, including Richard Albertini, Bruce Anderson, Melanie Anderson, Emma Ansara, Robert J. Baptista, Richard Bellis, Don Bennett, Esther Berezofsky, Michael Berry, Lois Bianchi, Eula Bingham, James Blumenstock, Lisa Boornazian, John Bucher, Bob Butler, Dan Carluccio, Judith Carluccio, Gary Casperson, Rajendra Chhabra, Dick Chinery, Angelo Cifaldi, Richard Clapp, Philip Cole, Craig Colten, Mark Cuker, Tom Curran, Bob De Sando, Michele Donato, John Paul Doyle, Michael Edelstein, James Etzel, Jerald Fagliano, Barry Finette, Heather Galick, Floyd Genicola, Robert Gialanella, Michael Gillick, Michael Gordon, Jon Gorin, Andrew Grange, Melvyn Greaves, John Groopman, Clark W. Heath Jr., Rich Henning, Peter Hibbard, Susan Hibbard, Bruce Hills, Jon Hinck, Kathleen Hughes, William Hyres, Donna Jakubowski, Laura Janson, Yu Jie, Steven Jones, Jeff Josephson, Allan Kanner, Charles Kauffman, John E. Keefe Jr., Judith Klotz, Alfred Knudson, Roden Lightbody, Liu Xiao-Mei, Liu Yu-Shu, Christine Livelli, Joanne Livelli, Gary Lotano, Harold Luker, Ray Lynnworth, David Malarkey, Jack Mandel, Jim Manuel, Morris Maslia, Nancy McGreevy, Sheila McVeigh, David Michaels, Bruce Molholt, Ernest Nagel, Gerald Nicholls, Terry Nordbrock, Kenneth Olden, Marian Olsen, J. Patrick O'Neill, David Ozonoff, Kim Pascarella, Richard Paules, Frederica Perera, Carole Peterson, Jerome Posner, Joseph Przywara, Dave Rapaport, Eric Rau, Bertha Reich, Juan Reyes, Herb Roeschke, Michael Rosenblum, John F. Russo Sr., Leona Samson, Jan Schlichtmann, Gale Scott, Nancy Menke Scott, William Skowronski, Martyn Smith, Wayne Smith, G. Wayne Sovocool, Tomm Sprick, Samuel Sprunt, Dennis Stainken, Shanna Swan, Jon Sykes, Jackie Talty, John Talty, Ray Talty, Deliang Tang, Ray Tennant, Anthony Travis, Pamela Vacek, William Warren, Dan Wartenberg, John Wauters, Stephanie Wauters, Dane Wells, Richard Wendel, Jorge Winkler, Steven Wodka, Carl Woodward III, George Woolley, and Mitchell Zavon.

Some of the people listed above provided assistance extending far beyond merely answering my questions. In their interactions with me, they showed the same commitment to unvarnished truth that they had demonstrated earlier in helping to uncover the secrets of Toms River. At the top of this list are those whose lives have been cleaved by cancer. The horror of losing a child, or almost losing one, is unfathomable to those of us who have not experienced it, yet Ray Lynn-worth, Kim Pascarella, and Bruce and Melanie Anderson tolerated my intrusive queries; so did Michael Gillick, one of the bravest people I have ever met. Others who provided extraordinary help include Don Bennett, Mark Cuker, Jerald Fagliano, Barry Finette, and Bruce Mol-holt. They are credits to their respective professions. I am also grate-ful to Frederica Perera, Deliang Tang, and Mariette DiChristina of *Scientific American* for making my trip to China possible.

My students at the Science, Health, and Environmental Reporting Program at New York University's Arthur L. Carter Journalism Insti-tute are a perpetual source of inspiration and delight. Several aided me directly as researchers. Susan Cosier and Alison Snyder did exten-sive document research, while Emily Elert, Kristina Fiore, Robert Grant, and Erica Westly also assisted. For help in tracking down sources and obtaining public records, newspaper clippings, and other documents, I owe thanks to Marcus Banks, Betsy Dudas, Stephen Greenberg, Jon Hinck, Cheryl Hogue, Robin Mackar, Mary Mears, Ralph Roper, Ellen Tracy, John Wauters, Stephanie Wauters, and Jorge Winkler as well as the reference librarians at the Ocean County Library, the Ocean County Historical Society, and the New York re-gional office of the U.S. Environmental Protection Agency.

Colleagues and friends vetted portions of this manuscript, provid-ing welcome suggestions and corrections. This group includes Marla Cone, Howard Frankel, Stephen S. Hall, Robert Lee Hotz, George Johnson, Ivan Oransky, Anthony Roisman, and Charles Seife. Any remaining errors are my own. I also want to thank my NYU colleague Brooke Kroeger, who was always supportive as I pursued this project and adjusted to academic life. It is a joy to work with such smart, dedicated colleagues and students, all of whom are committed to finding a sustainable future for deep journalism in service to democ-racy.

I have tested the patience of a succession of editors at Random House throughout the long gestation of this project. I am so grateful for their unflagging support, which began with the fabulous Ann Har-ris and continued with John Flicker, Beth Rashbaum, Susanna Porter, and finally Ryan Doherty, a terrific manuscript editor and a real champion of this book. My literary agent, Jane Dystel, and her business partner Miriam Goderich similarly never lost faith and have been sources of savvy counsel over the years.

My feelings about my family's contributions are impossible to commit to type. I hear the voice of my mother, the late poet and playwright Lois Levin Roisman, on every page. My father, Arnold Fagin, is my exemplar in many ways, even though, deep down, he still hopes I will become a lawyer. (Too late, Dad.) My wonderful daughters Anna and Lily tolerated my obsession with grace and good humor. As they grew into adulthood, so did this book; theirs is the more significant accomplishment by far. As for my wife, the legal journalist Alison Frankel, this book could not exist without her. She read and critiqued almost every word, but her contributions go much deeper than superb editing. Alison has listened to me talk incessantly about Toms River—and *Toms River*—for seven years, and about so many other stories for twenty-five years before that, and yet not once has she run screaming from the room (though there have been several close calls). With Alison's love, anything seems possible; without it, nothing would.

Notes

Chapter One

1. The origin story featuring Thomas Luker and Princess Ann was related in a twenty-nine-page book written in 1967 by Pauline S. Miller, who for many years held the officially designated title of Ocean County historian. Her self-published book, called *Early History of Toms River and Dover Township* and written to commemorate Dover's two-hundredth anniversary as a chartered township, also includes a version of the "Old Epic Poem" about Tom Luker and his Indian bride. In 1992, the town placed a plaque honoring Luker next to a footbridge in Huddy Park, near the ferry crossing he supposedly established in about 1712. Miller's brief chapter on Luker carries the optimistically definitive title "How Toms River Got Its Name." Skeptics of the story, however, note that Luker partisans, including his descendants, did not publicly champion their forebear until the 1920s. Only "Indian Tom" and William Tom (sometimes spelled William Toms) are suggested as likely namesakes in histories of the region written during the nineteenth century, including *A History of Monmouth and Ocean Counties, New Jersey* by Edwin Salter, published in 1890, and *Historical Collections of New Jersey* by John Warner Barber and Henry Howe, published in 1868. Casting a bit more doubt on the Luker story, the fight song of Toms River High School (now Toms River High School South) begins this way: "Oh! Old Indian Tom was the man who gave his name to our high school upon a hill!" Teams at the school, founded in 1891, are known as the Indians.

2. The description of the 1782 British raid on Toms River, first published in *Rivington's Royal Gazette,* a Loyalist newspaper in New York, is reprinted on pages 328 and 329 of *Historical Collections of New Jersey.*

3. The Dover Township name was finally formally discarded in 2006 when residents voted to change the name to Toms River Township. Fierce debate preceded the plebiscite, in which more than 40 percent of township voters voted against the switch.

4. Thomas A. Mathis's political career came to a startling end in 1958. Suffering from an undisclosed illness, he shot himself in the head in a second-floor room of his mansion, just a few days after he was discharged from a Philadelphia hospital.

While several local newspapers reported the suicide, the Mathis-owned *Ocean County Sun* was more circumspect, telling its readers that the eighty-eight-year-old party boss "succumbed at his capacitous Main Street residence Sunday morning after two weeks of illness." Today, most residents of Ocean County know Thomas A. Mathis only as the namesake of the older of two adjacent bridges that connect Toms River with the beach communities on the Barnegat Peninsula. The Thomas A. Mathis Bridge was built in 1950 and cost $6 million. Mathis and his son secured the state funds for the project.

5. Anthony S. Travis, "Perkin's Mauve: Ancestor of the Organic Chemical Industry," *Technology and Culture* 31 (January 1990): 51–82.

6. Working from recipe books, Egyptian dyers used reddish pigments from henna, madder, and safflower plants and blues from *Indigofera tinctoria,* or true indigo, to stain the linen cloth used to wrap mummies. The Greeks dyed their wool and linen tunics in vats of indigo. Centuries later, when the Romans invaded Britain, they were startled to discover that the Celts painted their bodies with a blue pigment that was probably made from woad. In fact, Britain may have gotten its name from the Greek *prettanoi,* or "tattooed people."

7. Arriving on the South American shore in 1500, Portuguese explorers were so excited to find dyewoods that they named the newly discovered territory *"Terra de Brasil"* because the trees there yielded a red that was the color of burning coals, or *braise* in medieval French.

8. Excerpted from "Beautiful Tar: Song of an Enthusiastic Scientist," *Punch,* September 15, 1888.

9. This chapter's account of Johann Jakob Müller-Pack's travails in Basel draws heavily on the work of a leading historian of the chemical industry, Anthony S. Travis of The Hebrew University of Jerusalem. See Anthony S. Travis, "Poisoned Groundwater and Contaminated Soil: The Tribulations and Trial of the First Major Manufacturer of Aniline Dyes in Basel," *Environmental History* 2:3 (July 1997): 343–65.

10. "Poisoned Groundwater and Contaminated Soil," 349. The quotation is originally from a speech by August Leonhardt, who worked for a British dye manufacturer in the 1860s. His speech, "Remarks on the Manufacture of Magenta," was prepared for delivery at the Philadelphia Centennial Exhibition of 1876.

11. The coloration of rivers was a mark of industrialization even before the rise of synthetic chemistry. In *Hard Times,* published in 1853, three years before William Perkin's discovery of aniline dyes, Charles Dickens described the book's fictional setting, Coketown, as "a town of machinery and tall chimneys, out of which interminable serpents of smoke trailed themselves for ever and ever, and never got uncoiled. It had a black canal in it, and a river that ran purple with ill-smelling dye, and vast piles of building full of windows where there was a rattling and a trembling all day long, and where the piston of the steam-engine worked monotonously up and down, like the head of an elephant in a state of melancholy madness." Almost seventy years later, in his modernist masterpiece *The Waste Land,* the poet T. S. Eliot wrote: "The river sweats/Oil and tar/The barges drift/With the turning tide."

12. The reference to the Rhine as a dumping ground for "effluents" and "rubbish" is on page 8 of *Society of the Chemical Industry in Basle, 1884–1934,* an official and

privately published corporate history published in Switzerland to commemorate Ciba's fiftieth anniversary.

13. Friedrich Goppelsröder reported his findings to the health committee of the Canton of Basel in a three-page handwritten letter dated June 8, 1864.

14. Markus Hammerle, *The Beginnings of the Basel Chemical Industry in Light of Industrial Medicine and Environmental Protection* (Schwabe & Co., 1995), 54. Translated from German.

15. The prosecution of Johann Jakob Müller-Pack in Basel was just the first of several highly publicized incidents involving water pollution by aniline dyes. There were also scattered reports of arsenic poisoning from eating artificially colored candies and even from wearing fuchsine-dyed stockings. The October 3, 1868, edition of *The Times* of London, for example, included two reports of "poisonous socks." One, a reprint from *The Lancet,* the medical journal, described a ballet dancer who had to be hospitalized with a severe rash after sweating through her "brilliant red" stockings. The other was a letter from a physician who reported that one of his patients had suffered similar symptoms after buying dyed socks. When the man returned to the store to protest, the storekeeper admitted that several buyers had lodged similar complaints.

16. Casimir Nienhas's experiment is described on page 56 of *The Beginnings of the Basel Chemical Industry in Light of Industrial Medicine and Environmental Protection.*

17. *Society of the Chemical Industry in Basle, 1884–1934,* 8, 76.

18. *Beginnings of the Basel Chemical Industry,* 44.

19. D. H. Killeffer, "Industrial Poisoning by Aromatic Compounds," *Industrial and Engineering Chemistry* 17:8 (August 1925): 820–22.

20. *Beginnings of the Basel Chemical Industry,* 41.

21. *Society of the Chemical Industry in Basle, 1884–1934,* 56.

22. M. W. Tatlock, "Industrial Waste Treatment: Cincinnati Chemical Works," *Proceedings of the Eleventh Purdue Industrial Waste Conference* (1956): 166–71.

23. The first comprehensive sampling of the entire Ohio River, from its source in western Pennsylvania to its terminus in the Mississippi River, was conducted September 18–29, 1950, by a newly created interstate agency called the Ohio River Valley Water Sanitation Commission. In the report that followed, "Pollution Patterns in the Ohio River—1950," the commission concluded that the most polluted section was at the Cincinnati waterfront and immediately downstream.

Chapter Two

1. The description of the facilities and dye-making processes at the Toms River plant during the first years of its operation is drawn from Kevin J. Bradley and Philip Kronowitt, "A Staff-Industry Collaborative Report: Anthraquinone Dyes," *Industrial and Engineering Chemistry* 46:6 (June 1954): 1145–56. At the time, Bradley was the assistant editor of the journal, a publication of the American Chemical Society. Kronowitt was a Ciba executive who helped to design and build the Toms River plant.

2. Robert J. Baptista, a former industry executive and historian of the dye in-

dustry, elaborated on why anthraquinone vat dyes were so useful in an interview with the author: "Vat dyes were a tremendous technology breakthrough because cotton fabrics would no longer fade. . . . The vat dyes were rugged molecules. They were built to last. They had just the right shape and weight to fit into the fiber structure and not get rubbed out or bleached out."

3. November 14, 1956, memo from Ciba executive Al Meier to five senior managers.

4. Julia E. Gwinn and David C. Bomberger, *Wastes from Manufacture of Dyes and Pigments: Volume 4, Anthraquinone Dyes and Pigments* (SRI International, 1984, under contract to the U.S. Environmental Protection Agency), 46.

5. Bradley and Kronowitt, "Staff-Industry Collaborative Report," 1153: "The capacity of four million pounds of dyestuffs requires about twenty-two million pounds of various raw materials. Chief among these are sulfuric acid, caustic soda, hydrochloric acid, phthalic anhydride, ammonia, benzene, nitrobenzene, glycerol and alcohol."

6. Two years after being chased out of Basel, having returned to a vagabond life, Paracelsus wrote of his critics: "Not one of you will survive, even in the most distant corner, where even the dogs will not piss. I shall be monarch and mine will be the monarchy." See page 73 of *Das Buch Paragranum* (The Book Against the Grain) by Paracelsus, as excerpted and translated from the original German in Nicholas Goodrick-Clarke, *Paracelsus: Essential Readings* (North Atlantic Books, 1999).

7. Scholars disagree on whether the infamous book-burning incident involving Paracelsus actually occurred or is an embellishment added much later by his followers. Some versions of the story have Paracelsus burning a tract by Galen instead of Avicenna. In any case, there is abundant evidence that Paracelsus antagonized the city's establishment during his brief stay in Basel. See, for example, Henry M. Prager, *Magic into Science—The Story of Paracelsus* (Sumner Press, 2007). The book was originally published in 1951, when Prager was a professor of history at City College of the City University of New York.

8. *Das Buch Paragranum*, excerpted in *Paracelsus: Essential Readings,* 74.

9. There is controversy over the wording of Paracelsus's famous handbill advertising the Basel lectures, since versions were published many years later by his acolytes and his enemies. The version quoted here is from *Magic into Science—The Story of Paracelsus,* 152–53.

10. Paracelsus's book on mining diseases, *Von der Bergsucht und anderen Bergkrankheiten* (On the Miners' Sickness and Other Miners' Diseases), was written in 1533 but not published until 1567, twenty-six years after his death. While the book is generally considered to be the first full-length work on occupational disease, there was at least one earlier effort. A German physician named Ulrich Ellenbog in 1473 wrote a short work for goldsmiths entitled *Von den gifftigen besen Tempffen und Reuchen* (On the Poisonous, Evil Vapors and Fumes). It was published as an eight-page pamphlet in 1524. In addition, another contemporary of Paracelsus, Georgius Bauer, better known by his pen name Agricola, published his *De Re Metallica* (On the Nature of Metals) in 1541. *De Re Metallica* was primarily a description of the mining industry, but Agricola also wrote about mining illnesses and preventative measures, including ventilation shafts.

11. According to Bernardino Ramazzini's *De Morbis Artificum Diatriba* (Diseases of Workers), published in 1700, Hippocrates also referred to mining diseases and noted that those affected by the "metallic pests" include not only miners but also "many others whose work is too near the mines." Thus Hippocrates, who died in approximately 375 B.C., may have been the first to recognize that pollution can harm more than just directly exposed workers.

12. This quotation is from the first chapter of the third tractate of *Von der Bergsucht und anderen Bergkrankheiten*. Henry G. Sigerist, ed., *Four Treatises of Theophrastus Von Hohenheim, Called Paracelsus* (Johns Hopkins Press, 1941), 68.

13. Robert Meerpol, *An Execution in the Family: One Son's Journey* (St. Martin's, 2003), 13–15.

14. "Governor Dedicates Ciba States Building," *Asbury Park Press,* June 5, 1953.

15. "'Effluent Pure,' Ciba Head Says," *Ocean County Sun,* June 11, 1953.

16. "Ciba Builds to Serve," a promotional booklet published by the company in 1953, 22.

17. Howard S. Pratt, "Rod and Gun," *Brooklyn Eagle,* March 24, 1953.

18. "Tour of Ciba Disposal Plant Dispels Rumors of Pollution," *New Jersey Courier,* April 2, 1953. Leading the tour that day was the Swiss company's public face in Toms River, a Hungarian émigré and chemical engineer named Philip Kronowitt. After World War II, the U.S. government sent Kronowitt back to Europe to study the advanced manufacturing techniques of the I. G. Farben cartel, which the Allies were dismembering with the enthusiastic support of competing firms in Switzerland and America. Returning home the following year, Kronowitt went to work for Ciba and took charge of the search for a new factory site in America. Moving to Toms River to supervise the plant's construction, he served as president of the Ocean County Boy Scouts Council and the Red Cross and led the drive to establish the county's first hospital. See "Philip Kronowitt," a profile in the Fall 1957 edition of *TRC Color,* a company newsletter.

19. Nicholas P. Cheremisinoff, *Handbook of Pollution Prevention Practices* (CRC Press, 2001), 238.

20. A 1953 article in *Chemical and Engineering News* typified the cheerleading coverage of the era. "Ciba Solves Water Pollution Danger at Toms River Dye Plant" was the headline. A relevant passage: "Actually, the water discharged into the Toms River will be a great deal more palatable for fish and humans than the river itself. The Toms River flows through such swampy land that by the time it reaches the sea it contains a high percentage of organic matter and has a fairly high pH." The article did not carry a byline but was written by an editor at the magazine, Kevin J. Bradley, who was also the coauthor—with Ciba's Philip Kronowitt—of a similarly glowing article on the plant published the following year in another publication of the American Chemical Society, *Industrial and Engineering Chemistry.* The initial article's final line noted that "Ciba's relations with its neighbors have been very cordial." See *Chemical and Engineering News* 31:36 (1953): 3691.

21. June 30, 1949, memo by an unnamed Ciba executive to senior managers in Basel, page 2. The complete quotation is: "The sewage treatment area is deliberately located at considerable distance from the river. The treated effluent will be carried in an open ditch to the river and the cost of such ditch is low. Long line will help aera-

tion of the effluent before it enters the river and we expect considerable seepage into adjacent area from ditch, which will reduce the volume of effluent carried into the river."

22. "Report: Consideration of Location of Vat Colors Project at St. Bernard and Toms River," May 2, 1949, page 4. It was written by an unidentified Ciba executive. Similarly, in a "Report on Toms River Property," dated December 30, 1948, Ciba executives Fritz Max and Philip Kronowitt wrote: "The acreage available is fairly level, sand and gravel soil, good for construction, easy to excavate, and level and very suitable for construction of sedimentation basins and filtration of such effluent which we do not want to dispose of into the stream."

23. January 19, 1949, letter to the New Jersey Department of Health from Philip Kronowitt. In a five-page report he prepared for the state two days earlier, Kronowitt asserted that "according to experience" from other dye factories, the Toms River plant's wastewater discharges would be "clear and transparent," "have no odor," and "be free of substances harmful to fish, animal or wild life."

24. Memo from Morris Smith to Philip Kronowitt, November 25, 1955. See also report entitled "Waste Disposal Plant Operations: Year 1955."

25. Anthony Travis, a leading chemical industry historian, told the author: "Back in '48 and '49, [Ciba] felt Toms River was a good site because the ground was sandy, they could dig a long trench and most of the wastewater would be gone before it even got to the river, and there were lots of trees they could hide behind."

26. For a detailed description of the wastewater treatment process at the Toms River plant, see John J. Baffa, "Treatment of Waste from Dye Manufacture at the Ciba States Ltd. Plant at Toms River, New Jersey," *Proceedings of the Ninth Industrial Waste Conference, May 10, 11 and 12, 1954* (Purdue): 560–66. A consulting engineer who helped design the treatment process in Toms River, Baffa presented the paper at the industrial waste conference hosted annually by Purdue University.

27. As late as the 1950s, "low costs and technological simplicity made lagoons a common, although when used in improper geological settings, highly inadequate chemical waste treatment." Craig Colten, "Creating a Toxic Landscape: Chemical Waste Disposal Policy and Practice: 1900–1960," *Environmental History Review* 18:1 (Spring 1994): 85–116, 105. Ciba's waste lagoon was designed by a professor of sanitary engineering at Rutgers University, Willem Rudolfs, who had written a textbook on waste treatment.

28. Opinion of New Jersey Superior Court Judge Lawrence Weiss, *Ciba-Geigy vs. Liberty Mutual Insurance Company et al.,* January 28, 1998, 1–36.

29. Ciba-Geigy's witness, Richard Dewling, a former chief of the New Jersey Department of Environmental Protection, testified that two industrial plants in New Jersey "provided a level of treatment in the '60s that were beyond what most industries were doing. One was American Cyanamid up in Bound Brook, and the other was at [CIBA's] Toms River facility." *Ciba-Geigy vs. Liberty Mutual Insurance et al.,* 20.

30. As far back as 1914, a New Jersey appeals court held a Newark gasworks liable for discharging coal tar wastes that contaminated groundwater used by a brewery next door. The case is *Ballantine & Sons vs. Public Service Corporation of New Jersey, New Jersey Law Reports* 86 (1914): 331.

31. The textbook is *Chemical Engineering Plant Design*, originally published in 1934 by Frank C. Vilbrandt, who chaired the department of chemical engineering at Virginia Polytechnic Institute. The relevant passage is on page 400 of the third edition, published in 1949 by McGraw-Hill: "Chemical plants often dispose of their waste by locating on a stream, river, or at tidewater. Disposal by tidewater is often satisfactory if there are no bathing beaches nearby. Disposal of waste into a stream or river is no longer satisfactory, for there is a growing list of states that have instituted legislation against such pollution by industrial wastes. Another method of waste disposal is by seepage through the ground. If such a method is to be used, soil tests should be made to determine whether the soil is porous enough to permit the disposal of considerable quantities of liquor without accumulation. It is also advisable to check the topography of the area to determine where the liquor will seep in order to avoid trouble from neighboring plants or the local authorities. Towns lower down the valley may draw their water supply from the drainage shed upon which the plant is situated."

32. In "A Historical Perspective on Industrial Wastes and Groundwater Contamination," *Geographical Review* 81:2 (April 1991): 215–28, author Craig Colten wrote: "By the early 1950s, governmental agencies, professional organizations, and industry-trade associations, drawing on three decades of experience, all publicly recognized the hazards posed by the surface disposal of liquid wastes. This generalization does not mean that the use of lagoons or infiltration ponds ceased, but it clearly illustrates the existence of sufficient understanding of hydrologic processes to argue that such waste-treatment methods were unsafe and irresponsible in most environmental settings."

33. This quotation is from the first chapter of the third tractate of *Von der Bergsucht und anderen Bergkrankheiten. Four Treatises of Theophrastus Von Hohenheim, Called Paracelsus*, 67.

Chapter Three

1. The close cooperation among the three major Swiss dye manufacturers in Cincinnati and then in Toms River helped set the stage for a series of later mergers. The Toms River factory was initially constructed by Ciba and in its initial years was known formally as the Toms River division of Ciba States Limited. In 1955, Geigy and Sandoz purchased minority shares, and the plant was renamed the Toms River–Cincinnati Chemical Corporation. (That name was rarely used in Toms River, where the company was known as Toms River Chemical or sometimes simply "Ciba" because Ciba always held a majority interest.) When the joint venture completely closed down its two Cincinnati factories in 1959, the Toms River Chemical Corporation name became official. In 1971, it would be changed again to Ciba-Geigy Toms River after those two companies merged. The last of the three major Swiss dye manufacturers joined in 1996 when Sandoz merged with Ciba-Geigy to form Novartis. The following year, Novartis spun off Ciba-Geigy's dye and chemical business into Ciba Specialty Chemicals, which was then acquired by BASF in 2009, marking the end of Ciba as an independent chemical manufacturer 150 years after Alexander Clavel began making synthetic dyes in Basel.

2. Among the beneficiaries of the Garden State Parkway project was the all-powerful Mathis family of Toms River, which received a lucrative state contract to insure a portion of the more than $300 million in bonds issued for the project.

3. This chapter's brief account of the contamination in Kimberton, Pennsylvania, relies on Environmental Protection Agency documents and an interview with James E. Etzel, a professor emeritus of environmental engineering at Purdue University who investigated it in 1957 and 1958 as a private consultant.

4. In 1981, twenty-four years after the initial discovery of carbolic acid and salts in streams near Kimberton, the Chester County Health Department found solvents in more than a dozen private water wells near the factory site. As a result, the site was placed on the federal Superfund list. After an investigation, the U.S. Environmental Protection Agency concluded that Ciba's abandoned waste lagoons were the source of the solvent contamination. In 1991 the agency negotiated a consent order with Ciba-Geigy and the owner at the time, The Monsey Company (an asphalt manufacturer), requiring both companies to pay the cost of pumping up and treating the contaminated groundwater and of extending public water lines to replace the contaminated wells. The companies also had to pay the EPA $200,000 to reimburse prior expenses. The pump-and-treat system began operating in 1993 and is likely to run for several decades before solvent concentrations decline to a level the EPA considers acceptable.

5. While its acute dangers to the lungs and skin were very well known even in the 1950s, epichlorohydrin was not officially considered a possible carcinogen until a 1976 assessment by the International Agency for Research on Cancer found that laboratory rats that inhaled the chemical developed nasal tumors. There has been inconclusive evidence since the early 1980s that epichlorohydrin may cause respiratory cancers in humans, based on studies of factory workers. The U.S. National Toxicology Program now classifies epichlorohydrin as a likely human carcinogen.

6. Until 2004, no government agency in the United States had officially classified anthraquinone as a carcinogen, although it had been suspected for many years. In a 2004 report, the National Toxicology Program found "clear evidence" of liver carcinogenicity in rats that ate anthraquinone-laced feed. The report also found evidence of noncancerous lesions in other organs and possible endocrine system effects.

7. J. Hohl, "Final Report on Explosion in Building 102 on December 22, 1960," January 30, 1961, 1, 5.

8. John A. Ross, "Percivall Pott 1714–1788," *Paraplegia* 24:5 (October 1986): 287–92.

9. Percivall Pott published his famous 725-word essay on "chimney sweeps' cancer" in 1775 as a chapter in his *Chirurgical Observations Relative to the Cataract, Polypus of the Nose, the Cancer of Scrotum, the Different Kinds of Ruptures and Mortification of the Toes and Feet.* The excerpt quoted here is from page 178 of *The Chirurgical Works of Percivall Pott,* by James Earle, Pott's son-in-law. It was originally published in London in 1808 by Wood and Innes.

10. Thanks in part to the public outrage engendered by Percivall Pott's description of "chimney sweeps' cancer," the British Parliament in 1778 restricted the job to older boys and in 1845 banned it altogether.

11. For more information about the plight of London's chimney sweeps in the eighteenth and nineteenth centuries, and Percivall Pott's identification of testicular cancer as an occupational disease of the "climbing boys," see H. A. Waldron, "A Brief History of Scrotal Cancer," *British Journal of Industrial Medicine* 40:4 (November 1983): 390–401. See also Meyer M. Melicow, "Percivall Pott (1713–1788)," *Urology* 4:6 (December 1975): 745–49.

12. Childless women and women who give birth at an advanced age are now thought to face a slightly higher risk of breast cancer because their lifetime exposure to estrogen is higher.

13. *Chirurgical Works of Percivall Pott,* 180–81.

14. A May 9, 1962, memo from Morris Smith described the accidental discharge of three thousand pounds of "arsenicals" into the sewer system seven days earlier. By quickly raising the pH level in the initial treatment lagoons, according to Smith's memo, the company managed "to prevent the discharge of [arsenic] into the Toms River in quantities which might otherwise have resulted in a fish kill."

15. "We have not been able to meet State standards with our present treatment plant," Emerich Varkony wrote in a March 14, 1962, memo to fellow Toms River Chemical manager Morris Smith.

16. The smelly drinking water at the factory was described in a memo Morris Smith wrote to his superiors on July 14, 1954; the well contamination was described in a memo from Smith dated September 27, 1954.

17. In a series of plaintive memos to his supervisors, including one dated August 14, 1956, and entitled "Seepage from Waste Disposal Basins," Morris Smith was explicit about the threat to the company's own wells and to those of its neighbors: "High seepage losses from the waste disposal basins, accompanied by continued heavy pumping from the deep wells, must inevitably result in contamination of all the deep wells. In addition, it is highly probable that ground water contamination will spread to foreign property to the east and southeast."

18. The consulting firm's report, "Ground-Water Conditions at the Toms River–Cincinnati Chemical Corporation Plant Near Toms River, New Jersey," was completed in February 1959. It was prepared for the company by the New York–based groundwater consultant Leggette, Brashears, and Graham and was not publicly released. Years later, Ciba-Geigy turned the thirty-six-page report over to the U.S. Environmental Protection Agency, from which the author obtained a copy via a Freedom of Information Act request.

19. O. B. Grant, "Design Considerations for the Expansion of Toms River Waste Disposal Facilities," August 31, 1955, 3.

20. "Seepage from Waste Disposal Basins," 2.

21. "Design Considerations," 5. The 1959 report was equally explicit about the purpose of relocating the seepage basins. Page 27 of that report recommends the following: "If the treatment basins are relocated, they should be constructed at a site near the Toms River so that infiltration from them will almost immediately be discharged into the river, rather than migrate into production wells or move beneath neighboring properties."

22. A profile of Morris Smith in the Spring 1961 edition of *TRC Color,* the com-

pany newsletter, begins this way: "The Toms River provides a handy receptacle for chemical wastes left over from our manufacturing processes. As a responsible neighbor, however, our Company cannot spoil its waters with unwholesome waste materials. Before it enters the river, the residue must be treated to remove potentially harmful substances. The man in charge of this operation is Morris Smith."

23. David C. Bomberger and Robert L. Boughton, *Wastes from Manufacture of Dyes and Pigments, Volume 1: Azo Dyes and Pigments* (SRI International, 1984, under contract to the U.S. Environmental Protection Agency), 34.

24. The new azo sludges were iron oxide sludge and a mixture of sulfuric acid and limestone called calcium sulfate.

25. "Creating a Toxic Landscape: Chemical Waste Disposal Policy and Practice: 1900–1960," 102–9.

Chapter Four

1. D. B. Dyche, "Alleged Pollution of the Toms River," September 9, 1963, memo, 4–5.

2. The minutes of the company's board meeting in November 1960 note that Toms River Chemical was getting complaints about the "distinct smell of the Toms River after our effluent has been added" as well as a "slimy deposit, containing solvents" just below the outfall pipe. "Report for the Board of Directors Meeting," Toms River Chemical Corporation, November 23, 1960, 4.

3. James Crane, "Progress Report on Liquid Waste Treatment Problems," March 14, 1961, memo.

4. J. A. Meier, "Waste Water Treatment Problems," August 9, 1961, memo.

5. Morris Smith, "Waste Disposal—Study of Fish in Toms River by Division of Fish and Game," April 15, 1963, memo.

6. Transcript of prepared remarks delivered by Robert Sponagel, general manager of Toms River Chemical, at a May 20, 1963, press conference.

7. In a September 30, 1964, memo to Toms River Chemical general manager Robert Sponagel, an engineer hired by the company named John C. Fellows Jr. recounts a meeting he had the previous day with Ernest Segesser, the supervising engineer of the state health department's stream pollution control program. In the memo, Fellows quotes Segesser as saying that New Jersey governor Richard Hughes "has become personally concerned" about public criticism of the company. "Off the record, he [Segesser] gave me his personal advice and opinion. He stated that apparently our news releases are only agitating people and supplying them with information. He felt that complete silence would improve our position."

8. Robert Shaw, director of environmental health at the New Jersey Department of Health, was quoted in a May 23, 1963, article in the *Ocean County Sun* headlined "Doubt Cast on River's Future as Development Seeps South."

9. James Crane and J. Zahuta, "Report on Analyses of T.R.C. Effluent and Waters of Toms River," a May 18, 1965, memo to ten senior managers at Toms River Chemical.

10. In Building 108, more than one thousand pounds of epichlorohydrin were "run to the sewer" every day, James Crane wrote in a January 14, 1964, report enti-

tled "Technical Report No. 2—Water-Borne Waste Treatment," sent to twenty-four senior managers at Toms River Chemical. In Building 102, more than two thousand pounds of nitrobenzene were going into the sewer daily—and had been since the 1950s.

11. James Crane, "Appropriation Request No. 2327-4: Force Main to Ocean—Justification," November 27, 1964, memo to Robert Sponagel.

12. Offshore dumping of sewage sludge by New Jersey municipalities ended in 1991, but as of 2012 there were still fourteen ocean outfalls in the state (three operated by the Ocean County Utilities Authority) discharging municipally treated wastewater into the ocean. Many cities around the world, from Miami to Melbourne, do the same.

13. "Toms River Chemical Pollution Suit," *Ocean County Sun,* August 12, 1965.

14. A. Bruce Pyle, "Toms River," September 23, 1965, memo to New Jersey Division of Fish and Game division director Lester G. MacNamara.

15. One of the local newspapers, the *New Jersey Courier,* on August 19, 1965, ran a cartoon showing a dartboard labeled "Toms River Chemical" being pierced by darts labeled "Philip Maimone" and "Beachfront Mayors." An accompanying editorial was sympathetic to the company it described as a "whipping boy" for opportunistic local politicians.

16. Kathy Wright, "Fischer Defends Flag of the United Nations," *New Jersey Courier,* June 24, 1965.

17. The term *epidemiology* was derived from three Greek words: *epi* (upon), *demos* (people), and *logos* (discourse). It could thus be defined as the "study of what is upon the people." Unsurprisingly, considering the primacy of infectious diseases in epidemiology, the new word was chosen because of its similarity to *epidemic,* which has a much older origin. Hippocrates used the Greek form, *epidemios* ("on the people"), to refer to groupings of similar symptoms or diseases occurring in a particular place and time.

18. John Ayrton Paris, *Pharmacologia, Volume 2* (W. Phillips, 1825), 96.

19. For a brief description of the contributions of Pierre Louis and Siméon Denis Poisson to epidemiology, see pages 358–59 of "Environment, Population and Biology: A Short History of Modern Epidemiology" by Alessandra Parodi, David Neasham, and Paolo Vineis, *Perspectives in Biology and Medicine* 49:3 (summer 2006): 357–68.

20. This chapter's account of the confidential negotiations between the Toms River Chemical Corporation and the Toms River Water Company over the contamination of the Holly Street wells and the subsequent actions of the two companies is based on memos and correspondence that were secret at the time they were written, in 1965, 1966, and 1967. Some of those documents became public more than thirty years later, after the two companies turned them over to the U.S. Environmental Protection Agency and the New Jersey Department of Health and Senior Services. Other documents became part of the public record in *Ciba-Geigy Corporation vs. Liberty Mutual Insurance Company et al.* Ciba-Geigy also described the Holly well episode in a December 3, 1993, letter from Ciba executive Daniel McIntyre to the EPA's Stephen Cipot.

21. According to the March 23, 1965, memo, "Water Analysis Record, Toms

River Water Company, Well No. 13," the water company added a huge amount of chlorine to the well to remove the color: eight parts per million. A standard level at the time would have been less than one part per million, which is also a generally accepted safety level for chlorine dioxide today. The 1965 "Water Analysis Record" is cited on page 15 of *Public Health Assessment: Ciba-Geigy Corporation,* prepared in 2001 by the New Jersey Department of Health and Senior Services.

22. Philip Wehner, "Water Supply Situation," March 21, 1966, confidential memo to Robert Sponagel.

23. "Toxicological Profile of Benzidine" (U.S. Agency for Toxic Substances and Disease Registry, 2001), 7.

24. James K. Mitchell, "Adjustment to New Physical Environments Beyond the Metropolitan Fringe," *Geographical Review* 66:1 (January 1976): 18–31, 26. "Ocean County residents act alike in at least one respect. Most regard the planting of trees, shrubs and flower gardens around individual homes as a prerequisite for living in the area. . . . Creating lawns and gardens, where they do not already exist, is an immediate priority for most newcomers."

25. The details of the talks between Toms River Water and Toms River Chemical are contained in a series of confidential letters from lawyers, consultants, and managers of the companies dated March 2, March 21, April 7, April 19, and September 6, 1966.

26. In its secret negotiations with Toms River Chemical in 1965 and 1966, the water company insisted on a backup plan in case chlorination did not work or the ocean pipeline was not finished in time. Toms River Water wanted a new water well immediately but also wanted the chemical company to dig it without applying for a permit from the state health department, which would have required public notice and thus broken the secrecy. Toms River Water's proposed solution was for the chemical company to drill the well on factory property without telling the state. Toms River Chemical refused, advising the water company instead either to drill an unpermitted well on its own property or to apply for a permit "immediately, but in a normal 'unhurried' manner" to minimize public attention. Toms River Chemical executive Philip Wehner, in a confidential April 19, 1966, memo, wrote that his company "thinks that it is unwise to apply for this permit on a 'rush' or 'emergency' basis." As it turned out, the backup well was not needed in 1966 because the chlorination worked and the ocean pipeline opened on schedule. Toms River Water got its new well in early 1967, without any public questions about why the water company needed a well in South Toms River at a location more than a mile south of Holly Street and the river.

27. James Crane, January 31, 1969, confidential memo to Philip Wehner. "Regular analyses of the Toms River since we started pumping to the ocean in 1966 show definitely that there is a discharge of our wastes to the Toms River to the extent of roughly 200,000 gals. per day. We have known of this seepage since 1960; it was of no importance until we started pumping to the ocean."

28. Page 3 of the legal agreement dated February 21, 1967, between the Toms River Chemical Corporation and the Toms River Water Company.

29. According a brief article on the front page of the October 12, 1967, edition

of the *New Jersey Courier,* the settlement included an acknowledgment by Philip Maimone's development company that "the Toms River is not in a polluted state by reason of past operations of TRC [Toms River Chemical] and that TRC is in no way altering the condition of the Toms River since its ocean outfall pipeline has been in operation."

Chapter Five

1. Frank Fernicola's quotation and the subsequent statements from both Fernicola brothers come from sworn depositions they gave in *Abbatemarco vs. Nicol,* a groundwater contamination case involving a housing development in Manchester Township called Pine Lake Park. The Fernicolas were deposed because Union Carbide was initially a defendant in the case, but the company was later dropped from the suit because there was no evidence that any of the Union Carbide waste that Nick Fernicola hauled ended up contaminating Pine Lake Park's drinking water supply. Instead, a local asphalt plant settled the case for $4 million. Although the Fernicola brothers were not implicated in the case, the *Abbatemarco* depositions are among the few surviving documents in which they spoke at length, and under oath, about their adventures in the waste-hauling business.

2. As a former Plumsted Township committeeman, Donald J. Knause, explained years later, "Goose Farm was the size of a football field, and the drums could be buried in layers. You could make a fortune." Knause was also a former investigator for the New Jersey Department of Environmental Protection. He is quoted in Caren Chesler, "Polluter of Plumsted Farm to Monitor Site," *Asbury Park Press,* October 11, 1997.

3. Don Bennett, "Much Work Left to Do at County Waste Sites," *Ocean County Observer,* February 15, 1991.

4. Manchester Township was a notorious location for dumping, especially for liquid waste. One site along Route 70 was owned by a drug company and was known to locals as the "Penicillin Dump," according to the *Abbatemarco* deposition of Bruce Egeland, who was a detective in the Manchester Police Department during the 1960s and 1970s.

5. *Abbatemarco* deposition of Nicholas Fernicola, 24.

6. February 16, 1971, memo from Nicholas Fernicola to Union Carbide Corporation. The signed note is on stationery with "Nick Fernicola Dealer in Barrels and Drums" printed in block letters on the top.

7. "We are starting to use Nick Fernicola to dispose of leaking, damaged drums. Has anyone taken a look at his disposal site, etc.," read the July 7, 1971, handwritten note from Joe Novak to Ed Moherek, both of Union Carbide's Bound Brook factory. Except for an initial visit to the town landfill, there is no indication that anyone from Bound Brook checked to see where Fernicola was dumping the company's waste drums.

8. Avicenna's insightful list of diseases spread by person-to-person contact included leprosy, smallpox, plague, tuberculosis, and the skin infections scabies and impetigo.

9. In the mid-seventeenth century, a London cloth merchant named John Graunt who had no training in mathematics or medicine started reading the Bills of Mortality as a hobby. Realizing that the reports were untapped founts of information about the lives of Englishmen, especially their health, Graunt compiled the first statistically derived estimates of life expectancy and tracked many other statistics, from fertility rates to suicides and murders. Published in 1662, his immensely popular *Natural and Political Observations Made upon the Bills of Mortality* helped to launch the modern science of demography.

10. The chapter by William Farr, "Vital Statistics," appeared in John Ramsay McCulloch, ed., *A Statistical Account of the British Empire,* published in 1837.

11. William Farr, "Tenth Annual Report of the Registrar General," cited in Elaine Freedgood, *Victorian Writing about Risk: Imagining a Safe England in a Dangerous World* (Cambridge University Press, 2000), 42.

12. John Snow, "The Cholera near Golden Square, and at Deptford" excerpted in Carol Buck et al., eds., *The Challenge of Epidemiology: Issues and Selected Readings* (Pan American Health Organization, 1988), 415–18.

13. John Snow also was one of the first disease investigators to recognize that the absence of illness could be just as informative as its presence. A brewery in the cholera outbreak zone, for instance, had no deaths, while an overcrowded workhouse had just five. Snow and Henry Whitehead investigated further, discovering that at the brewery everyone drank beer instead of water and that the workhouse had its own water well and did not rely on the Broad Street well. Snow scrutinized victims who appeared to be outliers, too, such as an older woman who several months earlier had moved from Soho to Hampstead, where there were no other cholera deaths. Interviewing the dead woman's relatives, Snow learned that she preferred the taste of Broad Street water and had a large bottle delivered to her every day.

14. John Snow learned of the basement cesspool from an interview with the mother of a baby who had died of cholera. She had cleaned her child's dirty diapers in the pool.

15. An Italian, Filippo Pacini, was the first to identify the cholera bacterium under the microscope as a likely source of the disease. But his 1855 observation was generally ignored; the bacterial hypothesis did not gain wide currency until Robert Koch independently reached the same conclusion thirty years later.

16. The first great champion of centralizing waste-handling was an English lawyer named Edwin Chadwick. Even before William Farr and John Snow confirmed the connection between filth and cholera, Chadwick vociferously argued that lack of sanitation was a root cause of poverty. He was the driving force behind the British Parliament's passage of the Public Health Act of 1848, which provided for the appointment of "inspectors of nuisances" (forerunners of today's public health officers), who were tasked with improving the safety of water distribution and sewage and garbage disposal.

17. Two statistics illustrate why urbanization in the developed world has vastly increased the amount of waste per capita. First, the average American generates almost twice as much garbage today as in 1960. Second, urban residents in wealthy countries are responsible for three times as much garbage as their counterparts in the cities of poorer nations. Franklin Associates, *Characterization of Municipal*

Solid Waste in the United States (U.S. Environmental Protection Agency, 1998), 141, table B-1; see also Sandra Cointreau, *Occupational and Environmental Health Issues of Solid Waste Management* (World Bank, 2006), 4, table 1.

18. Philip Wehner, "Waste Disposal/Effluent Problems," August 15, 1966, memo to Robert Sponagel, 1.

19. Nine days after the ocean pipeline opened, a man who lived near the plant, Ronald Clayton, complained at a meeting of the Ocean County Board of Chosen Freeholders that no one in his neighborhood could sleep because of a "terrible stench" that he believed was coming from the ten-acre holding pond at Toms River Chemical, according to an article that appeared on the front page of the July 21, 1966, edition of the *New Jersey Courier.* A front-page story that same day in the rival paper, the *Ocean County Sun,* noted that after Clayton voiced his complaints to the legislators, Mayor John G. Woods spoke up to defend the company, as did Robert Conti, secretary of the town sewer authority.

20. Morris Smith, "Solids and Solvent Waste Disposal," October 11, 1966, memo, 2: "Every 6–8 weeks, approximately 50 AQ [anthraquinone] laden filter socks must be disposed of. Because this material cannot be burned in the incinerator without generating huge quantities of black smoke, it is usually burned at night." See also September 7, 1967, letter from W. A. Helbig of Atlas Chemical Industries Inc. to James Crane of Toms River Chemical: "We understand it is the desire of Toms River Chemical Corporation that we refrain from making public the work on waste effluent treatment at your plant."

21. Raymond Simon, *Toms River Chemical Company Public Opinion Survey,* February 1, 1968, report to Toms River Chemical. Simon, a Utica College professor of public relations, and ten of his students designed the questions and conducted the poll.

22. "Public Opinion Survey Reveals TRC's Standing in the Community," *TRC Color* (Spring 1968).

23. In addition to the *Abbatemarco* deposition, this chapter's account of the events at Reich Farm in 1971 and 1972 relies on a 119-page report entitled, "Analysis of a Land Disposal Damage Incident involving Hazardous Waste Materials, Dover Township, New Jersey." It was prepared in May 1976 by Masood Ghassemi, an environmental engineer at TRW Inc. who was serving as a consultant to the U.S. Environmental Protection Agency.

24. Nick Fernicola "just kept digging holes all around the place. Digging holes and digging holes and digging holes, and we kept dumping the chemicals on the ground," recalled Richie Winton, one of the truck drivers Fernicola hired. *Abbatemarco* deposition of Winton, 73.

25. *Abbatemarco* deposition of Nicholas Fernicola, 37–38.

26. The state's "unfortunate" decision not to investigate further what Nick Fernicola was doing at Reich Farm is cited in an April 27, 1977, memo about the case written by New Jersey deputy attorney general Lawrence E. Stanley.

27. Twenty-two years after the 1971 dumping incidents at the town landfill and Reich Farm, Nick Fernicola claimed in a sworn deposition that drivers from Ciba-Geigy and the Lakehurst Naval Air Station "without a doubt" paid bribes in 1971 to get rid of drums at the town landfill. But he did not provide names or other

evidence to support this claim. See *Abbatemarco* deposition of Nicholas Fernicola, 83–84.

28. *Abbatemarco* deposition of Nicholas Fernicola, 43, 80.

Chapter Six

1. This chapter's description of the early life of Michael Gillick is drawn primarily from published sources, including numerous magazine and newspaper articles and Linda Gillick's book about his childhood, *For the Love of Mike* (WRS Publishing, 1994). The author also conducted a ninety-minute on-the-record interview with Michael Gillick and had several informal conversations with Linda Gillick at her home and office. Ultimately, however, Linda Gillick did not agree to a formal interview.

2. Bruce M. Rothschild, Brian J. Witzke, and Israel Hershkovitz, "Metastatic Cancer in the Jurassic," *The Lancet* 354 (July 31, 1999): 398. See also Bruce M. Rothschild et al., "Epidemiologic Study of Tumors in Dinosaurs," *Naturwissenschaften* 90 (2003): 495–500. In the second study, interestingly, although Rothschild and his collaborators used fluoroscopy to examine more than ten thousand fossilized bones from 708 dinosaurs representing seventy-seven species, they found tumors only in bones from the duck-billed hadrosaur family. The authors suggest that both genetic and environmental factors may explain the finding, since hadrosaurs had a unique diet that consisted largely of cone-bearing evergreen trees and shrubs.

3. James Henry Breasted, trans., *The Edwin Smith Surgical Papyrus* (University of Chicago Press, 1930), 404–5, 457–58.

4. W.H.S. Jones, trans., *Hippocrates, Volume IV, and Heraclitus On the Universe* (Heinemann, 1959), 189.

5. This chapter's discussion of Rudolf Virchow's contributions and character is based on Leon Eisenberg, "Rudolf Ludwig Karl Virchow, Where Are You Now That We Need You?" *American Journal of Medicine* 77 (September 1984): 524–32; J.M.S. Pearce, "Rudolf Ludwig Karl Virchow," *Journal of Neurology* 249 (2002): 492–93; Myron Schultz, "Rudolf Virchow," *Emerging Infectious Diseases* 14:9 (September 2008): 1480–81; and "Rudolf Virchow (1821–1902)—Anthropologist, Archeologist, Politician, and Pathologist," unsigned editorial, *Journal of the American Medical Association* 188:12 (June 22, 1964): 1080–81.

6. In another 1848 editorial in the same newspaper, Rudolf Virchow famously declared: "The physicians are the natural attorneys of the poor, and social problems fall to a large extent within their jurisdiction."

7. "The chief point," Rudolf Virchow wrote, is that "the cell is really the ultimate morphological element in which there is any manifestation of life, and that we must not transfer the seat of real action to any point beyond the cell." Rudolf Virchow, *Cellular Pathology as Based upon Physiological and Pathological Histology*, trans. Frank Chance (John Churchill, 1860), lecture 1, 3.

8. According to National Cancer Institute estimates for 2002–2006, the average yearly cancer incidence rate per year for people under age 65 was 221.1 cases per 100,000 people. For people over 65, the rate was 2,134.3 cases per 100,000.

9. William B. Ershler, "Cancer: A Disease of the Elderly," *Journal of Supportive Oncology* 1, Supp. 2 (November/December 2003): 5–10.

10. The estimate of thirty-eight newly diagnosed cases of childhood cancer per day in the United States is derived from the National Cancer Institute's incidence data for 2002–2006 for children under twenty years of age: 16.6 cases per 100,000 children per year. Also according to the institute, there were 12.9 new cases per 100,000 children and teenagers in 1975, compared to 17.6 in 2005, a 36 percent increase. Overall cancer incidence (for all ages) rose from 400.38 per 100,000 in 1975 to 459.94 in 2005, a 15 percent increase.

11. Because he was so young, Michael Gillick's chance of surviving was somewhat greater than the overall rate of 5 percent. For reasons that researchers do not yet understand, neuroblastoma tumors in infants are much more likely to undergo spontaneous regression compared to those in older children. Michael's tumors did not disappear, but their rate of growth slowed. Today, the overall five-year survival rate for late-stage metastatic neuroblastoma is greater than 50 percent and is nearly 90 percent for infants. Current treatments include a combination of surgery, drugs, radiation, and antibody therapies to stimulate the immune system.

12. James Crane and N. Morley, "Waste Clarification and Solids Separation Study," November 14, 1968, 2.

13. The 1970 merger of Ciba and Geigy allowed the Swiss to finally (if temporarily) surpass their longtime rivals in Germany and the United States and become the largest dye manufacturing company in the world. One of the merger's most important proving grounds was Toms River, where the two companies had worked closely together since 1955. Before the merger, Ciba owned 58 percent of Toms River Chemical and Geigy 21 percent. The other 21 percent was owned by Sandoz, which merged with Ciba-Geigy in 1996 to form Novartis.

14. The estimate of 8,800 waste drums per year dumped into the unlined landfill comes from Morris Smith in "Pre-Proposal Visit: Toms River Chemical," an August 6, 1968, memo from J. R. Lawson of Roy F. Weston Inc., page 4. The description of the "cliff" is in "Final Project: New Chemical Dump," an October 21, 1971, report from William Bobsein to Toms River Chemical executives, page 1. Describing practices at the old dump, Bobsein wrote: "At present, those waste chemicals which are suitable for burial are packed into 'scrap' steel drums. These drums are transported by truck to a dumping site at the southern side of the plant property, along the edge of an abandoned lagoon which formerly comprised part of the wastewater treatment facility. The drums are then dumped over the edge of the 'cliff.' Most of the contents are spilled. At intervals of three to four months, some drums are crudely rearranged and the pile consolidated by the use of a crane. Sand is then spread over the drums, and the working 'face' of the dump is thus advanced."

15. William Bobsein, "Disposal of Combustible Wastes Status Report," October 17, 1969, memo to Philip Wehner, 4.

16. "Pre-Proposal Visit: Toms River Chemical," 3.

17. William Bobsein, "Hazardous and Solid Wastes," a 1969 presentation, 1–10.

18. "TRC Teach-In on Control of Pollution Attracts 1,100-Plus," *TRC News*, December 1970.

19. What came to be known as the "Refuse Act" is actually a section of the Rivers and Harbors Act of 1899. More than twenty federal rivers and harbors laws were adopted between 1882 and 1970, their primary purpose being to keep the nation's waterways navigable to boat traffic, which was crucial to commerce. The "refuse" section in the 1899 version that would eventually become so important as a tool of environmental enforcement was an afterthought even to its sponsors, one of whom (Senator William P. Frye of Maine, the chairman of the Rivers and Harbors Committee) described it as a mere compilation of existing laws with "very slight changes to remove ambiguities." In the Refuse Act's heyday in the early 1970s, some legal scholars objected that Congress had never intended for its criminal penalties to apply to polluters, only to physical barriers to boat traffic such as dumped rocks and soil or illegal bridges. The controversy faded when new federal laws, including the Clean Water Act of 1977, became the primary tools for enforcing water-quality standards. For more information, see Diane D. Eames, "The Refuse Act of 1899: Its Scope and Role in Control of Water Pollution," *California Law Review* 48:6 (November 1970): 1444–73; and William H. Rodgers Jr., "Industrial Water Pollution and the Refuse Act: A Second Chance for Water Quality," *University of Pennsylvania Law Review* 119:5 (April 1971): 761–822.

20. Toms River Chemical had a 1965 construction permit from the Army Corps of Engineers for its ocean pipeline but not an operating permit, so the pipeline was operating with state, not federal, approval.

21. Philip Wehner, "Meeting with Representatives of the Federal Environmental Protection Agency on the Mercury Problem at TRC," September 15, 1971, memo, 1–3.

22. To settle the case, *U.S. vs. Asbury Park*, the offending towns agreed to barge their sludge to dump sites farther offshore; wastewater discharges through ocean pipelines continued unabated.

23. A January 18, 1977, quarterly report to Toms River Chemical's board of directors included this assessment of the company's talks with the U.S. Environmental Protection Agency, on page 15: "Considerable progress was made with EPA in regard to the effluent limitations for the new treatment plant. Litigation against EPA in regard to these limits will probably not be required. It appears that limitations will soon be adopted which are feasible and attainable by careful operation. Those limitations in the originally-proposed permit issued in October of 1976 could not have been attained."

24. J. Richard Pellington, "206 Count Indictment Astounds Officials," *Ocean County Observer*, July 17, 1972. See also "TRC and Pollution," unsigned editorial, *Ocean County Observer*, July 20, 1972.

25. The current Environmental Protection Agency standard for many industrial organic compounds is just five parts per billion, and New Jersey's current limit for some organic solvents is just one part per billion.

26. Bob De Sando, "State Finds Contaminants in Water Company Wells," *Asbury Park Press*, January 10, 1975. "A lot of people were very angry at me after that first story," recalled the reporter, Bob De Sando. "They thought it was inflammatory and would get people scared, and that then things would spiral out of control."

27. Bob De Sando, "Wells Held Safe: State Contends Phenol Levels Not Hazardous," *Asbury Park Press,* January 11, 1975. The article quotes Steven Corwin, a special assistant to the state environmental protection commissioner: "We feel there is no problem. . . . Phenols are not desirable by any means and could contribute to a taste and odor in some instances. But at those levels there is no physical detriment. . . . We sure wouldn't let people drink anything that would be dangerous." The article, which ran on the front page, essentially contradicted the story that appeared the previous day.

28. Gillick, *For the Love of Mike,* 94.

Chapter Seven

1. Don Bennett, "Lab Found More Problems," *Daily Observer,* September 16, 1984.

2. Among those who noticed occupational clusters of scrotal cancer was the Scottish surgeon Joseph Bell in 1876. A stickler for close observation, Bell was Arthur Conan Doyle's teacher and the chief inspiration for Sherlock Holmes. "A Brief History of Scrotal Cancer," 395.

3. The chapter's description of the Schneeberg mines and Walther Hesse and Friedrich Härting's "mountain sickness" studies is drawn from several sources: Walther Hesse, "Das Vorkommen von Primärem Lungenkrebs bei den Bergleuten der Consortschaftlichen Gruben in Schneeberg," translated as "The Occurrence of Primary Lung Cancer in the Miners of the Consortium Mines in Schneeberg," *Archiv der Heilkunde* 19 (1878): 160–62 [NIH Library Translation NIH-90-394]; Dieter H. M. Gröschel, trans., "Walther and Angelina Hesse—Early Contributors to Bacteriology," *American Society for Microbiology News* 58:8 (1992): 425–28; Margarethe Uhlig, "Schneeberg Lung Cancer," *Virchow's Archive for Pathological and Physiological Anatomy* 230 (1921): 76–98 [NIH Library Translation NIH-90-58]; and M. Greenberg and I. J. Selikoff, "Lung Cancer in the Schneeberg Mines: A Reappraisal of the Data Reporting by Härting and Hesse in 1879," *Annals of Occupational Hygiene* 37:1 (1993): 5–14.

4. Today, the term *mountain sickness* is used to describe an acute condition caused by exposure to low-density air at high altitudes; Hesse and Härting applied the term to a very different set of symptoms.

5. Before going to work for the German government, Walther Hesse took two trips to America as a steamship doctor. When he returned home, he married a wealthy, cultured woman he had met in New York and wrote one of the first papers on seasickness published in the medical literature.

6. "Das Vorkommen von Primärem Lungenkrebs," 160.

7. After leaving Schneeberg, Walther Hesse studied with the famed microbiologist Robert Koch, just as Koch was about to identify the bacteria responsible for tuberculosis and cholera. Hesse spent the rest of his life working on the control of bacterial diseases; he is best known for discovering, with his wife, that agar is an ideal surface for growing bacterial cultures. Angelina Hesse had used agar, a gelatinous material derived from seaweed, to keep her jellies and puddings solid.

8. Werner Schüttmann, "Schneeberg Lung Disease and Uranium Mining in the

Saxon Ore Mountains," *American Journal of Industrial Medicine* 23 (1993): 355–68, 361–62.

9. U.S. Environmental Protection Agency, *EPA Assessment of Risks from Radon in Homes* (June 2003), 64, table 20.

10. "Amendment to National Oil and Hazardous Substance Contingency Plan; National Priorities List," *Federal Register* 48:175, pages 40658–40673.

Chapter Eight

1. The staff reductions at the *Ocean County Observer* accelerated after 1998, when it was bought by Gannett Inc. On November 28, 2007, Gannett announced that it was switching the paper from daily to weekly publication, dropping all coverage outside of Toms River, merging it with a weekly newspaper and changing its name to the *Toms River Observer Reporter*. The paper had been known as the *Observer* since it was founded in 1850. Six months after Gannett's announcement, Don Bennett accepted a buyout offer and retired from the newspaper he had joined thirty years earlier. His career as a working journalist in Ocean County spanned forty-four years.

2. The first known attempt to induce tumor growth experimentally via transplantation occurred in 1775, when a Frenchman named Bernard Peyrilhe implanted a human breast tumor in a dog. The transplanted tumor did not grow, however. In 1889, a German scientist named Arthur Hanau finally managed to transfer squamous cell tumors from one rat to another. Researchers across Europe rushed to conduct their own transplants but were disappointed. Their experiments usually failed, and even successful transplants provided little useful information.

3. For a brief but cogent summary of the theories of cell irritation, embryonal rest, and dedifferentiation, see Sadhan Majumder, ed., *Stem Cells and Cancer* (Springer, 2009), 7–9. See also: Lorenzo Tomatis, "Cell Proliferation and Carcinogenesis: A Brief History and Current View Based on an IARC Workshop Report," *Environmental Health Perspectives* 101, Supp. 5: Cell Proliferation and Chemical Carcinogenesis (December 1993): 149–51; and Folke Henschen, "Yamagiwa's Tar Cancer and Its Historical Significance—From Percivall Pott to Katsusaboro Yamagiwa," *Gann: The Japanese Journal of Cancer Research* 58 (December 1968): 447–51, 447–48. Another useful source is William Seaman Bainbridge, *The Cancer Problem* (Macmillan, 1918), which is widely regarded as an authoritative source of information on competing theories of carcinogenesis of that era.

4. In addition to Henschen, "Yamagiwa's Tar Cancer," this chapter's account of Yamagiwa's life and accomplishments draws on the following sources: James R. Bartholomew, "Japanese Nobel Candidates in the First Half of the Twentieth Century," *Osiris, Second Series* 13 (1998): 238–84, 253–62; Murray J. Shear, "Yamagiwa's Tar Cancer and Its Historical Significance—From Yamagiwa to Kennaway," *Gann: The Japanese Journal of Cancer Research* 60 (April 1969): 121–27; "Katsusaboro Yamagiwa (1863–1930)," *CA: A Cancer Journal for Clinicians* 27:3 (May/June 1977): 172–73; and Katsusaburo Yamagiwa and Koichi Ichikawa, "Experimental Study of the Pathogenesis of Carcinoma," an excerpt of their original 1918 study, *CA: A*

Cancer Journal for Clinicians 27:3 (May/June 1977): 174–81. See also William Johnston, *The Modern Epidemic: A History of Tuberculosis in Japan* (Harvard University Asia Center, 1986), 202–4.

5. Waldron, "Brief History of Scrotal Cancer," 395.

6. At least two other researchers induced cancer under controlled experimental conditions before Katsusaburo Yamagiwa did in 1914, but their work was mostly ignored at the time. The French physician Pierre Edouard Jean Clunet in 1908 induced skin cancer in two rats by bombarding them with X-rays. The dosages were so high, however, that the other two rats in his experiment died, leading some scientists to question the validity of the experiment (though not the carcinogenicity of X-rays, which was already apparent from the illnesses that struck many pioneering radiation researchers). In 1911, the American pathologist Francis Peyton Rous identified the first cancer virus by demonstrating that cancer could be transmitted from one animal to another via injection. Working at what was then known as the Rockefeller Institute for Medical Research (now Rockefeller University) in New York City, Rous ground up a sarcoma tumor from a chicken, passed it through a filter, and injected the cell-free extract into a healthy chicken, which then developed cancer. Rous's conclusions were so far outside of the mainstream that few scientists tried to follow up on his work—including Rous, who shifted to other areas of research before returning later to cancer. In 1966, Rous was finally acknowledged with a Nobel Prize for describing what is now known as the Rous Sarcoma Virus, the first oncovirus. Today, viruses are thought to be responsible for 15 to 20 percent of all human cancer cases.

7. For more information about the Fibiger-Yamagiwa controversy, see Carl-Magnus Stolt, George Klein, and Alfred T. R. Jansson, "An Analysis of a Wrong Nobel Prize—Johannes Fibiger, 1926: A Study in the Nobel Archives," *Advances in Cancer Research* 92 (2004): 1–12. See also Bartholomew, "Japanese Nobel Candidates in the First Half of the Twentieth Century," 257–62.

8. Embarrassed by the 1926 fiasco, the Nobel Committee in Physiology or Medicine did not award another prize for cancer research until 1966, when Francis Peyton Rous was very belatedly acknowledged for his 1911 discovery that some cancers were virally transmitted.

9. Mysid shrimp are also known as opossum shrimp, for the thorax pouch in which a female carries her eggs. The species typically used in toxicity testing is *Americamysis bahia*. Despite the name and superficial resemblance, mysids are not actually shrimp.

10. Thomas Fikslin, biology section chief, New York regional office, U.S. Environmental Protection Agency, "Biomonitoring Inspection Report—Toms River Chemical Corporation," March 29, 1982, memo to the chiefs of the water enforcement and water facilities branches of the EPA regional office.

11. By the late 1970s, the mutagenicity test the state wanted Ciba to conduct had become an important tool for toxicologists. The Ames test was named for the man who had developed it just a few years earlier: Bruce Ames, a professor of biochemistry and microbiology at the University of California at Berkeley and a major figure in environmental cancer research. His inexpensive and relatively simple test helped

pave the way for the banning of many mutagenic chemicals in 1970s. By the 1990s, however, Ames was infuriating environmentalists by arguing that traditional tests on laboratory animals, in which they are dosed with high concentrations of chemicals and then checked for tumors, are not a good model for predicting whether chemicals cause cancer in humans. He also argued that many naturally occurring compounds posed at least as big a cancer risk as the synthetic chemicals produced by industry.

12. Leslie McGeorge and Tessie Wishart, Office of Cancer and Toxic Substances Research, New Jersey Department of Environmental Protection, "Chemical and Mutagenicity Analysis Report," February 9, 1983.

13. McGeorge and Wishart, "Chemical and Mutagenicity Analysis Report," 2: "It is suspected that the mutagenic activity may be due at least in part to unidentified nitrogen-containing compounds. Numerous nitrogenous compounds have been shown to be carcinogenic and/or mutagenic." See also Attachment A of "Biomonitoring Inspection Report—Toms River Chemical Corporation."

14. The series by Don Bennett appeared in the September 30 and October 1, 1984, editions of the *Ocean County Observer.*

Chapter Nine

1. Rose Donato died in 1996, at age eighty-three.

2. Almost thirty years later, no one involved in the events of 1984 remembers exactly how Greenpeace first found out that Ciba-Geigy was discharging industrial waste into the ocean off New Jersey. Dave Rapaport does not remember getting a letter from Rose Donato about it, although Rose's daughter, Michele, is certain her late mother sent one.

3. *Monkeywrenching* became a favored term of environmental activists after the publication of *The Monkey Wrench Gang,* by Edward Abbey (Lippincott, 1975). The darkly hilarious book, a counterculture classic, describes the adventures of four self-styled "environmental warriors" who blow up bridges, bulldozers, and other encroachments on wilderness in the American West. The phrase "throwing a monkey wrench" is much older, referring to an action that causes something to break down.

4. This chapter's description of Ernest Kennaway's life and work is based on the following sources: Ernest Kennaway, "The Identification of a Carcinogenic Compound in Coal Tar," *British Medical Journal* 4942 (September 24, 1955): 749–52; James W. Cook, "Ernest Laurence Kennaway, 1881–1958," *Biographical Memoirs of the Fellows of the Royal Society* 4 (November 1958): 139–54; David H. Phillips, "Fifty Years of Benzo(a)pyrene," *Nature* 303 (June 9, 1983): 468–72; Antoine Lacassagne, "Kennaway and the Carcinogens," *Nature* 191 (August 19, 1961): 743–47; and G. M. Badger, "Ernest Laurence Kennaway," *Journal of Pathology and Bacteriology* 78:2 (1959): 593–606.

5. Despite not being able to participate directly in the final identification of benzo(a)pyrene in 1932 due to his worsening Parkinson's symptoms, Ernest Kennaway lived until 1958 and was ultimately awarded a knighthood, among other high honors, for his work identifying carcinogens. He died in St. Bartholomew's Hospital, the place where, almost two hundred years earlier, Percivall Pott had first drawn attention to the carcinogenic potential of coal pollution through his observations of

chimney sweep boys. Kennaway was seventy-six when he died, having stayed an active researcher for almost his entire adult life thanks to the assistance of his wife, Nina, who helped him dissect animals, keep records, and perform other tasks requiring a steady hand.

6. Greenpeace's coming-out party actually began a few days before the press conference, thanks to an attempted ruse Dave Rapaport cooked up with William Skowronski, a founding member of Ocean County Citizens for Clean Water. Rapaport asked to appear on a local radio talk show to discuss the organization's pollution-fighting activities around the world. When it was time for outside callers, Skowronski dialed in and issued an invitation that he pretended was unrehearsed. "You don't have to go to the South Pacific to battle pollution," he told Rapaport live on the air, "you can battle it right here in Toms River." Rapaport responded enthusiastically, as if he were hearing about the Ciba-Geigy factory for the first time instead of having spent the previous two months investigating the plant and even sneaking onto the property twice.

7. The banner that Samuel Sprunt and Beverly Baker hung from the Ciba-Geigy water tower—"Reduce It, Don't Produce It"—referred to a key element of Greenpeace's toxics campaign that year. In addition to trying to shut down industrial discharges into oceans and lakes, the group was pressing companies to reassess their manufacturing processes to reduce the amount of waste they generated. In fact, as Dave Rapaport would later explain, the attempted shutdowns were just a dramatic way of attracting attention to Greenpeace's real goal: reducing the use of toxic chemicals in manufacturing.

8. Sam Sprunt and Beverly Baker did not have to return to Toms River to answer the trespassing charges because Greenpeace negotiated a plea bargain on their behalf. On October 24, 1984, the two activists pleaded guilty in absentia to four counts of trespassing. They were fined three hundred dollars each by a Dover Township municipal court judge, who agreed not to impose any jail time. Greenpeace paid the fines.

9. The attempted theft of Greenpeace's jar of effluent was mentioned in an August 5, 1984, editorial in *The Reporter,* a weekly newspaper in Ocean County. Its headline read, "The Greenpeace Show Ending Is Up to Us." An accompanying cartoon, captioned "The Sludge That Ate Lavallette (A Ciba-Geigy Production)," showed a monster emerging from a pipe. The jar incident was also described by Dave Rapaport, Nancy Menke Scott, Michele Donato, and Samuel Sprunt in interviews. The identity of the pilferer is unknown, although all sources agree he was affiliated with the Ocean County Chamber of Commerce.

10. Eugene Kiely, "Greenpeace Challenges Ciba to Water Tests," *Ocean County Observer,* August 20, 1984.

Chapter Ten

1. In a January 1980 letter to the state Department of Environmental Protection, a Ciba-Geigy manager estimated that the factory was burying 9,800 drums and 8,900 cubic yards of sludge in the landfill yearly.

2. Charles J. Trautman, environmental specialist, New Jersey Department of

Environmental Protection, "Toms River Chemical—Ciba-Geigy Chronology of Inspections from May 22, 1979 to May 14, 1984."

3. Burying a drum on site cost Ciba-Geigy just $33, compared to $136 to ship it to a hazardous waste landfill elsewhere.

4. Although the New Jersey Department of Environmental Protection allowed Ciba-Geigy to keep operating problem-plagued Cell Two of the factory landfill in the early 1980s, the DEP did negotiate a March 1981 consent order requiring the company to put a rainproof cap on Cell One after closing it and giving the state the right to close Cell Two if both the inner and outer liners were ever discovered to have leaked. The consent order also established a schedule of stepped-up state inspections of Cell Two.

5. John Cooney, the author of the *Environmental Crimes Deskbook* (Environmental Law Institute, 2004), bluntly describes the dilemma on page 9: "The reality is that environmental agencies cannot police the field alone; they must rely on corporations to regulate themselves."

6. David B. Spence, "The Shadow of the Rational Polluter: Rethinking the Role of Rational Actor Models in Environmental Law," *California Law Review* (July 2001): 917–98. See pages 924 and 925 for charts on federal environmental enforcement over the years.

7. In one infamous case, Kentucky health inspectors investigating a 1967 fire at a toxic dump there told the owner, A. L. Taylor, that he was operating illegally and needed a state permit. Taylor ignored them, and eleven years passed before a hearing officer finally recommended that he be fined three thousand dollars. By that time, in 1977, Taylor was dead and had left behind a toxic legacy of seventeen thousand half-buried drums and a web of contaminated waterways. The EPA-managed cleanup of the site, which came to be known as the "Valley of the Drums," cost more than $2 million. See *Superfund Third Five-Year Review Report for A.L. Taylor (Valley of the Drums)* (U.S. Environmental Protection Agency, 2003), 4.

8. The extent of organized crime's involvement in hazardous waste disposal has been the subject of heated debate, including before congressional committees. There is no doubt that some important people involved in waste-related crimes in New Jersey in the 1970s and 1980s had connections to crime syndicates, including the Lucchese, Genovese, and Gambino crime families. Several were even found murdered. But the authors of the most famous account of those connections, Alan A. Block and Frank R. Scarpitti, in the book *Poisoning for Profit: The Mafia and Toxic Waste in America* (William Morrow, 1985), were sued for libel and forced to acknowledge that some of their allegations were erroneous. In contrast, a later assessment of hazardous waste crimes in New Jersey and three other states found that "most commonly, the criminal dumper is an ordinary, profit-motivated businessman who operates in a business where syndicate crime activity may be present but [is] by no means pervasive." See Donald J. Rebovich, *Dangerous Ground: The World of Hazardous Waste Crime* (Transaction Publishers, 1992), xiv.

9. When New Jersey prosecutors did manage to get an environmental conviction, it was often because they resorted to using laws written for other purposes, such as mail fraud or creating a public nuisance. For example, on April 21, 1980, a suspicious fire at the Chemical Control Corporation in Elizabeth torched more than

five thousand drums of toxic waste, all of it stored illegally. Seven months later, the three owners of Chemical Control were indicted for storing the waste illegally and allowing it to seep into the Elizabeth River. They were also charged with mail fraud and conspiracy for falsely promising manufacturers to dispose of their waste legally. Two of the three men ended up going to prison for mail fraud and conspiracy, but the pollution charges mostly fizzled out. Two were acquitted, and the third was fined five thousand dollars and sentenced to three years' probation. The company, meanwhile, was fined just $23,500 for violating the state's water pollution control act and maintaining a public nuisance.

10. This account of the life and work of Wilhelm Hueper is based largely on an unpublished autobiography he completed in 1976 when he was eighty-two years old, three years before his death. He called it *Adventures of a Physician in Occupational Cancer: A Medical Cassandra's Tale*. The National Library of Medicine in Bethesda, Maryland, provided the author with a copy of the typewritten manuscript. The reference to the "stupid adventures" of war appears on page 44. The reference to the "orgy of mass murder" is on page 60. Two other important sources on Hueper's career are journal articles by David Michaels, an epidemiologist and author who in 2009 was appointed director of the federal Occupational Safety and Health Administration: "Waiting for the Body Count: Corporate Decision Making and Bladder Cancer in the U.S. Dye Industry," *Medical Anthropology Quarterly* 2:3 (September 1988): 215–32; and "When Science Isn't Enough: Wilhelm Hueper, Robert A. M. Case, and the Limits of Scientific Evidence in Preventing Occupational Bladder Cancer," *International Journal of Occupational Health* 1:3 (1995): 278–88.

11. In addition to aniline dyes, leaded gasoline was also manufactured at DuPont's Chambers Works complex starting in the 1920s. Tetraethyl lead was added to gas to reduce engine "knocking," but hundreds of workers suffered brain damage as a result of exposure to fumes. The poisonings were heavily publicized in the late 1920s, and the portion of the Chambers complex where the leaded gas was made came to be known as the "House of Butterflies" due to the insect hallucinations of its workers.

12. As Hueper put it on page 152 of his unpublished memoirs: "These and many other similar experiences with the often much delayed demonstration and admission of carcinogenic properties in industrial chemicals and wastes provide adequate and valid documentation incriminating industrial and commercial private parties as unsuitable media to be entrusted with safeguarding the health of their employees and of the general population."

13. Inserted into the urethra to examine the interior of the bladder, a cystoscope utilizes a thin tube equipped with lenses.

14. Hueper, *Adventures of a Physician in Occupational Cancer*, 156–57.

15. Michaels, "When Science Isn't Enough," 279.

16. Bailus Walker Jr. and Abbie Gerber, "Occupational Exposure to Aromatic Amines: Benzidine and Benzidine-Based Dyes," *National Cancer Institute Monograph* 58 (1981): 11–13, 11.

17. The true extent of bladder cancer at the Chambers Works—more than four hundred cases—would not be revealed until the 1980s. See Michaels, "When Science Isn't Enough," 280–81.

18. Wilhelm Hueper, "'Aniline Tumors' of the Bladder," *Archives of Pathology* 25 (1938): 855–99. See also W. C. Hueper, F. H. Wiley, and H. D. Wolfe, "Experimental Production of Bladder Tumors in Dogs with Beta-Naphthylamine," *Journal of Industrial Hygiene and Toxicology* 20 (1938): 46–84.

19. The headlines appeared in the *Ocean County Observer* on August 30, September 16, September 30, October 7, October 11, October 12, and October 17, 1984.

20. Don Bennett, "What's in Ciba's Waste?" *Ocean County Observer,* September 30, 1984.

21. Linda Gillick, "Help Fight for the Health of Our Families," letter to the editor, *Ocean County Observer,* October 25, 1984.

22. Bonnie Zukofski, "Ciba-Geigy Says Deadly Phosgene Gas Under Control," *Asbury Park Press,* December 16, 1984.

23. The men who led Greenpeace's campaigns in Ocean County in the summers of 1984 and 1985 moved on to colorful careers elsewhere. Jon Hinck became an environmental lawyer, with a client list that included Alaska fishermen harmed by the *Exxon Valdez* oil spill. He even briefly served as the acting attorney general of the newly independent South Pacific island of Palau. Hinck later became a member of the Maine House of Representatives and ran for Congress in 2012. His predecessor as Greenpeace's monkeywrencher-in-chief on the Jersey shore, Dave Rapaport, became a community activist in Vermont and then a manager at a company that sold environmentally sustainable consumer products and also worked with Walmart and other companies to "green" their supply chains. Both men would look back on their respective summers in Ocean County as high-spirited idylls in what was, for a while, an activist's Arcadia.

24. "An Open Letter from Ciba-Geigy," advertisement, *Ocean County Observer,* July 28, 1985.

25. One of the indicted executives, James McPherson, the plant's supervisor of solid waste processing, was a particular embarrassment to Ocean County legislators because they had appointed him chairman of the county's solid waste advisory council. Another, William Bobsein, the plant's manager of environmental technology, was on the state's hazardous waste advisory council. The third and youngest of the alleged conspirators, forty-four-year-old David Ellis, had a doctorate in chemistry and was the plant's assistant manager of environmental technology. He was also Jorge Winkler's partner in their side business, J. R. Henderson Labs, the water-testing firm that had caused controversy in 1984 for initially failing to recognize groundwater contamination in Oak Ridge. A fourth Ciba-Geigy executive, production manager Robert Fesen, was charged with illegal dumping, a lesser charge. Jorge Winkler, the suspended senior executive who had supervised all four men, was not charged.

26. On March 5, 1986, just hours after his conviction, Robert Marshall was sentenced to death. His appeals dragged on for eighteen years until 2004, when U.S. District Judge Joseph E. Irenas set aside the death sentence on the grounds that Marshall had had inadequate counsel during the sentencing phase of the trial. Two years later, Marshall was sentenced to life in prison, with the possibility of parole as early as 2014.

27. Joe McGinniss, *Blind Faith* (G.P. Putnam's Sons, 1989), 47.

Chapter Eleven

1. "Court Upholds Widow's Award," a United Press International story published May 4, 1984, described a case in which the widow of a worker at the Toms River plant was awarded $63,200 in death benefits and legal fees to compensate for her husband's death from lung cancer, even though he was a smoker. Ciba-Geigy appealed the award but lost. "We are convinced that the evidence proffered establishes within a reasonable probability that [the worker's] squamous carcinoma was aggravated, accelerated and exacerbated . . . by the chemicals to which [he] was exposed for a 23-year period in combination with cigarette smoking," the three-judge panel of the Appellate Division of Superior Court ruled.

2. For an insightful exploration of Wilhelm Hueper's conflicts with the environmental research establishment as it evolved in the mid-twentieth century, see Christopher Sellers, "Discovering Environmental Cancer: Wilhelm Hueper, Post–World War II Epidemiology and the Vanishing Clinician's Eye," *American Journal of Public Health* 97:11 (November 1997): 1824–35.

3. Sigismund Peller, "Mortality, Past and Future," *Population Studies* 1:4 (March 1948): 405–56, 445, table 16. The increase was at least partially due to better record keeping.

4. German prewar research into tobacco's carcinogenicity had an all-powerful sponsor in Adolf Hitler, whose mother had died of breast cancer. Hitler had smoked as a young man but later came to despise the habit, which he regarded as a plague foisted on the Aryan race. Citing its American origin, he called tobacco "the wrath of the Red Man against the White Man, vengeance for having been given hard liquor." Hitler made sure that the regime's antismoking propaganda included racist images associating cancer and cigarettes with Jews, Gypsies, and other undesirables.

5. The definitive book on cancer policy and research during the Third Reich is Robert Proctor's *The Nazi War on Cancer* (Princeton University Press, 1999). It begins with a chapter on Wilhelm Hueper's efforts to find a job in Nazi Germany after being fired (the first time) by DuPont Corporation. See also Devra Davis, *The Secret History of the War on Cancer* (Basic Books, 2007), ch. 3.

6. For Richard Doll's brief description of his 1936 visit to Nazi Germany, see Christopher Cook, "Oral History—Sir Richard Doll," *Journal of Public Health* 26:4 (2004): 327–36. Doll, who was Christian but not a churchgoer, described hearing a lecture by a famous German radiologist who used X-rays to treat cancer. The radiologist illustrated his talk with a drawing of X-ray "stormtroopers" annihilating cancer cells marked with Jewish stars. Later, sitting at a Frankfurt café with local medical students, Doll complained about the anti-Semitism and was immediately accused of being Jewish himself, which he "disproved" by showing his companions that he lacked the thick ankles they had insisted were characteristic of Jews. Doll added, "We didn't require many experiences of that sort to realize that there was something evil that had to be eliminated from the world."

7. Richard Doll and A. Bradford Hill, "Smoking and Carcinoma of the Lung," *British Medical Journal* 2:4682 (September 30, 1950): 739–48.

8. Ernest L. Wynder and Evarts A. Graham, "Tobacco Smoking as a Possible

Etiologic Factor in Bronchiogenic Carcinoma," *Journal of the American Medical Association* 143:4 (May 27, 1950): 329–36.

9. Hueper, *Adventures of a Physician,* 183.

10. Richard Doll and Richard Peto, "The Causes of Cancer: Quantitative Estimates of Avoidable Risks of Cancer in the United States Today," *Journal of the National Cancer Institute* 66:6 (June 1981): 1191–1308.

11. The trend toward case-control studies was approvingly documented in 1979 by epidemiologist Philip Cole, who would later play an important role in Toms River as a consultant to Ciba-Geigy. Cole looked at epidemiological studies published in 1957 and 1958 in the British journal *The Lancet* and found almost twice as many case series studies as case-control studies. But by 1976 and 1977, he found, the proportion had flipped. The trend in the *New England Journal of Medicine* was the same: many more case-controls, far fewer case series. See Philip Cole, "The Evolving Case-Control Study," *Journal of Chronic Disease* 32 (1979): 15–27.

12. Joseph Heller's description of Wilhelm Hueper's personality appears in Lester Breslow, *A History of Cancer Control in the United States* (National Institutes of Health, 1979), 139.

13. Wilhelm Hueper blamed DuPont for anonymous charges, lodged with the Federal Bureau of Investigation in 1948, that he was a secret Nazi. (Among the allegations: In 1934, while looking for work in Germany, Hueper had included the salutation "Heil Hitler" in some of his job application letters to Nazi Party representatives at German hospitals.) The FBI cleared him of being a Nazi, but his troubles were not over. In 1949, according to Hueper's unpublished memoirs, the National Cancer Institute received a letter from DuPont's chief of medicine asserting that Hueper had shown "communistic tendencies." The DuPont letter also objected to Hueper's plan to organize a joint government-industry investigation of cancer risk in dye manufacturing. With DuPont leading the opposition (the Swiss firms and others later joined in), the chemical industry's objections scuttled the planned joint investigation. Hueper reacted with characteristic belligerence, writing a letter to the New Jersey Department of Health urging it (unsuccessfully) to investigate the unusually high rate of bladder cancer in Salem County, home of the Chambers Works. Hueper clashed with other powerful entities, too, including the Atomic Energy Commission, which squelched his efforts to publicize the link between lung tumors and uranium mining in Colorado—a link that was first established in Schneeberg in 1878.

14. The forty-five-year-old man's case is one of many described in two journal articles written much later about the bladder cancer epidemic at the Cincinnati Chemical Works: Richard G. Wendel, Ulrich R. Hoegg, and Mitchell R. Zavon, "Benzidine: A Bladder Carcinogen," *Journal of Urology* 111 (May 1974): 607–10; and Mitchell R. Zavon, Ulrich Hoegg, and Eula Bingham, "Benzidine Exposure as a Cause of Bladder Tumors," *Archives of Environmental Health* 27 (July 1973): 1–7.

15. The century-long history of the link between dye manufacture and bladder cancer and the dye industry's reluctance to acknowledge it in Cincinnati and elsewhere are vividly described in David Michaels, "Waiting for the Body Count" and "When Science Isn't Enough." It is also briefly described in Michaels, *Doubt Is Their Product* (Oxford University Press, 2008), 25–28.

16. By 1952, British researchers had identified 341 dye workers in that country

who had developed bladder cancer since 1921—a cancer rate about thirty times higher than what would be expected among men in the general population. See Robert A. M. Case et al., "Tumors of the Urinary Bladder in Workmen Engaged in the Manufacture and Use of Certain Dyestuff Intermediates in the British Chemical Industry," *British Journal of Industrial Medicine* 11:75 (1954): 75–104.

17. In one oft-cited story, a scientist from a British dye company visiting DuPont's Deepwater factory asked his DuPont counterpart why he was so sure benzidine was not the source of the bladder cancers at the plant. The DuPont scientist reportedly replied: "We here know very well that benzidine is causing bladder cancer, but it is company policy to incriminate only the one substance, beta-naphthylamine." The 1948 conversation between the British and DuPont scientists was initially recounted in a column published in the opinion section of *The Washington Post* on July 15, 1979. The author was Barry Castleman, an environmental consultant and frequent critic of chemical industry practices, and the headline was "Another View: DuPont's Business Ethics." Castleman did not name the DuPont scientist, but he identified the British scientist as Michael Williams of Imperial Chemical Industries and claimed that Williams, who died in 1961, frequently retold the story to friends. According to Castleman, the conversation took place in a car and was witnessed by another scientist, an academic, who was thought to be dozing in the backseat but was actually awake.

Two weeks after Castleman's column appeared, the eminent British epidemiologist Robert A. M. Case wrote a lengthy letter to the editor of the *Post* identifying himself as the person in the backseat and confirming Castleman's account of the 1948 conversation. In his letter, Case explained that he was writing because, after Castleman's column was published, a DuPont representative contacted Case and asked him to confirm or deny the story. "I informed the man from Duponts [*sic*] that the tale as recounted was absolutely true," Case wrote. Castelman's column and Case's letter to the editor later caught the attention of congressional investigators. Both were republished on pages 100–103 of the official transcript of hearings on "Corporate Criminal Liability" conducted in 1979 and 1980 by the Subcommittee on Crime of the Judiciary Committee of the U.S. House of Representatives. (The Congressional Information Service Index Number for the transcript is 81-H521-54.)

18. T. S. Scott, "The Incidence of Bladder Tumors in a Dyestuffs Factory," *British Journal of Industrial Medicine* 9 (1952): 127–32.

19. Sophie Spitz, W. H. Maguigan, and Konrad Dobriner, "The Carcinogenic Action of Benzidine," *Cancer* 3:5 (September 1950): 789–804, 804.

20. Zavon, Hoegg, and Bingham, "Benzidine Exposure," 2.

21. Thomas F. Mancuso and Elizabeth Jackson Coulter, "Methods of Studying the Relation of Employment and Long-Term Illness—Cohort Analysis," *American Journal of Public Health* 49:11 (November 1959): 1525–36. In 1967, Mancuso's follow-up study of workers at the Cincinnati plant produced similar results. The incidence of fatal bladder cancer had jumped again and was now eighteen times higher than the statewide rate for men of the same age. See Thomas F. Mancuso and Anas A. El-Attar, "Cohort Study of Workers Exposed to Betanaphthylamine and Benzidine," *Journal of Occupational Medicine* 9:6 (June 1967): 277–85.

22. In an interview many years later, Mitchell Zavon explained that he and Ar-

thur Wendel did not try to publish their 1958 findings because there was already convincing evidence from Europe about benzidine's carcinogenicity and also because their study was small and lacked a control group. Thomas Mancuso's study, which used a control group, was published in 1959 but did not identify the Cincinnati Chemical Works by name. "My job was to correct the situation at that plant and make the best health possible for the workers there, and I did that," Zavon told the author. "Anybody [else] who wanted to know could have read the earlier European studies and known about benzidine and bladder cancer."

23. The two articles were Wendel, Hoegg, and Zavon, "Benzidine," and Zavon, Hoegg, and Bingham, "Benzidine Exposure."

24. In one of the most sweeping actions in its history, the Occupational Safety and Health Administration in 1973 issued tough rules on the use of fourteen carcinogenic chemicals, including benzidine and BNA. The rules were so strict—forbidding benzidine manufacturers from selling the chemical to other companies and requiring that it be handled in enclosed containers (no more open piles)—that they effectively halted benzidine use in the United States.

25. Kevin H. Ferber, William J. Hill, and Donald A. Cobb, "An Assessment of the Effect of Improved Working Conditions on Bladder Tumor Incidence in a Benzidine Manufacturing Facility," *American Industrial Hygiene Association Journal* 37 (1976): 61–68.

26. In one of her last official acts as the director of the Occupational Safety and Health Administration in 1980, Eula Bingham, who knew benzidine well from her involvement in the Cincinnati study more than twenty years earlier, issued a special "health hazard alert" stating that ortho-tolidine and ortho-dianisidine, as well as benzidine, cause cancer in lab animals and that "exposure of workers to the dyes should be reduced to the lowest feasible levels. This should include discontinuing use of the dyes where possible." See "Health Hazard Alert—Benzidine-, o-Tolidine-, and o-Dianisidine-Based Dyes," *NIOSH Publication 81-106* (December 1980), 1.

27. William Ehart, "Ciba-Geigy to Participate in Health Hazard Studies," *Ocean County Observer,* August 18, 1985.

28. Bruce W. Hills, "Industrywide Studies Report of Industrial Hygiene Surveys at the Ciba-Geigy Corporation, Toms River, New Jersey," Report 93.26 (National Institute for Occupational Safety, 1987).

Chapter Twelve

1. Don Bennett, "Ocean Groups Blame Ciba for Shelter Cove Pollution," *Ocean County Observer,* June 25, 1986.

2. The dialogue between Maria Pavlova and Frank Livelli is from *Chemical Town USA,* the WNET-Thirteen television documentary that aired November 19, 1986.

3. Don Bennett, "Ciba Doing What It Thought Was Impossible a Year Ago," *Ocean County Observer,* June 10, 1986.

4. *Status Update: Ciba-Geigy's Toms River Plant,* eight-page brochure from Ciba-Geigy, June 2, 1986.

5. John Simas, Victor Baker's successor as plant manager, later explained the decision to end chemical production this way: "We saw the writing on the wall and decided that Toms River was not a suitable site for heavy chemicals manufacture." William Lichtenstein, "The Toms River Experience," *Chemical Engineering* (April 1991): 46.

6. According to the state cancer registry, 31,920 New Jersey residents were diagnosed with cancer in 1979, the first year of mandatory reporting. By 2000, the total had risen to 45,641 cases, a 43 percent increase, even though the state's population had risen just 15 percent during that time. Improved diagnoses explain at least some of the increase in cancer incidence, though how much is unknown. The cancer totals did not include non-invasive, *in situ* cancers, in which tumors had not spread and were confined to one site in the body because those types of tumors were not reported to the registry.

7. Title 26, Section 2-104, New Jersey Statutes.

8. Donald M. Parkin, "The Evolution of the Population-Based Cancer Registry," *Nature Reviews Cancer* 6 (August 2006): 603–12, 605.

9. Lisa Roche et al., "Trends in Cancer Incidence and Mortality in New Jersey 1979–2002," New Jersey Department of Health and Senior Services (October 2005), 46–47, tables 5 and 6.

10. Marlene Monfiletto, "Toxic Waste, Cancer Incidence Raising Unanswered Questions," *Asbury Park Press,* November 16, 1986.

11. On the night before the Channel Thirteen broadcast, a local woman named April Stopa brought her copy of *TV Guide* to the meeting of the Dover Township Committee and complained that the town was about to get a black eye. "New Jersey may be thought of as one big chemical plant, but that's not what Toms River is like," she said. Mayor W. Thomas Renkin admitted that he, too, was worried and sought to downplay Ciba-Geigy's importance to the fast-growing town. See Donna E. Flynn, "Dover Officials Wary of TV Show's Publicity," *Asbury Park Press,* November 19, 1986; and Anthony A. Gallotto, "Ciba TV Show Worries Panel," *Ocean County Observer,* November 19, 1986.

Chapter Thirteen

1. Lawrence Schmidt, "Impacts and Implications of the Summer of 1987, New Jersey Floatable Incidents," published in conference proceedings of *Oceans '88: A Partnership of Marine Interests, October 31–November 2, 1988,* 790–93.

2. Sara Rimer, "After Summer of Sun, A Silent Protest," *New York Times,* September 8, 1987.

3. Janet Picknally, "1,000 Protest Ciba Pipeline," *Ocean County Observer,* September 6, 1987. A few months later, Frank Livelli took his bullhorn to New York City and picketed the Swiss consulate.

4. For more about Lawrence Bathgate, see Ronald Brownstein, "Raising Bucks for Bush," *New York Times Magazine,* May 17, 1987; Peter Overby, "N.J. Mogul Revs Up the Money Machine, Bathgate Is Buddy to Bush, Kean," *Bergen Record,* August 18, 1988; and Chris Conway, "Bank Lowers the Boom on GOP's Top Financier," *Philadelphia Inquirer,* June 17, 1991.

5. Richard McDonald, "Lead Readings May Force Dover to Shut Family's Well," *Ocean County Observer,* January 25, 1988.

6. Among the eighty-nine Cincinnati veterans who had moved to the Toms River plant in 1960, there were seventeen cancer deaths—twice as many as expected.

7. A. J. McMichael, an epidemiologist at the University of North Carolina School of Public Health, first coined the term "healthy worker effect" in 1975 in a study of mortality rates at a rubber factory. For more about the healthy worker effect, see A. J. McMichael, "Standardized Mortality Ratios and the 'Healthy Worker Effect': Scratching Beneath the Surface," *Journal of Occupational Medicine* 18:3 (March 1976): 165–68; C. Y. Lee and E. C. Sung, "A Review of the Healthy Worker Effect in Occupational Epidemiology," *Occupational Medicine* 49:4 (1999): 225–29; and Timothy Wilcosky and Steve Wing, "The Healthy Worker Effect," *Scandinavian Journal of Work, Environment, and Health* 13 (1987): 70–72.

8. Elizabeth Delzell, Maurizio Macaluso, and Philip Cole, "A Follow-Up Study of Workers at a Dye and Resin Manufacturing Plant," *Journal of Occupational Medicine* 31:3 (March 1989): 276–77.

9. Richard McDonald, "Union Won't Sit Still for Outbursts," *Ocean County Observer,* March 20, 1988.

10. In addition to being hazardous, trichloroethylene is also devilishly difficult to clean up. The usual ways of cleaning up underground spills, by digging up contaminated soil or pumping up tainted groundwater, work poorly for compounds like TCE, perchloroethylene, and vinyl chloride. These dense solvents dissolve very slowly in groundwater but sink to the bottom of the water table, like the oil at the bottom of an unshaken bottle of salad dressing. As a result, the solvents, which can travel for miles in groundwater without degrading much, are treacherously difficult to pump up through recovery wells. Hydrologists call them dense non-aqueous phase liquids, or DNAPLs. Wherever DNAPLs turn up, including in Toms River, multiyear, multimillion-dollar cleanups follow.

11. See three stories by Anthony A. Gallotto in the *Ocean County Observer:* "Taint Seen in Well of TR Water," November 11, 1987; "Water Co. Followed Procedures," November 15, 1987; and "DEP Has Done Little to Find Source of TR Water Taint," November 17, 1987.

12. Lauren Ascione, "Ciba Indicted Again by Grand Jury," *Ocean County Observer,* February 17, 1988.

13. Gillick, *For the Love of Mike,* 109–12.

14. U.S. Environmental Protection Agency, *EPA Superfund Record of Decision: Ciba-Geigy Corp., Operating Unit 1,* April 24, 1989, 31.

15. Don Bennett, "DEP Rejects Ciba's Plant Plans," *Ocean County Observer,* October 25, 1988.

16. Patricia A. Miller, "Ciba-Geigy Has No Plans to Expand Its Dover Plant," *Asbury Park Press,* April 25, 1991.

Chapter Fourteen

1. This chapter's brief sketch of the life and work of Theodor Boveri is drawn from the following articles: Fritz Baltzer, "Theodor Boveri," *Science* 144 (May 15,

1964): 809–15; Ulrich Wolf, "Theodor Boveri and His Book *On the Problem of the Origin of Malignant Tumors,*" in *Chromosomes and Cancer,* ed. James German (John Wiley and Sons, 1974), 1–20; and Thomas Ried, "Homage to Theodor Boveri (1862–1915): Boveri's Theory of Cancer as a Disease of the Chromosomes, and the Landscape of Genomic Imbalances in Human Carcinomas," *Environmental and Molecular Mutagenesis* 50 (2009), 593–601.

2. "Taking everything into consideration, I believe that the essential point can finally be approached," Boveri wrote in a 1901 letter to a scientific colleague. "I feel beyond any doubt that the individual chromosomes must be endowed with different qualities, and that only certain combinations permit normal development." See Wolf, "Theodor Boveri and His Book," 7.

3. Theodor Boveri's book on cancer, published in 1914, was called *Zur Frage der Entstehung Maligner Tumoren* (On the Problem of the Origin of Malignant Tumors).

4. As cancer geneticist Allan Balmain pointed out almost ninety years later, if a modern reader substitutes "gene" for "chromosome" (Boveri had no way of seeing the genetic machinery tucked inside chromosomes), Boveri in 1914 managed to foretell the identification of oncogenes, tumor-suppressor genes, and even telomeres, the protective "caps" at the ends of chromosomes that shorten each time a cell divides in an aging organism, increasing the risk of cancer. See Allan Balmain, "Cancer Genetics: From Boveri and Mendel to Microarrays," *Nature Reviews Cancer* 1 (October 2001): 77–82.

5. "The connection between cancer and certain chemical irritants is ever clearer than it is between cancer and the physical agents I have mentioned. I need only refer to the cancers of paraffin works," Theodor Boveri wrote. Boveri, *Zur Frage der Entstehung Maligner Tumoren,* trans. Henry Harris, reprinted in *Journal of Cell Science* 121, Supp. 1 (2008): 1–84.

6. This chapter's brief description of Hermann J. Muller and his work is drawn from the following articles: Guido Pontecorvo, "Hermann Joseph Muller, 1890–1967," *Biographical Memoirs of the Fellows of the Royal Society* 14 (November 1968): 349–89; Daniel J. Kevles, "Hermann J. Muller," *Science* 214 (December 11, 1981): 1232–33; and Tove Mohr, "Hermann J. Muller, 1890–1967: An Appreciation by a Friend," *Journal of Heredity* (1972): 132–34.

7. Hermann Muller once attempted suicide, via an overdose of sleeping pills, after hearing that a rival was awarded a Nobel Prize. The next day a search party found him, dazed, in the woods near the University of Texas.

8. Thomas Edison, who developed one of the first X-ray imaging machines, or fluoroscopes, was also one of the first to notice that long-term, direct exposure to X-rays often led to cancer. One of his assistants, a glassblower named Clarence Madison Dally who had helped Edison invent the incandescent lightbulb, developed tumors and severe radiation burns on his hands from demonstrating the fluoroscope. His left hand and four right fingers were amputated, prompting Edison to declare that he would no longer work with X-rays. "Don't talk to me about X-rays, I am afraid of them," he told the *New York World* in an article published August 31, 1903. Dally died the following year from metastatic cancer.

9. Hermann J. Muller, "Artificial Transmutation of the Gene," *Science* 66: 1699 (July 22, 1927): 84–87.

10. Hermann J. Muller, "Time Bombing Our Descendants," *American Weekly,* January 3, 1948.

11. Hermann J. Muller, "Radiation Damage to the Genetic Material," *American Scientist* 38:1 (January 1950): 35–59.

12. Peter Armitage and Richard Doll, "The Age Distribution of Cancer and a Multi-Stage Theory of Carcinogenesis," *British Journal of Cancer* 8:1 (March 1954): 1–12.

13. Alfred G. Knudson, "Mutation and Cancer: Statistical Study of Retinoblastoma," *Proceedings of the National Academy of Sciences* 68:4 (April 1971): 820–23.

14. Alfred Knudson provides an insightful historical overview of research into mutation and cancer in "Two Genetic Hits (More or Less) to Cancer," *Nature Reviews Cancer* 1 (November 2001): 157–62. See also Ezzie Hutchinson, "Alfred Knudson and His Two-Hit Hypothesis," an interview of Knudson, *Lancet Oncology* 2 (October 2001): 642–45.

15. The Ames test works by using Salmonella bacteria that have mutated and lost their ability to produce the amino acid histidine, without which the bacteria cannot grow. The tester applies the mutated Salmonella to a medium that includes a small amount of histidine as well as the chemical being tested. The bacteria will grow for a short while until the histidine is consumed but then will stop—unless the test chemical causes the bacteria's DNA to mutate back into a form that can produce its own histidine. If the chemical is highly mutagenic, the Salmonella will mutate and thrive even in a histidine-free medium. Some versions of the test involve using bacteria with defective outer coats, which makes them more vulnerable to the chemicals being tested. Sometimes liver enzymes are added to the culture medium to simulate the effects of human metabolism, which often alters the molecular structure of synthetic chemicals after ingestion.

16. The identification of mutagenic compounds is not always a clear-cut process and is heavily dependent on the assay being used. The compounds identified as mutagenic in this chapter are listed in two government sources: Walter W. Piegorsch and David G. Hoel, "Exploring Relationships between Mutagenic and Carcinogenic Potencies," National Institute of Environmental Health Sciences, *Mutation Research* 196 (1988): 161–75, 167–69, table 2; and "2007 Right to Know Special Health Hazardous Substance List: Mutagens," available online from the New Jersey Department of Health and Senior Services.

17. United States General Accounting Office, *Superfund Program: Current Status and Future Fiscal Challenges,* July 31, 2003.

18. The author was able to identify and interview Lisa Boornazian thanks to Ellen Tracy, the director of oncology nursing at the Children's Hospital of Philadelphia. In 2007, the author contacted Tracy, who had worked in the oncology ward for many years, to ask if she knew the identity of the nurse who had anonymously initiated the Toms River investigation more than twelve years earlier. She did. Tracy agreed to find the nurse and to ask her if she was willing to consent to an interview. Lisa Boornazian gave her consent, having decided there was no longer any reason to remain anonymous because so much time had passed and she no longer worked at the hospital. Her sister-in-law, Laura Janson, who still works in the EPA's Philadelphia office, felt the same way.

Chapter Fifteen

1. In America, the New York tycoon John Jacob Astor III built the country's first cancer hospital in 1887 after his earlier effort to fund a cancer wing at an existing hospital was rejected on the grounds that it might threaten other patients with infection. See James T. Patterson, *The Dread Disease: Cancer and Modern American Culture* (Harvard University Press, 1987), 23.

2. The famous member of the Webb family was Thomas's younger brother, Captain Matthew Webb, who in 1875 became the first person to swim the English Channel without floats or other artificial aids. Victorian-era matchboxes displayed his picture. Webb drowned in 1883 while trying to swim across the swirling rapids just below Niagara Falls in New York. His brother Thomas Law Webb, despite his contribution to the field of biostatistics, earns but one line in his brother's two-page entry in the 2001 edition of the *Dictionary of National Biography* (60:104–5).

3. For more about the life and work of Karl Pearson, see Helen W. Walker, "The Contributions of Karl Pearson," *Journal of the American Statistical Association* 53:281 (March 1958): 11–22; and M. Eileen Magnello, "Karl Pearson and the Origins of Modern Statistics: An Elastician Becomes a Statistician," *Rutherford Journal* 1 (December 2005).

4. For a concise, if dated, recounting of the history of significance testing and the long-running argument over its importance, see Ronald N. Giere, "The Significance Controversy," *British Journal for the Philosophy of Science* (May 1972): 170–81.

5. Karl Pearson, "On 'Cancer Houses,' from the Data of the late Th. Law Webb, M.D.," *Biometrika* (January 1912): 430–35. Pearson cofounded *Biometrika* in 1901 after the Royal Society refused to publish articles on biostatistics, deeming the field insufficiently scientific. To test how likely it was that the 377 cancer deaths in Madeley were distributed randomly, Karl Pearson conceived of the following experiment: If a bag were filled with two thousand balls, numbered one through two thousand, and someone then drew a ball out of the bag 377 times, recording the ball's number each time before returning it to the bag and picking again, at the end of the experiment how many numbers would be written down twice, or even three or more times? Writing numbers on two thousand balls would be a tedious task, so Pearson designed some less burdensome experiments to answer the question through the use of random numbers and the drawing of playing cards. He conducted five such experiments, and each time found that the distribution was very close to what he had predicted. In only one of the five experiments was a number selected more than twice, and it was never picked four times. But in Madeley, there were *six* houses with three cancer cases and one with four, an extremely unlikely result for a chance distribution.

6. Karl Pearson wrote, "Dr. Law Webb's data provide sufficient evidence to justify a demand for a thorough investigation of the subject, such as is not feasible in the case of the individual medical man. They do not finally demonstrate that cancer is more frequent in one house than a second, but they do justify a complete inquiry into the possibility that 'cancer-houses' are not wholly a myth, in other words, that immediate environment is in the long run a factor of the frequency of cancer." See Pearson, "On 'Cancer Houses,'" 434.

7. In 1932, two French statisticians, Auguste Lumière and Paul Vigne, analyzed 6,703 cancer deaths in 5,027 homes in the city of Lyon over a twenty-year period, including one home in which there had been eight deaths. Instead of trying to use a statistical formula to estimate a normal distribution of cases, as Pearson had, Lumière and Vigne instead obtained 6,703 birth, marriage, and death certificates for the city and studied the distribution of addresses on the certificates. What they found was that births, marriages, and deaths in Lyon clustered just as much as cancer did. Their discovery strongly suggested, the Frenchmen concluded, that mere chance was at work, not a hidden cause. A description of Lumière and Vigne's work in Lyon appears in Percy Stocks, "The Frequency of Cancer Deaths in the Same House and in Neighbouring Houses," *Journal of Hygiene* (February 1935): 46–63. The Lyon study is also described on page 10 of the May 5, 1933, edition of *Science* entitled "Cancer House Disproved Statistically."

8. Stocks, "Frequency of Cancer Deaths," 46–63.

9. At the time, the CDC was known as the Communicable Disease Center. Since then, as its portfolio has expanded, its name has changed four times: National Communicable Disease Center in 1967, Center for Disease Control in 1970, Centers for Disease Control in 1980, and Centers for Disease Control and Prevention in 1992.

10. Clark W. Heath Jr. and Robert J. Hasterlik, "Leukemia Among Children in a Suburban Community," *American Journal of Medicine* 34 (June 1963): 796–812.

11. Sister Mary Viva, the school principal in Niles, Illinois, did give investigator Clark Heath one bit of tantalizing information: During the same years the leukemia cases were diagnosed, several waves of a contagious fever had swept through the school. There were no reliable records of the outbreaks, but Heath's conversations with the school nurses suggested that they coincided with the months in which several of the leukemia cases were diagnosed. The researchers also found antibodies that might be related to leukemia in the blood samples of several relatives of the victims. To Heath, this was a hint that the same virus that had spread the fever might also have triggered leukemia in a few genetically susceptible children. As Alfred Knudson would have put it, the infection might have provided the second "hit."

Fifty years later, researchers are still searching for a virus associated with acute lymphocytic leukemia in children. Their efforts have been spurred by two closely related hypotheses suggested by studies of childhood leukemia clusters. The "population mixing" hypothesis was first proposed in 1988 by University of Oxford epidemiologist Leo Kinlen, who was seeking to explain a cluster of childhood leukemia near a nuclear power plant. Kinlen suggested that ALL may arise from an extreme immune response to a viral infection. Though very rare, these extreme reactions would most likely occur in fast-growing communities in which infected populations mix with nonimmune children who are highly susceptible. Another eminent British researcher, biologist Melvyn Greaves of the Institute of Cancer Research in London, has expanded on Kinlen's ideas, as well as Knudson's, by proposing the "delayed infection" hypothesis. Greaves and his frequent collaborator, Joseph Wiemels of the University of California at San Francisco, believe that the first "hit" occurs in utero, while the second involves a communicable infection that most children easily suppress but may trigger leukemia between the ages of two and five in a very small number of highly susceptible children. (These rare children, Greaves suggests, are less likely to have been in day care, where exposure to other children would have built up their immune systems.)

In an interview with the author, Greaves said that it is possible that the initial hit during pregnancy could be from exposure to a carcinogenic compound but noted that the dose would have to be large enough to trigger genetic damage. (Greaves is especially interested in a translocation called TEL-AML1 often found in ALL patients but also present in many healthy children.) For a current assessment of the population mixing and delayed infection hypotheses, see Kevin Urayama et al., "A Meta-Analysis of the Association between Day-Care Attendance and Childhood Lymphoblastic Leukaemia," *International Journal of Epidemiology* 39 (2010): 718–32.

12. Clark W. Heath, "Community Clusters of Childhood Leukemia and Lymphoma: Evidence of Infection?" *American Journal of Epidemiology* 162:9 (November 1, 2005): 817–22. After leaving the Centers for Disease Control, Heath was a vice president of the American Cancer Society, where his generally skeptical views about environmental causation of cancer would raise the ire of activists.

13. By the mid-1980s, thirty-seven states had established cancer registries or were in the process of doing so. See Stephanie C. Warner and Timothy E. Aldrich, "The Status of Cancer Cluster Investigations Undertaken by State Health Departments," *American Journal of Public Health* 78:3 (March 1988): 306–7.

14. Michael Greenberg and Daniel Wartenberg, "Communicating to an Alarmed Community about Cancer Clusters: A Fifty State Survey," *Journal of Community Health* 16:2 (April 1991): 71–82. Greenberg and Wartenberg, of Rutgers University, surveyed all fifty state health departments in 1989 and concluded that they were getting a total of 1,300 to 1,650 requests for cluster investigations per year.

15. For more about "occult multiple comparisons," see Raymond Richard Neutra, "Counterpoint from a Cluster Buster," *American Journal of Epidemiology* 132:1 (July 1990): 1–8.

16. This kind of *post hoc* hypothesis generation is sometimes called Texas sharpshooting, after a joke in which a cowboy wildly fires shots into the side of a barn, then looks to see where the bullet holes have randomly clustered and draws a target centered on the densest cluster, declaring himself a sharpshooter.

17. There is, arguably, at least one exception to the universal failure of residential cluster analysis in identifying new carcinogens. The mineral erionite was identified as a lung carcinogen in the 1980s based on studies of two villages in central Turkey in which there were very high rates of mesothelioma and also high concentrations of airborne erionite fibers. As Raymond Neutra points out in "Counterpoint from a Cluster Buster," however, the erionite example is not a conventional time-limited cluster because lung cancer has apparently been endemic in those villages for centuries.

18. Glyn G. Caldwell, "Twenty-Two Years of Cancer Cluster Investigations at the Centers for Disease Control," *American Journal of Epidemiology* 132, Supp. 1 (1990): S43–S47.

19. From 1978 to 1984, the National Institute for Occupational Safety and Health investigated sixty-one alleged workplace cancer clusters and identified a

plausible cause in just five, none of which could be scientifically confirmed. In forty-six of those clusters, NIOSH found, rates were not actually higher than expected. See Paul A. Schulte et al., "Investigation of Occupational Cancer Clusters: Theory and Practice," *American Journal of Public Health* 77:1 (January 1987): 52–56.

20. Phil Brown, "Popular Epidemiology and Toxic Waste Contamination: Lay and Professional Ways of Knowing," *Journal of Health and Social Behavior* 33 (September 1993): 267–81.

21. Kenneth Rothman's keynote address to the "Cluster Buster" conference in February of 1989 was later published in a special issue of the *American Journal of Epidemiology* that also included twenty-five other papers presented at the same conference. Rothman's contribution, entitled "A Sobering Start for the Cluster Busters' Conference" is on pages S6–S13 of the special issue, which was published in July of 1990 (vol. 132, Supp. 1). The entire issue serves as an excellent guide to the perils and promise of investigating disease clusters, including cancer clusters.

22. Atul Gawande, "The Cancer Cluster Myth," *The New Yorker,* February 8, 1998, 34–37.

23. In addition to the 75 percent who were satisfied after an explanatory telephone conversation or letter, another 20 percent of cluster callers in Minnesota, according to Alan Bender, could be mollified by supplying them with preexisting data about cancer rates and environmental conditions. "Relatively modest" data collection and analysis took care of another 4 percent. Only 1 percent of the time did Minnesota get to step four: an in-depth investigation. Bender's department conducted only five of those between 1981 and 1988, a period during which it had fielded 420 telephone calls or letters about clusters. See Alan P. Bender et al., "Appropriate Public Health Reponses to Clusters: The Art of Being Responsibly Responsive," *American Journal of Epidemiology* 132, Supp. 1 (July 1990): S48–S52.

24. In its "Guidelines for Investigating Clusters of Health Events," published in final form on July 27, 1990, the Centers for Disease Control declared: "The unofficial consensus among workers in public health is that most reports of clusters do not lead to a meaningful outcome."

25. Craig W. Trumbo, "Public Requests for Cancer Cluster Investigations: A Survey of State Health Departments," *American Journal of Public Health* 90:8 (August 2000): 1300–1302.

26. The 1994 report, entitled *Childhood Cancers in New Jersey: 1980–1988,* is no longer available on the state health department's website. The revised report, *Childhood Cancer in New Jersey: 1979–1995,* was published in 1999, in the midst of the Toms River cancer furor, and does not include the county-to-county comparisons that incited local activists.

Chapter Sixteen

1. The terms of the 1992 legal settlement that resolved the criminal charges against Ciba-Geigy also required the company to put $50 million into a trust fund for environmental cleanup at the factory site—money it was already obligated to pay under the Superfund program.

2. Patricia A. Miller, "Ciba Guilty of Pollution," *Asbury Park Press,* February 29, 1992.

3. From the river, the Toms River "core zone" extended north for two miles to State Road 571. Its eastern border was Vaughn Avenue, its western edge an irregular line that included parts of the Garden State Parkway and Old Freehold Road.

4. The cancer incidence rate for United States children under age twenty was fourteen cases per one hundred thousand per year in 1995, according to the National Cancer Institute's Surveillance, Epidemiology, and End Results (SEER) program. By 2005, the rate had risen to sixteen cases per one hundred thousand.

5. This small-numbers problem was less severe in the township, where there were about twenty thousand children, and in Ocean County, where there were one hundred thousand.

6. The data tables in this chapter are based on tables produced by Michael Berry for his 1995 incidence analysis. They were later reproduced in Appendix 1 of *Childhood Cancer Incidence Health Consultation: A Review and Analysis of Cancer Registry Data, 1979–1995 for Dover Township,* New Jersey Department of Health and Senior Services (December 1997).

7. Michael Berry's cancer rates for all New Jersey children were based on the average of three years of statewide incidence data, 1986 through 1988. He did not have access to statewide data for any other years. Berry also needed to account for the fact that the year-by-year age distributions of the childhood population of the county, town, and core zone differed slightly from each other and from the state as a whole. In the fast-growing township, for example, the average age of the total population of children under age twenty was slightly lower than the statewide average. That was significant because certain cancers were much more common in younger children than teenagers, while other kinds tended to be found in older children. So, in order to calculate an accurate "expected" number of cases, Berry had to adjust his estimates to reflect the local age distributions. He did so by subdividing the childhood population into groupings of ages zero to four, five to nine, ten to fourteen, and fifteen to nineteen. Then he calculated what percentage of the overall childhood population in the Toms River core, Dover Township, Ocean County, and the entire state fell into each of those subgroups. Finally, he adjusted the statewide figures so that the "expected" number of cases reflected the specific age distributions of the county, the town, and the core zone.

8. Fabio Barbone's doctoral thesis, *A Nested Case-Control Study of Lung Cancer and Central Nervous System Neoplasms Among Chemical Workers* (1989), was later republished in the form of two journal articles: Fabio Barbone et al., "A Case-Control Study of Lung Cancer at a Dye and Resin Manufacturing Plant," *American Journal of Industrial Medicine* 22 (1992): 835–49; and Fabio Barbone et al., "Exposure to Epichlorohydrin and Central Nervous System Neoplasms at a Resin and Dye Manufacturing Plant," *Archives of Environmental Health* 49:5 (September 1, 1994): 355–58.

9. Don Bennett, "Widow Sues Ciba over Cancer," *Ocean County Observer,* July 27, 1995.

10. The reinjection plan ended up costing tens of millions of dollars less than Ciba had initially feared, thanks to an EPA decision not to require extra treatment to

remove salts from the treated wastewater since it was not being directly discharged into the river—even though the reinjection zone was so close to the river that much of the treated water seeped into the river anyway. See Don Bennett, "EPA Flip-Flops on Ciba Water," *Ocean County Observer,* May 14, 1993; and Richard Peterson, "Ciba Proposal Gets Praise from Activists, Neighbors," *Ocean County Observer,* June 18, 1993.

11. The "Ex-Pariah" headline was published May 25, 1995, in the *Observer* and referred to an award Ciba received from the United Nations–affiliated World Environment Center. The editorial, published June 6, 1995, included this line: "Fortress Ciba, where officials once arrogantly insisted that everything was pristine, is now openly encouraging participation by neighbors in its affairs."

12. Craig Wilger, January 7, 1993, letter to Gary Adamkiewicz of the U.S. Environmental Protection Agency, 2.

13. U.S. Environmental Protection Agency, *EPA Superfund Record of Decision: Reich Farms,* September 30, 1988, 14.

14. The name change from Toms River Water to United Water Toms River was prompted by a 1994 merger between its parent company, Philadelphia-based General Waterworks, and one of the largest water companies in the United States: United Water Resources, based in Harrington Park in northeastern New Jersey.

15. U.S. Environmental Protection Agency, "Responsiveness Summary: Reich Farms Superfund Site" (1995). See also "Dover Water Cleanup Plan May Be Altered," *Asbury Park Press,* August 17, 1995.

16. Michael Berry's August 31, 1995, letter to Steven Jones was later reproduced as Appendix I of *Childhood Cancer Incidence Health Consultation: A Review and Analysis of Cancer Registry Data, 1979–1995 for Dover Township, New Jersey* (December 1997), 67.

17. Though it was a key moment in the Toms River saga, Michael Berry and Robert Gialanella have different recollections as to why Berry sent Gialanella a copy of the 1995 analysis Berry conducted of childhood cancer incidence in Toms River. Berry thinks that Gialanella called him to express renewed concern about childhood cancer in Toms River, prompting Berry to send him the newly completed analysis. Gialanella, however, says he has a clear memory of the analysis being sent to him without his having requested it, and says that by 1995 he had given up on asking for help from the state. (Gialanella had requested a similar investigation in 1991, with ambiguous results.) Berry, who was involved in dozens of cluster investigations during his long tenure at the state health department, acknowledges that he is not sure.

Chapter Seventeen

1. Gale Scott, "Cluster of Concern: Kids' Cancer Rate Alarms County," *Star-Ledger,* March 10, 1996.

2. Brett Pulley, "State to Study Ocean County over Cancer," *New York Times,* March 12, 1996.

3. Elin Gursky was quoted in a March 21, 1996, broadcast on *The CBS Evening News.*

4. The description of the March 15 protest at the Ocean County Health Department is drawn from three sources: a story broadcast on March 21 on *The CBS Evening News;* Richard Peterson, "High Anxiety; Local Cancer Scare Spins Out of Control; Parents Demand Answers," *Ocean County Observer,* March 17, 1996; and Carol Gorga Williams, "Officials Call in Vain for Patience," *Asbury Park Press,* March 16, 1996.

5. As physician Henry M. Vyner explained in his influential 1988 book *Invisible Trauma,* chemical pollutants were singularly terrifying because they were invisible environmentally and medically: They could not be detected without special equipment, and even the best-informed experts could not reliably predict their health effects. As a result, residents of a singled-out community could never know whether they were in a truly dangerous situation and whether the appearance of an illness like cancer was due to a past toxic exposure. See Henry M. Vyner, *Invisible Trauma* (D.C. Heath, 1988), ch. 3.

6. Vyner, *Invisible Trauma,* 190–91.

7. The description of the March 21, 1996, meeting at Toms River High School North is based on interviews with people who were there as well as taped footage broadcast May 5 on *The NBC Nightly News* and three newspaper accounts: Michael S. Yaple, "Residents Air Fears About Cancer Cluster," *The Press of Atlantic City,* March 22, 1996; Carol Gorga Williams, "1,000 Hear State's Cancer-Study Plan," *Asbury Park Press,* March 22, 1996; and Gale Scott and John Hassell, "Fear and Recrimination Engulf Cancer Forum," *Star-Ledger,* March 22, 1996.

8. Paul Mulshine, "Ignorance Is the Real Cancer in Phony Ocean County Scare," *Star-Ledger,* March 31, 1996.

9. U.S. Environmental Protection Agency, *Twenty-Five Years of the Safe Drinking Water Act: History and Trends* (December 1999), 5, fig. 2. By 2011, the number of contaminants mandated for testing under the Safe Drinking Water Act had risen only slightly, to eighty-eight.

10. New Jersey Department of Health and Senior Services, New Jersey Department of Environmental Protection, U.S. Agency for Toxic Substances and Disease Registry, *Public Health Consultation: Drinking Water Quality Analyses, March 1996 to June 1999, United Water Toms River* (2000), 110–11, table 9a. Drinking water at ten schools exceeded the five picocuries per liter regulatory standard, and four more were over it if margins of error were included.

11. *Public Health Consultation: Drinking Water Quality Analyses, March 1996 to June 1999, United Water Toms River,* 113–16, table 9b.

12. A New Jersey Department of Environmental Protection scientist, Bahman Parsa, eventually realized that United Water's privately tested water samples typically sat for weeks before reaching the laboratory—enough time for the chief culprit, Radium-224, to decay to low levels. (Radium-224 has a half-life of just three and a half days.) Later, the DEP would discover similarly high levels of Radium-224 in groundwater almost everywhere it looked in central and southern New Jersey, as long as it analyzed the sample within two days. The state would institute a two-day turnaround requirement for radiation testing, and several utilities in the region—including United Water Toms River—would spend years and millions of dollars trying to bring down radioactivity levels in their drinking water.

13. Ted Sherman, "2d Water Test Shows Acceptable Radium Reading," *Star-Ledger,* April 19, 1996.

Chapter Eighteen

1. In addition to the radiation issues, the state's initial checks of Toms River's water system showed that six schools had relatively high levels of lead or copper, a common problem in buildings with old pipes. As expected, the state also found trichloroethylene in three Parkway wells at levels slightly over the safety limit, but the air stripper was removing almost all of it before the water was distributed to residents.

2. "Water Tests Find No Link with Cancer," *New York Times,* May 9, 1996.

3. Another half-century would pass before Danish physicist Niels Bohr explained why atoms and molecules have unique spectral signatures. Bohr recognized that spectral lines were generated by the release or absorption of photons—particles of light with distinctive, measurable wavelengths—as the electrons orbiting an atomic nucleus jumped predictably from one discrete orbit (or energy level) to another in atoms that had been put into an excited state by the flame of a Bunsen burner or some other catalyst. When an atom absorbs energy from the burner, one or more of its electrons will jump to a higher orbit. And when the atom returns to an unexcited ground state, it will release that energy in the form of photons at specific spectral wavelengths that can be measured on a detector as lines of light.

4. This chapter's account of the sequence of events that led to the identification of styrene acrylonitrile trimer in Toms River's water is based on interviews with, and documents provided by, Floyd Genicola and Jerald Fagliano of the New Jersey Department of Health and Senior Services; Gerald Nicholls and Eric Rau, formerly of the New Jersey Department of Environmental Protection; and G. Wayne Sovocool of the U.S. Environmental Protection Agency. Although they had many disagreements over how best to proceed in 1996, their recollections many years later of what had transpired were quite consistent. Readers interested in a detailed scientific account of the identification process should consult Andrew H. Grange et al., "Identification of Pollutants in a Municipal Well Using High Resolution Mass Spectrometry," *Rapid Communications in Mass Spectrometry* 12 (1998): 1161–68. The process is also explained in Susan D. Richardson et al., "Identification of Drinking Water Contaminants in the Course of a Childhood Cancer Investigation in Toms River, New Jersey," *Journal of Exposure Analysis and Environmental Epidemiology* 9 (1999): 200–216.

5. In 1805, the British Quaker chemist John Dalton published a primitive but highly accurate table of six elements—including hydrogen, the lightest, to which he assigned a weight of one. He gave oxygen a weight of 16 because its mass was 16 times greater than hydrogen. More than a century later, using a very similar principle, Francis Aston helped develop the modern scale of atomic and molecular mass, which is now named after Dalton. The subtle difference is that the fundamental unit of the modern scale, one dalton, is equivalent to one-twelfth of the mass of a carbon-12 atom, which turns out to be nearly identical but slightly lighter than the

mass of a hydrogen atom (1.00794 daltons). The same subtle differences apply to other elements, which is why oxygen is now considered to have an atomic weight of 15.9994 daltons.

6. Years later, Gerald Nicholls would remember his battles with Floyd Genicola in intricate detail: "Floyd had a history with the department," Nicholls recalled. "He was largely viewed by his colleagues as very intelligent and a good analytical chemist. In fact, I can't remember a time when he was ever proven wrong about an analysis. But he also had the reputation of being a loose cannon, and of pursuing things well beyond what we could argue for scientifically and logically."

7. Sociologist Phil Brown coined the term *popular epidemiology* to describe the very different process by which nonexperts gather information about environmental risks to their health and seek solutions. Phil Brown, "Popular Epidemiology and Toxic Waste Contamination: Lay and Professional Ways of Knowing," *Journal of Health and Social Behavior* 33 (September 1992): 267–81.

8. New Jersey State Department of Health and Senior Services, *Cancer Incidence in New Jersey, 1995–1999* (September 2001), tables 1, 2.

9. Agency for Toxic Substances and Disease Registry, *Public Health Consultation: Toms River General Post Office* (December 1998).

10. The spectrogram actually showed 211 daltons, not 210, but a quirk of isobutane chemical ionization is that it yields a result with one extra hydrogen ion, with an atomic mass of 1.

11. The fax was Floyd Genicola's last interaction with anyone at Radian International. As soon as Union Carbide managers heard about his call to the contractor, they complained to Genicola's supervisors, who ordered him to stop communicating with Radian.

12. For a brief summary of styrene acrylonitrile's uses and environmental properties, see the National Toxicology Program's background documents on styrene acrylonitrile trimer, available at the NTP website. In consumer products, styrene acrylonitrile polymer was eventually replaced by acrylonitrile butadiene styrene, or ABS plastic, in which the same two compounds—styrene and acrylonitrile—are bonded in the presence of polybutadiene, a synthetic rubber. ABS plastic is less toxic and is widely used in products such as Lego toy blocks, plastic clarinets, and certain auto parts and plumbing pipes.

13. Acrylonitrile is classified as "reasonably anticipated to be a human carcinogen" in the *Eleventh Report on Carcinogens,* published in 2005 by the National Toxicology Program, because it causes stomach and central nervous system cancers in rats. It has also been associated with lung cancer in textile workers in some studies. The *Twelfth Report on Carcinogens,* issued in 2011, for the first time listed styrene as "reasonably anticipated to be a human carcinogen" for causing lung tumors in mice and increasing incidence of leukemia and lymphoma in factory workers in the plastics and rubber industries. The International Agency for Research in Cancer classifies both styrene and acrylonitrile as possibly carcinogenic to humans based on animal studies and lists acrylonitrile as a genotoxic mutagen that damages DNA in animal studies. For styrene, the evidence of mutagenicity is mixed, according to the IARC.

14. Later, styrene acrylonitrile trimer would also be found in soil at the town landfill, which made sense because Nick Fernicola had briefly dumped Union Carbide waste there, too.

Chapter Nineteen

1. Jerald Fagliano, "A Case-Control Study of Childhood Brain Cancer and Drinking Water Contamination," unpublished dissertation (The Johns Hopkins University, 1998).

2. Peter J. Murphy, *Exposure to Wells G and H in Woburn Massachusetts,* report produced under contract to the Massachusetts Health Research Institute, the Massachusetts Department of Public Health, and the U.S. Centers for Disease Control (August 1990).

3. Kevin Costas, Robert S. Knorr, and Suzanne K. Condon, "A Case-Control Study of Childhood Leukemia in Woburn, Massachusetts: The Relationship between Leukemia Incidence and Exposure to Public Drinking Water," *Science of the Total Environment* 300 (December 2002): 23–35, 29–30, tables 3, 4.

4. Suzanne Condon of the Massachusetts Department of Public Health, quoted in Richard Saltus, "Woburn Poisons' Route Mapped; Study Blames Water for Leukemia," *Boston Globe,* May 10, 1996.

5. Gale Scott, "Dover Twp. Childhood Cancers Alarmingly High to Feds; Study Will Check Link to Water Supply Toxins," *Star-Ledger,* January 24, 1997.

6. The third defendant in the Woburn case, a uniform launderer called UniFirst, had less potential liability than W.R. Grace or Beatrice Foods and settled early in the case for slightly more than $1 million.

7. Patricia A. Miller, "Officials Quizzed on Water in Dover," *Asbury Park Press,* July 15, 1997.

8. "Requests to Conserve Water, Power Ignored by Many in N.J.," *Associated Press,* July 16, 1997.

9. Jean Mikle, "Cancer Cluster Confirmed; Federal, State Studies Verify High Rates in Dover," *Asbury Park Press,* April 8, 1997.

10. Paul Kix, "In the Shadow of Woburn," *Boston Magazine,* September 22, 2009.

11. Paula Span, "One Man's Poisons; Jan Schlichtmann's Last Pollution Case Turned into a Toxic Obsession. Now He's Digging for the Truth from a Different Angle," *Washington Post,* February 22, 1999.

12. Michael Gillick, who was diagnosed in the spring of 1979, barely met the criteria for inclusion in the legal case. Randy Lynnworth did not because he had been diagnosed in 1982; his parents had reached a separate legal settlement with Ciba.

Chapter Twenty

1. Patricia A. Miller and Jean Mikle, "Lawyer Vows Answers for Families Hit by Cancer," *Asbury Park Press,* December 12, 1997.

2. In an interview, Robert Butler said his memory of the confrontation differed from Esther Berezofsky's account but, citing the confidentiality of the discussions, would not provide specifics.

3. Lois Gibbs, the onetime leader of the Love Canal protests who was now a full-time environmental activist in Washington, D.C., had recommended Richard Clapp to Linda Gillick.

4. Frederick Kunkle, "Activist Cites Five New Kid Cancers," *Star-Ledger,* May 14, 1998. See also Patricia A. Miller, "Utility's Aim: Keep Two Wells Off Line," *Asbury Park Press,* May 19, 1998.

5. Patricia A. Miller, "Cancer Scare 'Hysteria' Irks C of C Leader," *Asbury Park Press,* May 21, 1998.

6. Angela Stewart, "State to 'Aggressively' Pursue Report of New Child Cancer Cases," *Star-Ledger,* May 15, 1998.

7. Michael R. Edelstein, *Contaminated Communities: Coping with Residential Toxic Exposure* (Westview Press, 2004), ch. 4.

8. Nancy Lorenzo, quoted in Mark Fritz, "Little League Giants Put Town on Map, For Good," *Los Angeles Times,* August 31, 1998.

9. The second and third quotations are from Laurie Baiamonte Amato and Kevin Turner in Nichole M. Christian, "What Tropical Storm? Only Bliss in Toms River," *New York Times,* August 29, 1998.

10. Eugene Gessler, Ciba Toms River site manager, quoted in Jean Mikle, "Cancer Death Rate Normal among Ex-Ciba Workers," *Asbury Park Press,* July 2, 1998.

11. Nalini Sathiakumar and Elizabeth Delzell, "An Updated Mortality Study of Workers at a Dye and Resin Manufacturing Plant," *Journal of Occupational and Environmental Medicine* 42:7 (July 2000): 762–71.

12. Sathiakumar and Delzell, "Updated Mortality Study," 762.

13. Delzell, Macaluso, and Cole, "Follow-Up Study of Workers," 276–77. See also Barbone, *Nested Case-Control Study,* 39–40.

14. Jean Mikle, "Third Toms River Well Has Traces of Pollutants," *Asbury Park Press,* September 10, 1998.

15. John Jenks et al., *Characterization of Non-Target Substances in the Groundwater near the Reich Farm Superfund Site, Dover Township, New Jersey,* New Jersey Department of Environmental Protection (April 10, 2000), 5.

Chapter Twenty-One

1. For a list of Paracelsus's attributed works, see Franz Hartmann, *Life and Doctrines of Philippus, Theophrastus Bombast of Hohenheim Known as Paracelsus* (Kessinger, 1992), 31–36.

2. The quotation, "Alle Ding sind Gift und nichts ohn Gift; allein die Dosis macht, das ein Ding kein Gift ist," appears in Paracelsus's *Seven Defensiones,* first published in 1538. See *Four Treatises of Theophrastus Von Hohenheim, Called Paracelsus,* 22.

3. Laura N. Vandenberg et al., "Hormones and Endocrine-Disrupting Chemicals: Low-Dose Effects and Nonmonotonic Dose Responses," *Endocrine Reviews*

33:3 (June 1, 2012): 378–455. See also Dan Fagin, "Toxicology: The Learning Curve," *Nature* 490:7421 (October 25, 2012): 462–65.

4. Morris Maslia faced many other complications, too, in constructing the water model for Toms River. The old records of the Toms River Water Company were sparse, and he could not conduct a real-world tracer test to confirm the accuracy of his model. In Woburn, Peter Murphy had arranged to add a fluoride tracer to each well, one at a time, and then collect water samples all over town to see if each well's water was distributed as the model predicted. But after the trauma Toms River had been through since 1996, no one wanted to seek permission to add anything to the water system—even table salt, which Maslia considered using as a tracer. Fortunately, during the emergency water tests of 1996, Jerry Fagliano had noticed that concentrations of naturally occurring barium varied significantly from well to well and thus could serve as a natural tracer Maslia could use to verify his model.

5. New Jersey Department of Health and Senior Services, *Public Health Assessment: Reich Farm* and *Public Health Assessment: Ciba-Geigy Corporation* (March 12, 2001). Said Bruce Molholt, whose work experience included four years at the EPA: "I've seen at least one hundred, maybe two hundred of these health consultations, and they never find anything. I was astounded when ATSDR came right out and said we do find reason to believe there was harm caused by both air and water contamination."

6. U.S. Environmental Protection Agency, *Final Source Control Remedial Investigation Report: Ciba-Geigy Site, Volume 4* (December 1994).

7. William P. Eckel, Disposal Safety Incorporated, *Tentatively Identified Compounds at Ciba-Geigy Site, Toms River, NJ* (May 28, 1997), 4. This report was produced by a consultant hired by the Ocean County Citizens for Clean Water, funded by Ciba to provide independent oversight of the factory cleanup.

8. Radiation is the only conclusively proven environmental risk factor for pediatric leukemia and nervous system cancers. For more about risk factors, see National Cancer Institute Cancer Statistics Branch, *Cancer Incidence and Survival Among Children and Adolescents: United States SEER Program, 1975–1995* (1999).

9. The azo dye ingredient benzidine, for example, was not just a bladder carcinogen; it also caused nervous system cancers in mice. Acrylonitrile was one of the few chemicals known to cause brain tumors in rats, and factory studies suggested that it might cause brain cancer in humans, too, although there was conflicting evidence. Other research found that parents exposed to nitrobenzene were more likely to have children with brain cancer and that parental exposure to TCE and benzene raised the risk of childhood leukemia.

10. National Toxicology Program, *Draft NTP Technical Report on the Toxicology and Carcinogenesis Study of Styrene-Acrylonitrile Trimer in F344/N Rats* (December 2010), 22–24.

11. James Huff, "Long-Term Chemical Carcinogenesis Bioassays Predict Human Cancer Hazards: Issues, Controversies, and Uncertainties," *Annals of the New York Academy of Sciences* 895 (December 1989): 56–79.

12. To be placed into the National Toxicology Program's highest category ("clear evidence" of carcinogenicity in animals), a compound generally needed to

cause tumors in multiple organs or in both mice and rats. Only about 25 percent of chemicals tested by the NTP met this higher standard. See Victor A. Fung et al., "Predictive Strategies for Selecting 379 NCI/NTP Chemicals Evaluated for Carcinogenic Potential: Scientific and Public Health Impact," *Fundamental and Applied Toxicology* 20 (1993): 413–36.

13. In 1999, the National Toxicology Program was just completing a study of acrylonitrile that would find "clear evidence" that it caused cancer in mice. See National Toxicology Program, *NTP Technical Report on the Toxicology and Carcinogenesis Studies of Acrylonitrile in B6C3F1 Mice (Gavage Studies),* Report TR-506 (October 2001).

14. D. J. Caldwell, "Review of Mononuclear Cell Leukemia in F-344 Rat Bioassays and Its Significance to Human Cancer Risk: A Case Study Using Alkyl Phthalates," *Regulatory Toxicology and Pharmacology* 30:1 (August 1999): 45–53.

15. Only a few chemicals had ever been found to trigger brain tumors in rats, but one was acrylonitrile, one of SAN trimer's two ingredients, in a test involving Fisher 344 rats. See D. D. Bigner et al., "Primary Brain Tumors in Fischer 344 Rats Chronically Exposed to Acrylonitrile in Their Drinking Water," *Food and Chemical Toxicology* 24:2 (1986): 129–37.

16. There was one fewer control family than planned, because one had been chosen errantly.

17. One of the only truly bizarre results from the Toms River interview study, almost certainly attributable to chance, was that children with brain cancer were 70 percent *less* likely to have been born to mothers who ate cured meat at least once a week during pregnancy, compared to mothers who followed their doctors' advice and avoided foods that were high in sodium and nitrates. See New Jersey Department of Health and Senior Services, *Case-Control Study of Childhood Cancers in Dover Township (Ocean County), New Jersey. Volume III: Technical Report Tables and Figures* (January 2003), 14, table 34.

18. *Childhood Cancers in Dover Township (Ocean County), New Jersey. Volume III: Technical Report Tables and Figures,* 1–6, table 34.

19. In addition to Ciba, the class-action lawsuit, *Kramer et al. vs. Ciba-Geigy et al.* also named United Water and the two former Ciba employees who pleaded guilty to illegal dumping back in 1992, James McPherson and William Bobsein, as defendants. The lawsuit was resolved in August of 2002 without any admission of liability and with modest but undisclosed payments to the injured adults. For the uninjured, the companies agreed only to pay for ten years' worth of screening for bladder cancer. Anyone who had drunk Toms River water in 1965 and 1966 was eligible. The settlement was the closest Ciba and United Water would ever come to acknowledging that their predecessors, Toms River Chemical and Toms River Water, had kept the public in the dark in 1965 and 1966 when dye chemicals seeped into the riverside wells on Holly Street and then into the homes of thousands of water customers in Oak Ridge and elsewhere. See Ana Alaya, "Toms River Families Press Fight," *Star-Ledger,* December 18, 2002.

20. The property-damage class-action lawsuit, *Janes et al. vs. Ciba-Geigy Corp. et al.,* stagnated in the court system for more than ten years before it was finally re-

solved in June of 2011. Under the terms of the court-approved settlement, Oak Ridge homeowners who bought their homes before March of 2000 and sold them before 2009 were awarded $4,000 each (up to a total of $900,000). Ciba also agreed to create a permanent, sixteen-acre buffer zone on the southeastern edge of the property (abutting homes along Cardinal Drive) and to spend $25,000 to landscape the buffer. By far the largest cash payment went to the plaintiffs' attorneys; Ciba agreed to pay them $3 million.

21. For a mid-1990s overview of the progress and potential of molecular epidemiology, see Frederica P. Perera, "Molecular Epidemiology: Insights into Cancer Susceptibility, Risk Assessment, and Prevention," *Journal of the National Cancer Institute* 88:8 (April 17, 1996): 496–509; and Frederica P. Perera et al., "Molecular Epidemiology in Environmental Carcinogenesis," *Environmental Health Perspectives* 104, Supp. 3 (May 1996): 441–43.

Chapter Twenty-Two

1. B. A. Finette et al., "Gene Mutations with Characteristic Deletions in Cord Blood T Lymphocytes Associated with Passive Maternal Exposure to Tobacco Smoke," *Nature Medicine* 4:10 (October 1998): 1144–51.

2. In an email to the author, Jack Mandel wrote that he "has no recollection of this discussion with Clapp. I can't believe I would have said something like this." For his part, Richard Clapp says he is certain that Mandel made the remark: "I'll never forget it."

3. For detailed accounts of how the water model was constructed, see Morris Maslia et al., *Historical Reconstruction of the Water-Distribution System Serving the Dover Township Area, New Jersey: January 1962–December 1996* (Agency for Toxic Substances and Disease Registry, 2001), and also Morris Maslia et al., *Analysis of the 1998 Water-Distribution System Serving the Dover Township Area, New Jersey: Field-Data Collection Activities and Water-Distribution System Modeling* (Agency for Toxic Substances and Disease Registry, 2000).

4. Because there was so much uncertainty about when pollution from Reich Farm first reached the Parkway wells, the state's case-control studies also included exposure scenarios that started as early as 1978 or as late as 1986, although the health department considered 1982 to be the best guess of when the first contamination occurred. In any case, moving the exposure window forward or back a few years did not materially change the results of the case-control studies (odds ratios were highest for a 1984 start date). For the Holly well contamination, which was assumed to have ended by 1975, the state also tested an alternate scenario in which the contamination did not end until 1980. Again, there was little change in the study results.

5. To quantify exposure to contamination from private wells, the state identified eleven neighborhoods where private wells had been contaminated by industrial chemicals, including Pleasant Plains and Oak Ridge Estates. A mother or child who lived in one of those eleven areas and relied on a private well during the months and years before diagnosis would be classified as highly exposed.

6. State cancer registries were crucial to identifying the forty-eight case children in the birth record study. Investigators looked for name and birth date matches

between Toms River birth certificates and registries in New Jersey and nine other states. They then identified cancer-free children for the control group by looking for birth certificates that matched case children, in groups of ten, by age and sex.

7. Since they had no information about the birth record study children other than their birth certificate and diagnosis, investigators had to make two assumptions in order to estimate a mother's exposure to contaminated air and water during pregnancy: first, that she had not changed addresses while pregnant and, second, that if she lived near a water line, she got her water from the public supply system, not from a backyard well or from bottled water.

8. To avoid potentially biasing his calculations, Morris Maslia did not know which addresses were for children with cancer and which were for healthy controls. He also did not know which months during the study period were relevant for each child.

9. A prenatal water exposure score of 50 percent or higher was classified as high, less than 10 percent was low, and everything in between was medium.

10. Twenty-five case children in the interview study were also in the birth record study, but the overlap was limited because the two studies defined cases differently. The interview study cases included only children diagnosed with leukemia, brain or nervous system cancers (the birth record study included any form of childhood cancer), and only children who lived in Toms River at the time of diagnosis (the birth record study included children who had moved away). In one way, the interview study was more inclusive: It included some children who were not born in Toms River, while the birth record study, based on birth certificates, included only natives.

11. As an example of how these adjustments worked, a boy who lived in a home that got an average of 30 percent of its water from the Parkway wells during his mother's pregnancy and during his life before a cancer diagnosis would normally be classified as having a "medium" exposure to Parkway water, but if the boy's mother reported that during her pregnancy she had drunk an unusually high number of glasses of water each day, the boy would be bumped up into the "high" category.

12. The total number of children—702—was fewer than 727 because of overlap between the 528 children in the birth record study and the 199 children in the interview study.

13. Frederica P. Perera and I. Bernard Weinstein, "Molecular Epidemiology and Carcinogen-DNA Adduct Detection: New Approaches to Studies of Human Cancer Causation," *Journal of Chronic Diseases* 35 (1982): 581–600.

14. Frederica Perera and her collaborators found that mice and rabbits injected with benzo(a)pyrene had more DNA adducts than control animals did. But dogs exposed to cigarette smoke for twenty months did not. In humans, they found high adduct concentrations in the lung tissue of five lung cancer patients, all heavy smokers. But twenty-two other smokers did not have more adducts, and there was no dose-response pattern: Heavier smokers did not have more adducts. Frederica P. Perera et al., "A Pilot Project in Molecular Cancer Epidemiology: Determination of Benzo(a)pyrene-DNA Adducts in Animal and Human Tissues by Immunoassays," *Carcinogenesis* 3:12 (1982): 1405–10.

15. Mutated TP53 genes are in about half of all cancers, including neuroblastomas, melanomas, and many other tumors. Timothy E. Baroni et al., "A Global Sup-

pressor Motif for P53 Cancer Mutants," *Proceedings of the National Academy of Sciences* 101:14 (April 6, 2004): 4930–35.

16. Richard Albertini and R. B. Hayes, "Somatic Cell Mutations in Cancer Epidemiology," *IARC Scientific Publications 142* (International Agency for Research on Cancer, 1997), 159–84.

17. Barry Finette's mentor at the University of Vermont, Richard Albertini, had pioneered the use of the HPRT gene in cancer research in the late 1980s. One of the gene's many advantages as a biomarker is that it is located on the X chromosome. Since there is normally only one active X chromosome in a human cell, just one mutation event is enough to guarantee a heritable mutation, which makes HPRT a more sensitive biomarker for genetic damage than genes on paired chromosomes. Albertini also identified mutational mechanisms in HPRT that seemed to apply to many other genes, supporting the idea that HPRT really was a good surrogate for genome-wide damage.

18. Barry Finette's mutant frequency study was published as Pamela M. Vacek et al., "Somatic Mutant Frequency at the HPRT Locus in Children Associated with a Pediatric Cancer Cluster Linked to Exposure to Two Superfund Sites," *Environmental and Molecular Mutagenesis* 34 (2005): 339–45.

Chapter Twenty-Three

1. To understand how Jerry Fagliano calculated adjusted odds ratios in the Toms River case-control studies, consider this example: Nine case children and twenty-three healthy controls in the interview study were in the "high" exposed category because their mothers had gotten more than 50 percent of their water during pregnancy from the Parkway well field. Twenty-six cases and 110 controls were in the "low" exposure category because their mothers had gotten less than 10 percent of their water from the Parkway wells during pregnancy. Calculating a "crude" odds ratio is simply a matter of comparing the odds that high-exposed women would have a child with cancer, versus low-exposed women. The simple equation looks like this: (exposed cases/unexposed cases)/(exposed controls/unexposed controls) = odds ratio. Applied to the Parkway data for high-exposure mothers, the math is (9/26)/(23/110), yielding an odds ratio of 1.66. Jerry Fagliano then adjusted that result in a way that helped to account for the influence of possible confounding factors such as the age and sex of each study child. After this final calculation, known as conditional logistical regression, the adjusted odds ratio is slightly higher: 1.68. In other words, the odds of a case child having been highly exposed to Parkway water were 68 percent greater than the odds that a healthy child had been exposed.

2. All of the odds ratios described in this chapter come from data tables in *Case-Control Study of Childhood Cancers in Dover Township (Ocean County), New Jersey, Volume III: Technical Report Tables and Figures.*

3. New Jersey Department of Health and Senior Services, *Case-Control Study of Childhood Cancers in Dover Township (Ocean County), New Jersey, Volume II: Final Technical Report* (January 2003), 23. Like many of the others in the study, this odds ratio fell short of statistical significance because the lower bound of its very wide confidence interval—0.79 to 274—was below 1.0.

4. The same small-numbers problem stymied the health department's attempts to assess the consequences of the groundwater pollution that tainted hundreds of backyard wells in the neighborhoods like Oak Ridge Estates and Pleasant Plains in the 1960s and early 1970s, before water mains were fully extended to those areas. Eleven such neighborhoods had been identified by investigators, but so much time had passed that they could track down just two exposed cases from that era—too few to generate a credible odds ratio.

5. In order to create high-, medium-, and low-exposure categories for the air pollution model, its developers at Rutgers first analyzed air exposures for the residences of the control families. The family at the midpoint was assigned a 1.0 rating on a scale of "relative air impact units." Under the same scale, the top 25 percent of controls in the interview study were at or above a rating of 2.18. They then applied the same scale to the case families, placing cases and controls into three categories: low (below 1.0), medium (1–2.17), and high (2.18 and above). For the birth record study, the "cut point" separating low from medium was 1.5; for separating medium from high exposure, the cut point was 3.04.

6. The size of the Toms River legal settlement has never been disclosed but, based on court filings, it is very likely that the total was somewhere between $35 and $40 million, before deductions for the lawyers' 25 percent fee and other reimbursable expenses. Records from judicial hearings, held only for the twenty-nine families with children still under age eighteen, show that those families were awarded approximately $11.4 million in gross payments. The payments went to children and siblings and also to parents for reimbursable costs (such as lost wages, uninsured medical bills, and trips to distant hospitals). After deductions for legal fees and expenses, the net payout was about $8.4 million, or an average of about $290,000 per family. Assuming that those twenty-nine families were roughly representative of the forty families not subject to the judicial hearings, then the total gross payout was approximately $27.1 million. In addition, there was a "recurrence fund" of $5 million ($3.75 million after legal fees) to compensate any surviving children who relapsed before 2016, at which time the proceeds would be distributed to all the surviving children. There were also additional payments to parents beyond expense reimbursement. Thus, the gross amount of the settlement in 2001 (not including subsequent interest on funds held in trust for minors) was probably somewhere between $35 million and $40 million. The court records also show that some families with children under eighteen received as much as $501,570, after deductions for fees and expenses, while others got as little as $91,954. The lawyers donated $150,000 from their fees to TEACH, the environmental advocacy group formed by the families.

7. Dow, Ciba, and United Water have never revealed what percentage of the overall settlement each company paid, but Dow (as Union Carbide's corporate successor) likely paid the largest share because the Parkway well contamination was so well documented.

8. Like the Toms River settlement, the Woburn settlement was supposed to be secret. Unlike in Toms River, however, its $8 million size was immediately leaked to the press. Each of the eight Woburn families ultimately received about $435,000, while the payouts to the sixty-nine Toms River families varied widely but probably averaged about $290,000 per family. In Toms River, families of children with leuke-

mia received some of the largest settlements—in some cases, larger than what the Woburn families received—because the scientific evidence of environmental causation was stronger for leukemia than for other cancers. The Woburn lawsuit included only leukemia cases.

9. New Jersey Department of Health and Senior Services, *Case-Control Study of Childhood Cancers in Dover Township, Volume I* (January 2003), 18.

10. For the four nervous system cancer cases diagnosed before age five, the 9.00 odds ratio for high exposure to Ciba air emissions came close to achieving statistical significance because the lower bound of the 95 percent confidence interval was 0.86, just under 1.0. The upper bound was an extremely high 94.2.

11. For birth record study children diagnosed with cancer before age five, the odds ratio for high prenatal exposure to Ciba air emissions was 2.95, with a 95 percent confidence interval of 0.92 to 9.45.

12. Tom Feeney and Mark Mueller, "Crusading Mom Shrugs Off Vindication," *Star-Ledger,* December 19, 2001.

Chapter Twenty-Four

1. Carol S. Rubin et al., "Investigating Childhood Leukemia in Churchill County, Nevada," *Environmental Health Perspectives* 115:1 (January 2007): 151–57.

2. Craig Steinmaus et al., "Probability Estimates for the Unique Childhood Leukemia Cluster in Fallon, Nevada, and Risks near Other U.S. Military Aviation Facilities," *Environmental Health Perspectives* 112:6 (May 2004): 766–71. The estimate that a childhood cancer cluster as large as Fallon's would occur randomly only once every 22,000 years in a country the size of the United States (which had about 77 million children in 2003) is based on a case count of eleven childhood leukemias diagnosed in Fallon and the surrounding area from 1999 to 2001. The authors' estimate that there was only a 1-in-232 million chance that the Fallon cluster was random is a sharp contrast to the Woburn childhood leukemia cluster, which the authors calculate had a 1-in-120 probability of occurring by chance. The authors did not attempt to estimate the odds that the Toms River cluster was random.

3. U.S. Centers for Disease Control and Prevention, *Biosampling Case Children with Leukemia (Acute Lymphocytic and Myelocytic Leukemia) and a Reference Population in Sierra Vista, Arizona: Final Report* (November 30, 2006).

4. Florida Department of Health, *Acreage Cancer Review, Palm Beach County* (August 28, 2009), subsequently revised as *Acreage SIR Recalculation Population Estimate Methods and Results.* Both are available on the Palm Beach County Health Department website. Because The Acreage grew so fast during the years at issue, the state calculated incidence rates using four different methods of estimating the local population. For the four cases of pediatric brain cancer diagnosed from 2005 to 2007, the 95 percent confidence intervals for the four methods were (1.7–11), (1.5–9.8), (1.3–8.2), and (1.3–8.2). Thus, by any of those methods, the cluster was statistically significant. Since three of the four cases were in girls, the incidence ratios and confidence intervals for girls were higher still.

5. Stephen P. Hunger et al., "Improved Survival for Children and Adolescents with Acute Lymphoblastic Leukemia Between 1990 and 2005: A Report from the

Children's Oncology Group," *Journal of Clinical Oncology,* published online March 12, 2012. Gleevec, the Novartis trade name for the drug imatinib, is especially effective in treating children with ALL who carry the Philadelphia chromosomal translocation, the article notes.

6. The biggest dosimeter success story so far is for aflatoxin, a naturally occurring fungal secretion that is carcinogenic and often contaminates peanuts and other grains, especially in Africa and China. In the early 1990s, John Groopman of The Johns Hopkins University confirmed that aflatoxin-DNA adducts can be readily detected in urine and are a superb measure for identifying populations at high risk for liver cancer. See John D. Groopman et al., "Molecular Epidemiology of Aflatoxin Exposures: Validation of Aflatoxin-N7-Guanine Levels in Urine as a Biomarker in Experimental Rat Models and Humans," *Environmental Health Perspectives* 99:3 (March 1993): 107–13.

7. One clue about the SUOX gene is that it produces an enzyme that is inhibited by tungsten exposure. That may be significant because unusually high levels of the naturally occurring metal have been found in the blood of Fallon children, both those with cancer and matched healthy controls. While case-control studies have not associated tungsten exposure with cancer in Fallon or anywhere else, it is possible that the gene variation may make certain children more vulnerable to tungsten-related toxicity. See Karen K. Steinberg et al., "Genetic Studies of a Cluster of Acute Lymphoblastic Leukemia Cases in Churchill County, Nevada," *Environmental Health Perspectives* 115:1 (January 2007): 158–64.

8. Barry Finette's Toms River research did uncover some suggestive evidence, but he was not sure what it meant, if anything. He found that small changes in three genes—a detoxifier on Chromosome One known as EPHX1, another detoxifier on Chromosome Eleven called GSTP2, and a DNA repairer on Chromosome Fourteen called APEX1—more than doubled cancer risk among children who did not live in Toms River. For Toms River children, however, the presence of the three genetic alterations did not significantly change their cancer risk. A more intriguing result concerned the gene ABCB1, on the long arm of Chromosome Seven. A drug transporter gene, it affects a cell's ability to expel invading molecules that are capable of crossing complex bodily barriers, including the placenta that separates mother from child. Finette found that Toms River children were more likely to have one version of the ABCB1 gene than out-of-town children. But those differences in genotype did not match up with differences in cancer risk—at least, not that could be elucidated from such a small sample.

9. Barry Finette had been relatively optimistic about the chromosome damage study because there was already evidence from other studies that it was a useful biomarker for certain cancers and for exposure to some mutagenic chemicals. See, for example, Lars Hagmar et al., "Chromosomal Aberrations in Lymphocytes Predict Human Cancer: A Report from the European Study Group on Cytogenetic Biomarkers and Health," *Cancer Research* 58:18 (September 15, 1998): 4117–21.

10. Another powerful motivation for manufacturers is that compounds deemed carcinogenic in National Toxicology Program rodent tests usually end up listed in the NTP's *Report on Carcinogens.* The 2011 edition included 246 compounds—a blacklist manufacturers try to avoid at all cost.

Notes

11. National Toxicology Program, *Draft NTP Technical Report on the Toxicology and Carcinogenesis Study of Styrene-Acrylonitrile Trimer in F344/N Rats* (December 2010), 81, table 19.

12. For more about the limitations of 400-animal bioassays (50 per sex at four doses), see David Gaylor, "Are Tumor Incidence Rates from Chronic Bioassays Telling Us What We Need to Know about Carcinogens?" *Regulatory Toxicology and Pharmacology* 41 (2005): 128–33.

13. The Fischer 344 rat is a poor model for human leukemia because it is naturally prone to mononuclear cell leukemia, found only in rats. See Caldwell, "Review of Mononuclear Cell Leukemia."

14. Compiled by Jerry Fagliano, data for the two charts in this chapter are from the New York State Cancer Registry and the New Jersey Department of Health and Senior Services. Incidence rates are diagnoses per 100,000 children per year and are age-adjusted to the U.S. population, as measured by the 2000 Census.

15. Daniel Wartenberg, "Investigating Disease Clusters: Why, When and How?" *Journal of the Royal Statistical Society A* 164:1 (2001): 13–22, 15.

16. Erika Bolstad et al., "Congress Urged to Track Cancer Clusters Better," McClatchy Newspapers (*Bradenton Herald*), March 29, 2011. For a brief overview of proposed federal legislation, as well as descriptions of forty-two disease clusters reported in thirteen American states since 1977, see Kathleen Navarro et al., *Health Alert: Disease Clusters Spotlight the Need to Protect People from Toxic Chemicals,* a report of the Natural Resources Defense Council and the National Disease Clusters Alliance (March 2011).

17. The author interviewed parents and physicians at Chongqing Children's Hospital in October of 2007.

18. As of 2007, the five-year survival rate for acute lymphoblastic leukemia among children who received at least some chemotherapy at Chongqing Children's Hospital—one of the top facilities in China—was about 70 percent, compared to more than 95 percent for leukemia patients in the United States.

19. Lee Liu, "Made in China: Cancer Villages," *Environment* (March/April 2010).

20. Xie Chuanjiao, "Pollution Makes Cancer the Top Killer," *China Daily,* May 21, 2007. Published in China's largest English-language publication, the article quotes Chen Zhizhou, a cancer researcher affiliated with the Chinese Academy of Medical Sciences: "The main reason behind the rising number of cancer cases is that pollution of the environment, water and air is getting worse by the day."

21. Chemical production figures are drawn primarily from the "Facts and Figures of the Chemical Industry" survey published every summer by *Chemical & Engineering News*.

22. BASF, "BASF Chongqing MDI Project Approved," news release, March 25, 2011.

23. For more about Frederica Perera and Deliang Tang's research in China, see Frederica Perera et al., "Benefits of Reducing Prenatal Exposure to Coal-Burning Pollutants to Children's Neurodevelopment in China," *Environmental Health Perspectives* 116:10 (October 2008): 1396–1400; and Dan Fagin, "China's Children of Smoke," *Scientific American* (August 2008): 72–79.

24. The state cancer registry's inadequacies, exposed during the 1996 crisis in Toms River, have not been completely resolved. But the current lag time of about eighteen months is less than half as long as in 1995, when Michael Berry had to abandon the idea of incorporating recent statistics into his cluster analysis. The diminished lag time is a direct result of pressure from Linda Gillick and the other TEACH families.

25. During the dry summer of 2005, United Water's two senior executives in Toms River, General Manager George Flegal and Operations Manager Richard Ottens Jr., were so worried they might have to take wells offline due to radium contamination that they faked a safety test and filed a false report, according to state regulators. An audit by the water company's home office discovered that the men had failed to report seven instances of elevated radium levels, prompting the state to fine United Water $64,000. The two were ousted from their jobs and fined $5,000 each. The company was also fined $104,000 for exceeding state-mandated pumping limits.

United Water's new management in Toms River, however, has received a strong endorsement from Linda Gillick, who even gave the water company Ocean of Love's 2010 Public Service Award. "I could not have imagined five years ago that I would be presenting this award to United Water," she said at the ceremony. "The people at the company have earned my trust and respect through their commitment to honesty and integrity in everything they do."

26. Like its corporate predecessor, Ciba, BASF has not been willing to spend the tens of millions of dollars it would cost to empty out Cell One of the lined landfill. The town has filed suit to try to force a Cell One cleanup, with the support of the TEACH families. Bruce and Melanie Anderson and their two sons have even picketed at the front gate of the shuttered plant to press for a cleanup. However, Ciba and now BASF have not relented, and the state Department of Environmental Protection, which is responsible for regulating the landfill, has not joined efforts to compel the company to remove the buried drums.

27. The excavation of the older dumps began in 2003, when workers pulled more than forty-seven thousand crumbling drums from the leaky pit of the old drum dump, which was on the site of an even older settling lagoon from the 1950s. Excavators then moved on to the other very old dumps scattered around the property. The original EPA plan had been for the soil from those dumps to be heat-treated in a thermal desorption machine, as had happened at Reich Farm in 1995, but the agency changed course after TEACH and other groups protested that the desorption process would generate toxic air emissions. However, the EPA refused to order Ciba to truck all of the chemical-laced soil to an off-site landfill, as TEACH and other groups wanted. Instead, for all but the most severely contaminated earth, the EPA allowed Ciba to use an experimental composting process. The tainted soil was mixed with straw, wood chips, nutrients, and water and then aerated to spur the growth of microbes that break down complex pollutants. After eight weeks of composting—first inside a shed the size of an airplane hangar and then outside, on an asphalt slab the size of a football field—the soil was tested. If contaminant levels were low enough, the dirt was then buried elsewhere on the factory grounds and another load of contaminated soil was brought in for composting.

28. Novartis, which inherited Ciba's pharmaceutical business in the 1997 merger that broke up Ciba, still uses the Ciba name in its unit that makes contact lenses and lens care products, known as Ciba Vision. Ciba's chemical products have been manufactured under the BASF corporate brand since 2009, when the German company bought what was left of the old Ciba.

Index

About the Author

DAN FAGIN is an associate professor of journalism and the director of the Science, Health, and Environmental Reporting Program at New York University's Arthur L. Carter Journalism Institute. For fifteen years, he was the environmental writer at *Newsday*, where he was twice a principal member of reporting teams that were finalists for the Pulitzer Prize. His articles on cancer epidemiology were recognized with the Science Journalism Award of the American Association for the Advancement of Science and the Science in Society Award of the National Association of Science Writers.

About the Type

This book was set in Sabon, a typeface designed by the well-known German typographer Jan Tschichold (1902–74). Sabon's design is based upon the original letter forms of Claude Garamond and was created specifically to be used for three sources: foundry type for hand composition, Linotype, and Monotype. Tschichold named his typeface for the famous Frankfurt typefounder Jacques Sabon, who died in 1580.